Generalized Concavity

Books in the Classics in Applied Mathematics series are monographs and textbooks declared out of print by their original publishers, though they are of continued importance and interest to the mathematical community. SIAM publishes this series to ensure that the information presented in these texts is not lost to today's students and researchers.

Classics in Applied Mathematics

C. C. Lin and L. A. Segel, *Mathematics Applied to Deterministic Problems in the Natural Sciences*

Johan G. F. Belinfante and Bernard Kolman, *A Survey of Lie Groups and Lie Algebras with Applications and Computational Methods*

James M. Ortega, *Numerical Analysis: A Second Course*

Anthony V. Fiacco and Garth P. McCormick, *Nonlinear Programming: Sequential Unconstrained Minimization Techniques*

F. H. Clarke, *Optimization and Nonsmooth Analysis*

George F. Carrier and Carl E. Pearson, *Ordinary Differential Equations*

Leo Breiman, *Probability*

R. Bellman and G. M. Wing, *An Introduction to Invariant Imbedding*

Abraham Berman and Robert J. Plemmons, *Nonnegative Matrices in the Mathematical Sciences*

Olvi L. Mangasarian, *Nonlinear Programming*

Carl Friedrich Gauss, *Theory of the Combination of Observations Least Subject to Errors: Part One, Part Two, Supplement*. Translated by G. W. Stewart (*First time in print.*)

Richard Bellman, *Introduction to Matrix Analysis*

U. M. Ascher, R. M. M. Mattheij, and R. D. Russell, *Numerical Solution of Boundary Value Problems for Ordinary Differential Equations*

K. E. Brenan, S. L. Campbell, and L. R. Petzold, *Numerical Solution of Initial-Value Problems in Differential-Algebraic Equations*

Charles L. Lawson and Richard J. Hanson, *Solving Least Squares Problems*

J. E. Dennis, Jr. and Robert B. Schnabel, *Numerical Methods for Unconstrained Optimization and Nonlinear Equations*

Richard E. Barlow and Frank Proschan, *Mathematical Theory of Reliability*

Cornelius Lanczos, *Linear Differential Operators*

Richard Bellman, *Introduction to Matrix Analysis, Second Edition*

Beresford N. Parlett, *The Symmetric Eigenvalue Problem*

Richard Haberman, *Mathematical Models: Mechanical Vibrations, Population Dynamics, and Traffic Flow*

Peter W. M. John, *Statistical Design and Analysis of Experiments*

Tamer Başar and Geert Jan Olsder, *Dynamic Noncooperative Game Theory, Second Edition*

Emanuel Parzen, *Stochastic Processes*

Petar Kokotović, Hassan K. Khalil, and John O'Reilly, *Singular Perturbation Methods in Control: Analysis and Design*

Classics in Applied Mathematics (continued)

Jean Dickinson Gibbons, Ingram Olkin, and Milton Sobel, *Selecting and Ordering Populations: A New Statistical Methodology*

James A. Murdock, *Perturbations: Theory and Methods*

Ivar Ekeland and Roger Témam, *Convex Analysis and Variational Problems*

Ivar Stakgold, *Boundary Value Problems of Mathematical Physics, Volumes I and II*

J. M. Ortega and W. C. Rheinboldt, *Iterative Solution of Nonlinear Equations in Several Variables*

David Kinderlehrer and Guido Stampacchia, *An Introduction to Variational Inequalities and Their Applications*

F. Natterer, *The Mathematics of Computerized Tomography*

Avinash C. Kak and Malcolm Slaney, *Principles of Computerized Tomographic Imaging*

R. Wong, *Asymptotic Approximations of Integrals*

O. Axelsson and V. A. Barker, *Finite Element Solution of Boundary Value Problems: Theory and Computation*

David R. Brillinger, *Time Series: Data Analysis and Theory*

Joel N. Franklin, *Methods of Mathematical Economics: Linear and Nonlinear Programming, Fixed-Point Theorems*

Philip Hartman, *Ordinary Differential Equations, Second Edition*

Michael D. Intriligator, *Mathematical Optimization and Economic Theory*

Philippe G. Ciarlet, *The Finite Element Method for Elliptic Problems*

Jane K. Cullum and Ralph A. Willoughby, *Lanczos Algorithms for Large Symmetric Eigenvalue Computations, Vol. I: Theory*

M. Vidyasagar, *Nonlinear Systems Analysis, Second Edition*

Robert Mattheij and Jaap Molenaar, *Ordinary Differential Equations in Theory and Practice*

Shanti S. Gupta and S. Panchapakesan, *Multiple Decision Procedures: Theory and Methodology of Selecting and Ranking Populations*

Eugene L. Allgower and Kurt Georg, *Introduction to Numerical Continuation Methods*

Leah Edelstein-Keshet, *Mathematical Models in Biology*

Heinz-Otto Kreiss and Jens Lorenz, *Initial-Boundary Value Problems and the Navier-Stokes Equations*

J. L. Hodges, Jr. and E. L. Lehmann, *Basic Concepts of Probability and Statistics, Second Edition*

George F. Carrier, Max Krook, and Carl E. Pearson, *Functions of a Complex Variable: Theory and Technique*

Friedrich Pukelsheim, *Optimal Design of Experiments*

Israel Gohberg, Peter Lancaster, and Leiba Rodman, *Invariant Subspaces of Matrices with Applications*

Lee A. Segel with G. H. Handelman, *Mathematics Applied to Continuum Mechanics*

Rajendra Bhatia, *Perturbation Bounds for Matrix Eigenvalues*

Barry C. Arnold, N. Balakrishnan, and H. N. Nagaraja, *A First Course in Order Statistics*

Charles A. Desoer and M. Vidyasagar, *Feedback Systems: Input-Output Properties*

Stephen L. Campbell and Carl D. Meyer, *Generalized Inverses of Linear Transformations*

Alexander Morgan, *Solving Polynomial Systems Using Continuation for Engineering and Scientific Problems*

I. Gohberg, P. Lancaster, and L. Rodman, *Matrix Polynomials*

Galen R. Shorack and Jon A. Wellner, *Empirical Processes with Applications to Statistics*

Richard W. Cottle, Jong-Shi Pang, and Richard E. Stone, *The Linear Complementarity Problem*

Rabi N. Bhattacharya and Edward C. Waymire, *Stochastic Processes with Applications*

Robert J. Adler, *The Geometry of Random Fields*

Mordecai Avriel, Walter E. Diewert, Siegfried Schaible, and Israel Zang, *Generalized Concavity*

Rabi N. Bhattacharya and R. Ranga Rao, *Normal Approximation and Asymptotic Expansions*

Generalized Concavity

Mordecai Avriel
Technion – Israel Institute of Technology
Haifa, Israel

Walter E. Diewert
University of British Columbia
Vancouver, British Columbia, Canada

Siegfried Schaible
Chung Yuan Christian University
Chung-Li, Taiwan

Israel Zang
The Academic College of Tel-Aviv-Yaffo
Tel-Aviv-Yaffo, Israel

siam®
Society for Industrial and Applied Mathematics
Philadelphia

This SIAM edition is an unabridged republication of the work first published by Plenum Press, 1988.

10 9 8 7 6 5 4 3 2 1

Library of Congress Cataloging-in-Publication Data

Generalized concavity / Mordecai Avriel ... [et al.].
p. cm. -- (Classics in applied mathematics ; 63)
Originally published: New York : Plenum Press, 1988.
Includes bibliographical references and index.
ISBN 978-0-898718-96-6
1. Concave functions. I. Avriel, M.
QA353.C64G46 2010
515'.94--dc22

2010035259

To my wife, Hava,
our children, Dorit and Yoram,
Ron and Ayelet, Tal and Dror,
and our grandchildren,
Or, Shir, Yael, Yuval, Sara, and Gal.

Mordecai Avriel

To my wife, Virginia Diewert.

Erwin Diewert

To my late wife, Ingrid,
our children, Sue, John, and Rickie,
and our grandchildren,
Hannah, Joel, Elise, and Cody.

Siegfried Schaible

To my wife, Miri,
our children, Ben, Shira, and Avi,
and our grandchildren, Dan and Orr.

Israel Zang

Contents

Preface to the Classics Edition

With the republication of our book, first published in 1988, we would like to recall how it all started.

Mordecai Avriel worked on nonlinear programming, where convexity plays a major role. If the nonlinear program is minimizing a convex function (or maximizing a concave function) over a convex set, every local optimum is a global one. Consequently, searching for a global optimum of a convex nonlinear program is much easier than seeking the optimum in a general nonlinear program. In other words, convexity is a sufficient condition for local minima to be a global one, but it is definitely not a necessary one. Therefore, Mordecai began studying functions that are more general than convex and in special cases can be reduced to convex ones. He published his first results in 1972, in the paper "r-Convex Functions."

Siegfried Schaible was working at about the same time on various forms of generalized convexity in his doctoral dissertation in Germany. Siegfried and Mordecai met at the Department of Operations Research of Stanford University.

At about the same time Israel Zang started his doctoral research at the Technion in Haifa, Israel, under the supervision of Mordecai. The dissertation dealt with convex transformable functions and programs. The concept of r-convex functions is concerned with a particular range transformation of the function under discussion. Israel's work created the framework to handle general range and/or domain transformations. In addition, this dissertation discusses the characterization of functions whose local minima are global through continuity properties of their level sets mappings.

The next phase of collaboration started when Mordecai spent a sabbatical at the University of British Columbia (UBC) in Vancouver in 1974–1975, where he met Erwin Diewert, who was interested in generalized convexity from an economist's viewpoint. It turns out that convexity (and concavity) have important implications in economic theory. There are three important constrained maximization or minimization problems that play a fundamental role in economics: (i) the cost minimization problem of a producer; (ii) the expenditure minimization problem of a consumer; and (iii) the profit maximization problem of a producer. The first two optimization problems generate a cost function and an expenditure function, respectively, and it turns out that these two functions must be concave functions in the input prices that the producer faces and the commodity prices that the consumer faces. The third optimization problem generates a profit function, which must be a convex function in the prices that the producer faces. Thus the theory of convex and concave functions plays a very important role in economics. Another area of economics where generalized concave functions play an important role is concerned with the following question: What type of generalized concavity is required to guarantee that the satisfaction of first order necessary conditions for maximizing a differentiable function will also be sufficient? The answer is that pseudoconcavity will do the job.

Israel Zang completed his doctoral studies in 1974, and following a one-year visit to the Center for Operations Research and Econometrics (CORE), he joined the Faculty of Management at Tel Aviv University and was subsequently invited to UBC in Vancouver in 1980–1982, where he continued his cooperation with Erwin. During this period the JET paper on nine kinds of quasi concavity and concavity, written jointly with Mordecai and Erwin, was finalized, and first versions of chapters of this book were written.

The four researchers decided to organize a conference on the topic of generalized convexity. It was decided to ask for the support of NATO, and indeed the conference was held as a NATO Summer School in Vancouver in August 1980. At the conference, the four researchers also decided on writing the first book of its kind on the subject of generalized convexity. Erwin thought that in order to interest economists in the book it should be called *Generalized Concavity*. It took 8 years until the book was finally published in 1988.

Since the book's original publication, three of the coauthors have turned to other fields.

Mordecai took early retirement from Technion and became Director of Analytic Development at Bank Hapoalim in Israel. He is head of a multidisciplinary group working on financial engineering.

Erwin turned to more traditional subjects in economics. Up until 1985, he worked in various areas associated with microeconomic theory such as duality theory, the pure theory of international trade, the measurement of waste in an economy, and nonparametric producer and consumer theory. Duality theory does have close connections with generalized concavity; i.e., one looks for types of generalized concave production functions such that the dual cost function can provide a complete representation of the primal production function.

After 1985, Erwin focused on economic measurement issues. Over the past two decades, Erwin has been working with international agencies such as the OECD, the IMF, and the World Bank in writing manuals that address some of the above measurement issues.

Israel turned to higher education administration. After serving as Dean of the Recanati Faculty of Management at Tel Aviv University, he became Vice Provost of the university, retiring when his term was completed. He is currently the President of the Academic College of Tel-Aviv-Yaffo.

Siegfried has remained in the field of generalized convexity. In 1994, he organized a session cluster for the International Symposium of the Mathematical Programming Society (ISMP). He had chosen as the focus for the cluster the topic of generalized convexity. It turned out to be the largest cluster on that topic up to that point, with well over 10 sessions.

By that point in time, Siegfried was also aware that researchers in applied mathematics, including optimization, usually worked as individuals behind closed doors. This seemed to him to be less efficient than a more community-based, team-oriented work style such as that found in the natural sciences. In fact, he had read several biographies and found positive examples of those scientists who had favored a team approach. So he set out to start a collaborative effort in generalized convexity; thus, the international Working Group of Generalized Convexity (WGGC) was born. More information on the group can be found at www.genconv.org. Since its inception the group has organized the triennial symposia on generalized convexity.

In time the WGGC also started to explore the related area of generalized monotonicity. The most notable event of the group was the publication of the *Handbook of Generalized Convexity and Generalized Monotonicity* (Springer, 2005). Users of the *Handbook* soon learned that this book, *Generalized Concavity*, was needed to benefit fully from the *Handbook*. Since *Generalized Concavity* was by then out of print, finding copies to use as reference texts was quite difficult. Therefore, the authors welcomed the efforts of SIAM to return the original monograph to print in the SIAM Classics in Applied Mathematics series. Together with the *Handbook* and the proceedings from each GCM conference, the community is now well served with key references on the topic.

We thank Sara Murphy and the other SIAM staff who worked on the book for the great support in preparing the SIAM Classics edition of our book.

Mordecai Avriel
Walter E. Diewert
Siegfried Schaible
Israel Zang

Preface

Concavity plays a central role in mathematical economics, engineering, management science, and optimization theory. The reason is that concavity of functions is used as a hypothesis in most of the important theorems concerning extremum problems. In other words, concavity is usually a sufficient condition for satisfying the underlying assumptions of these theorems, but concavity is definitely not a necessary condition. In fact, there are large families of functions that are nonconcave and yet have properties similar to those of concave functions. Such functions are called generalized concave functions, and this book is about the various generalizations of concavity, mainly in the context of economics and optimization.

Although hundreds of articles dealing with generalized concavity have appeared in scientific journals, numerous textbooks have specific chapters on this subject, and scientific meetings devoted to generalized concavity have been held and their proceedings published, this book is the first attempt to present generalized concavity in a unified framework. We have collected results dealing with this subject mainly from the economics and optimization literature, and we hope that the material presented here will be useful in applications and will stimulate further research.

The writing of this book constituted a unique experience for the authors in international scientific cooperation—cooperation that extended over many years and at times spanned three continents. It was an extremely fruitful and enjoyable experience, which we will never forget.

We are indebted to our respective home universities—the Technion-Israel Institute of Technology, the University of British Columbia, the University of Alberta, and Tel Aviv University—for including the other authors in their exchange programs and for the technical assistance we received. Thanks are also due to the Center for Operations Research and Econometrics, Université Catholique de Louvain, for the hospitality extended to one of the authors.

The writing of this book was partially supported by the Fund for the Advancement of Research at Technion, the Natural Sciences and Engineering Research Council of Canada, the Deutsche Forschungsgemeinschaft (West Germany), the U.S. National Science Foundation, and the Israel Institute of Business Research at Tel Aviv University.

Mordecai Avriel
Haifa, Israel
Walter E. Diewert
Vancouver, British Columbia, Canada
Siegfried Schaible
Edmonton, Alberta, Canada
Israel Zang
Tel Aviv, Israel

1

Introduction

In this introductory chapter, we provide a brief and mostly nonmathematical description of the contents of this book on generalized concavity. Formal mathematical definitions of the various types of concavity may be found in subsequent chapters.

The first question we must attempt to answer in this chapter is: why do concave functions occupy such an important position in economics, engineering, management science, and applied optimization theory in general? A real-valued function of n variables defined over a convex subset of Euclidean n-dimensional space is *concave* iff (if and only if) the line segment joining any two points on the graph of the function lies on or below the graphs; a set is convex iff, given any two points in the domain of definition of the set, the line segment joining the two points also belongs to the set.

Returning to the question raised above, we suggest that the importance of concave functions perhaps rests on the following two properties: (i) a local maximizer for a concave function is also a global maximizer, and (ii) the usual first-order necessary conditions for maximizing a differentiable function f of n variables over an open set [i.e., x^* is a point such that the gradient vector of f vanishes so that $\nabla f(x^*) = 0$] are also sufficient to imply that x^* globally maximizes f if f is a concave function defined over a convex set. Various generalizations of concavity (studied in Chapter 3) preserve properties (i) and (ii), respectively. In Chapter 2, we also study two classes of functions that are more restrictive than the class of concave functions: *strictly concave* and *strongly concave* functions. Strictly concave functions have the following useful property, which strengthens property (i) above: (iii) A local maximizer for a strictly concave function is also the unique global maximizer. A function is *strictly concave* iff the line segment joining any two distinct points on the graph of the function lies below the graph of the function (with the obvious exception of the end points of the line segment). A function f is *strongly concave* iff it is equal to the sum of a concave function and a negative definite quadratic form, i.e., $f(x) = g(x) - \alpha x^T x$ for every x belonging to the convex domain of definition set, where g is a concave function, $\alpha > 0$ is a positive scalar, and $x^T x = \sum_{i=1}^{n} x_i^2$. Strongly concave functions also have a property (iii) above, and, in addition, in a neighborhood of a local maximizer, a strongly concave twice continuously differentiable function will have the curvature of a negative definite quadratic form. This property is useful in proving convergence of certain optimization algorithms, and it is also useful in enabling one to prove comparative statics theorems in economics; see Section 4.9 in Chapter 4.

We now describe the contents of each chapter.

Chapter 2 deals with concave functions and the two classes of functions that are stronger than concavity, namely, strictly and strongly concave functions. The first three sections of Chapter 2 develop alternative characterizations of concave functions. In addition to the definition of a concave function, there are three additional very useful characterizations of concavity: (i) the hypograph of the function (the graph of the function and the set in $(n + 1)$-dimensional space lying below the graph) is a convex set; (ii) the first-order Taylor series approximation to the function around any point in the domain of definition lies on or above the graph of the function (this characterization requires the existence of first-order partial derivatives of the function); (iii) the Hessian matrix of second-order partial derivatives of the function evaluated at each point in the domain of definition is a negative semidefinite matrix (this characterization requires the existence of continuous second-order partial derivatives of the function).

Section 2.3 of Chapter 2 also develops some composition rules for concave functions; e.g., a nonnegative sum of concave functions is a concave function or the pointwise minimum of a family of concave functions is a concave function, and so on. Additional composition rules are developed in Chapter 5. Section 2.3 also provides characterizations for strictly and strongly concave functions.

Section 2.4 derives the local-global maximizer properties of concave functions referred to earlier. As we stated before, these properties are probably the main reason for the importance of the concavity concept in applied optimization theory.

Section 2.5 deals with another extremely important topic from the viewpoint of applications, namely, concave mathematical programming problems. A *concave program* is a constrained maximization problem, where (i) the objective function being maximized is a concave function; (ii) the functions used to define equal to or greater than zero inequality constraints are concave functions; and (iii) the functions used to define any equality constraints are linear (or affine). If we have a concave program with once-differentiable objective and constraint functions, then it turns out that certain conditions due to Karush (1939) and Kuhn and Tucker (1951) involving the gradient vectors of the objective and constraint functions evaluated at a point x^* as well as certain (Lagrange) multipliers are *sufficient* to imply that x^* solves the concave programming problem; see Theorem 2.30. In addition, if a relatively mild constraint qualification condition is satisfied, then these same Karush-Kuhn-Tucker conditions are also *necessary* for x^* to solve the constrained maximization problem; see Theorem 2.29. The multipliers that appear in the Karush-Kuhn-Tucker conditions can often be given physical or economic interpretations: the multiplier (if unique) that corresponds to a particular constraint can be interpreted as the incremental change in the optimized objective function due to an incremental relaxation in the constraint. For further details and rigorous statements of this result, see Samuelson (1947, p. 132), Armacost and Fiacco (1974), and Diewert (1984). Another result in Section 2.5, Theorem 2.28, shows that a concave programming problem has a solution iff a certain Lagrangian saddle point problem (which is a maximization problem in the primal variables and a minimization in the dual multiplier variables) has a solution. This theorem, due originally to Uzawa (1958) and Karlin (1959), does not involve any differentiability conditions; some economic applications of it are pursued in the last section of Chapter 4.

Chapter 3 deals with generalized concave functions; i.e., functions that have some of the properties of concave functions but not all.

Section 3.1 defines the weakest class of generalized concave functions, namely, the class of quasiconcave functions. A function (defined over a convex subset of Euclidean n-dimensional space—throughout the book we make this domain assumption) is *quasiconcave* iff the values of the function along the line segment joining any two points in the domain of definition of the function are equal to or greater than the minimum of the function values at the end points of the line segment. Comparing the definition of a quasiconcave function with the definition of a concave function, it can be seen that a concave function is quasiconcave (but not vice versa). Recall that concave functions played a central role in optimization theory because of their extremum properties. Quasiconcave functions also have a useful extremum property, namely: every strict local maximizer of a quasiconcave function is a global maximizer (see Proposition 3.3). Quasiconcave functions also play an important role in the *generalized concave mathematical programming problem* (see Section 3.6), where the concave inequality constraints that occurred in the concave programming problem of Section 2.5 become quasiconcave inequality constraints. Finally, quasiconcave functions play a central role in economic theory since the utility functions of consumers and the production functions of producers are usually assumed to be quasiconcave functions (see Chapter 4 below).

Sections 3.1 and 3.2 provide various alternative characterizations of quasiconcavity in a manner that is analogous to the alternative characterizations of concavity that were developed in the opening sections of Chapter 2. Three alternative characterizations of quasiconcavity are as follows: (i) the upper level sets of the function are convex sets for each level (Definition 3.1); (ii) if the directional derivative of the function in any feasible direction is negative, then function values in that direction must be less than the value of the function evaluated at the initial point (this is the contrapositive to Theorem 3.11); and (iii) the Hessian matrix of second-order partial derivatives of the function evaluated at each point in the domain of definition is negative semidefinite in the subspace orthogonal to the gradient vector of the function evaluated at the same point in the domain of definition (Corollary 3.20). Characterization (ii) above requires once differentiability of the function, while characterization (iii) requires twice continuous differentiability over an open convex set *and* the existence of a nonzero gradient vector at each point in the domain of definition. The restriction that the gradient vector be nonzero can be dropped (see Theorem 3.22), but the resulting theorem requires an additional concept that probably will not be familiar, namely, the concept of a *semistrict local minimum*, explained in Definition 3.3. This concept is also needed to provide a characterization of semistrictly quasiconcave functions in the twice-differentiable case; see Theorem 3.22. On the other hand, the familiar concept of a *local minimum* is used to provide a characterization of strictly quasiconcave functions in the twice-differentiable case; see Theorem 3.26. In fact, all of the different types of generalized concave functions can be characterized by their local minimum or maximum behavior along line segments; see Diewert, Avriel, and Zang (1981) for the details.

Section 3.4 deals with the properties and uses of the class of semistrictly quasiconcave functions. A function is *semistrictly quasiconcave* iff for every two points in the domain of definition such that the function has unequal values at those two points, then the value of the function along the interior of the line segment joining the two points is greater than the minimum of the two end-point function values; see Definition 3.11. If the function is continuous (or merely upper semicontinuous so that its upper level sets are closed), then a semistrictly quasiconcave function is also quasiconcave (Proposition 3.30). It is easy to verify that a concave function is also semistrictly quasiconcave. Hence, in the continuous (or upper semicontinuous) case, the class of semistrictly quasiconcave functions lies between the concave and quasiconcave classes. An alternative characterization of the concept of semistrict quasiconcavity for continuous functions in terms of level set properties is given by Proposition 3.35: the family of upper level sets must be convex and each nonmaximal level set must be contained in the boundary of the corresponding upper level set (a maximal level set obviously must coincide with the corresponding upper level set). Semistrictly quasiconcave functions have the same extremum property that concave functions had, namely: any local maximizer for a semistrictly quasiconcave function is a global maximizer (Theorem 3.37). Semistrictly quasiconcave functions also play a role in consumer theory; see Section 4.5.

A more restrictive form of generalized concavity than semistrict quasiconcavity is strict quasiconcavity, discussed in Section 3.3 A function is *strictly quasiconcave* iff for every two distinct points in the domain of definition of the function the value of the function along the interior of the line segment joining the two points is greater than the minimum of the two end-point function values; see Definition 3.8. It is easy to verify that a strictly concave function is strictly quasiconcave and that a strictly quasiconcave function is semistrictly quasiconcave and quasiconcave. Strictly quasiconcave functions have the same extremely useful extremum property that strictly concave functions had: any local maximum is the unique global maximum. Strictly quasiconcave functions also play an important role in economics; see Section 4.6. Continuous strictly quasiconcave functions have strictly convex upper level sets (Proposition 3.28).

Section 3.5 deals with three new classes of generalized concave functions: (i) pseudoconcave, (ii) strictly pseudoconcave, and (iii) strongly pseudoconcave. These classes of functions are generalizations of the class of concave, strictly concave, and strongly concave functions, respectively. The three new classes of functions are usually defined only in the differentiable case (although nondifferentiable definitions exist in the literature and are referred to in the text).

A *pseudoconcave function* may be defined by the following property (the contrapositive to Definition 3.13): if the directional derivative of the function in any feasible direction is nonpositive, then function values in that direction must be less than or equal to the value of the function evaluated at the initial point. Pseudoconcave functions have the same important extremum property that concave functions had: if the gradient vector of a function is zero at a point, then that point is a global maximizer for the function (Theorem 3.39). A characterization of pseudoconcave functions in the twice continuously differentiable case is provided by Theorem 3.43.

A *strictly pseudoconcave function* may be defined by the following property (the contrapositive to Definition 3.13): if the directional derivative of the function in any feasible direction is nonpositive, then the function values in that direction must be less than the value of the function evaluated at the initial point. Strictly pseudoconcave functions have the same important extremum property that strictly concave functions had: if the gradient vector of a function is zero at a point, then that point is the unique global maximizer for the function (Theorem 3.39). A characterization of strictly pseudoconcave functions in the twice continuously differentiable case is provided by Theorem 3.43.

Strongly pseudoconcave functions are strictly pseudoconcave functions with the following additional property: if the directional derivative of the function in any feasible direction is zero, then the function diminishes (locally at least) at a quadratic rate in that direction. Recall that in the twice differentiable case, a strongly concave function could be characterized by having a negative definite Hessian matrix of second-order partial derivatives at each point in its domain of definition. In the twice differentiable case, a function is strongly pseudoconcave iff its Hessian matrix is negative definite in the subspace orthogonal to the gradient vector at each point in the domain of definition (Proposition 3.45). The property of strong pseudoconcavity is sometimes called *strong quasiconcavity* in the economics literature, and some economic applications of this concept are developed in Section 4.7.

It should be noted that all of our concavity and quasiconcavity concepts have *convex* and *quasiconvex* counterparts: a function f is convex (quasiconvex) iff $-f$ is concave (quasiconcave).

Chapter 3 is concluded by Section 3.6, which deals with generalizations of the concave programs studied in Section 2.5. An example shows that the Karlin–Uzawa Saddle Point Theorem for (not necessarily differentiable) concave programming problems cannot be readily generalized. However, for differentiable programs, the sufficiency of the Karush–Kuhn–Tucker conditions for concave problems can be generalized to programming problems involving objective and constraint functions that satisfy some type of generalized concavity property: Theorem 3.48 shows that the objective function need only be pseudoconcave, the equal to or greater than inequality constraint functions need only be quasiconcave, and the equality constraint functions need only be quasimonotonic. A function is *quasimonotonic* iff it is both quasiconcave and quasiconvex (inequality 3.35). Thus these pseudoconcavity, quasiconcavity, and quasimonotonic properties replace the earlier concavity and linearity properties that occurred in Theorem 2.30.

Chapter 4 deals with economic applications. We consider four models of economic behavior: (i) a producer's cost minimization problem, (ii) a consumer's utility maximization problem, (iii) a producer's profit maximization problem, and (iv) a model of national product maximization for an economy that faces world prices for the outputs it produces and is constrained by domestic resource availabilities. In the context of the above four models, we show how each of the types of generalized concavity studied in Chapters 2 and 3 arises in a natural way.

Chapter 4 also proves some economics *duality theorems*. Many problems in economics involve maximizing or minimizing a function subject to another functional constraint. If either the objective function or the constraint function is linear (or affine), then the optimized objective function may be regarded as a function of the parameters or coefficients (these are usually prices) of the linear function involved in the primal optimization problem. This optimized objective function, regarded as a function of the prices appearing in the primal problem, is called the *dual function*. Under certain conditions, this dual function may be used to reconstruct the nonlinear function that appeared in the primal optimization problem. The regularity conditions always involve some kind of generalized concavity restrictions on the nonlinear primal function. Some applications of these economics duality theorems are provided in Chapter 4.

Chapters 5 and 6 deal with the following important question: how can we recognize whether a given function has a generalized concavity property?

In the first part of Chapter 5, *composition rules* for the various types of generalized concave functions are derived. Suppose we know that certain functions have a generalized concavity property (or are even concave). Then under what conditions will an increasing or decreasing function of the original function or functions have a generalized concavity property?

In the second part of Chapter 5, we apply these composition rules to derive conditions under which a *product* or *ratio* of two or more functions has a generalized concavity property, provided that the original functions are concave or convex. Special attention is given to the case of products and ratios of only two functions. The material in this chapter draws heavily on the work of Schaible (1971, 1972).

Chapter 6 deals with the generalized concavity properties of an important class of functions, namely, the class of *quadratic* functions. It turns out that restricting ourselves to the class of quadratic functions simplifies life somewhat: quasiconcave and semistrictly quasiconcave quadratic functions cannot be distinguished. Furthermore, strictly and strongly pseudoconcave quadratic functions cannot be distinguished. However, even with these simplifications, the characterization of the generalized concavity properties of quadratic functions proves to be a rather complex task. Chapter 6 develops all known results using a unified framework (based on the composite function criteria developed in Chapter 5) on the generalized concavity properties of quadratic functions. Furthermore, many of the criteria are expressed in alternative ways using eigenvalues and eigenvectors or determinantal conditions. The material in this chapter summarizes and extends the work of Schaible (1981a, b).

Chapter 7 provides a brief survey of *fractional programming* and indicates how generalized concavity concepts play a role in this important applied area. A *fractional program* is a constrained maximization problem where the objective function is a ratio of two functions, say $f(x)/g(x)$, and the decision variables x are restricted to belong to a closed convex set S in finite-dimensional Euclidean space. In a *linear fractional program*, the functions f and g are both restricted to be linear or affine. In a *concave fractional program*, the numerator function f is restricted to be nonnegative and concave and the denominator function is restricted to be convex and positive over the constraint set S. In a *generalized fractional program*, we maximize a sum of ratios or we maximize the minimum of a finite number of ratios.

In Section 7.1, we show that the objective function in a concave fractional programming problem is semistrictly concave. Hence, a local maximum for the problem is a global maximum. If, in addition, the objective function in a concave fractional program is differentiable, we show that the objective function is pseudoconcave. In this latter case, the Karush–Kuhn–Tucker conditions are sufficient (and necessary if a constraint qualification condition is satisfied) to characterize the solution to the fractional programming problem.

Section 7.2 surveys a number of applications of fractional programming.

Business and economics applications of fractional programming include the following:

1. *Maximization of productivity.* The productivity of a firm, enterprise, or economy is usually defined as a function of outputs produced divided by a function of the inputs utilized by the firm.
2. *Maximization of the rate of return on investments.*
3. *Minimization of cost per unit of time.*
4. *Maximization of an economy's growth rate.* This problem originates in von Neumann's (1945) model of an expanding economy. The overall growth rate in the economy is the smallest of certain sectoral growth rates. Maximizing the minimum of the sectoral growth rates leads to a generalized fractional programming problem.
5. *Portfolio selection problems in finance.* Here we attempt to maximize the expected return of a portfolio of investments divided by the risk of the portfolio.

Applied mathematics applications of fractional programming include the following:

1. *Finding the maximal eigenvalue.* The maximal eigenvalue λ of a symmetric matrix A can be found by maximizing the ratio of two quadratic forms, i.e., $\lambda = \max_x \{x^T Ax/x^T x: x \neq 0\}$.
2. *Approximation theory.* Some problems in numerical approximation theory generate generalized fractional programs.
3. *Solution of large-scale linear programs.* Using decomposition methods, the solution to a large linear program can be reduced to

the solution of a finite number of subproblems. These subproblems are linear fractional programs.

4. *Solving stochastic programs.* Certain stochastic linear programming problems lead to fractional programming problems. This class of applications includes the portfolio selection problem mentioned above.

The above applications of fractional programming (and additional ones) are discussed in Section 7.2 and references to the literature are provided there.

In Section 7.3, we indicate how a concave fractional program may be transformed into a family of ordinary concave programs using a separation of variable technique. However, an even more convenient transformation is available. Propositions 7.2 and 7.3 show how concave fractional programs can be transformed into ordinary concave programs using a certain change of variables transformation. We also derive the (saddle point) dual programming problems for a concave fractional program in this section.

Section 7.4 concludes Chapter 7 by outlining some possible algorithmic approaches to the solution of concave fractional programs.

The material in Chapter 7 draws heavily on Schaible (1978, 1981c).

Chapter 8 introduces two new classes of generalized concave functions: transconcave functions and (h, ϕ)-concave functions.

A function f defined over a convex subset C of Euclidean n-dimensional space is *transconcave* (or *G-concave*) iff it can be transformed into a concave function by means of a monotonically increasing function of one variable G; hence f is G-concave iff $h(x) \equiv G[f(x)]$ is a concave function over C.

Transconcave functions are used in at least two important areas of application. The first use is in numerical algorithms for maximizing functions of n variables; if the objective function f in the nonlinear programming problem can be transformed into a concave function $G[f(x)]$ by means of an increasing function of one variable G, then the original objective function $f(x)$ may be replaced by the concave objective function $G[f(x)]$ and one of many concave programming algorithms may be used to solve the problem. A second use is in the computation of general equilibria in economic models where the number of consumers in the model is smaller than the number of commodities. In order to compute a general equilibrium (see Debreu, 1959, for a formal definition and references to the literature), an algorithm is required that will compute a fixed point under the hypotheses of the Kakutani (1941) Fixed Point theorem. Scarf (1967) has constructed such an algorithm, but it is not efficient if the number of commodities exceeds 50. However, if the preferences of all consumers in the general equilibrium model can be represented by means of concave utility functions, then Negishi (1960) showed how a general equilibrium could be computed by solving a sequence of concave programming problems, where the objective function in each problem was a weighted sum of individual utility functions. Diewert (1973) later showed how the Negishi framework could be simplified to a problem where it is necessary to find a fixed point in the space of utility weights. Hence, for an economic model where the number of consumer classes is smaller than the number of commodity classes, it will be more efficient to compute the fixed point over the space of nonnegative utility weights rather than over the space of nonnegative commodity prices.

It is obvious that transconcave functions must be at least quasiconcave. In fact, Proposition 8.1 shows that transconcave functions must be semistrictly quasiconcave. An example shows that the property of semistrict quasiconcavity is not sufficient to imply transconcavity. In the case of differentiable functions, Proposition 8.7 shows that transconcave functions must be pseudoconcave. Furthermore, Theorem 8.25 shows that a twice continuously differentiable strongly pseudoconcave function is transconcave. Hence, in the twice differentiable case, necessary and sufficient conditions for transconcavity lie between the properties of pseudoconcavity and strong pseudoconcavity.

In Section 8.2, we derive necessary and sufficient conditions for transconcavity in the twice differentiable case. Even in this differentiable case, the conditions are somewhat complex. [We do not attempt to treat the nondifferentiable case, which has been treated by Kannai (1977, 1981) in some detail.] The first necessary condition for transconcavity is pseudoconcavity. The next necessary condition is the existence of a negative semidefinite augmented Hessian matrix (Condition B). Let us explain what this condition means when $n = 1$, i.e., when $f(x)$ is a function of one variable. In this case, $G[f(x)]$ is a concave function over the convex set C iff the second derivative $G'[f(x)]f''(x) + G''[f(x)][f'(x)]^2 \leqq 0$ for every x belonging to C. Since we restrict ourselves to functions G such that $G'[f(x)] > 0$, we require $f''(x) - r(x)[f'(x)]^2 \leqq 0$, where $r(x) \equiv -G''[f(x)]/G'[f(x)]$. In this $n = 1$ case, Condition B becomes $f'(x) = 0$ implies $f''(x) \leqq 0$; i.e., if the first derivative of f is zero at a point x belonging to C, then the second derivative is nonpositive. This condition is already implied by Condition A, pseudoconcavity of f, when $n = 1$. For a general n, Condition B becomes: there exists a scalar $r(x)$ such that $\nabla^2 f(x) - r(x)\nabla f(x)\nabla f(x)^T$ is negative semidefinite for each x belonging to C, where $\nabla^2 f(x)$ is the $n \times n$ matrix of second-order partial derivatives of f evaluated at $x \equiv (x_1, \ldots, x_n)^T$ and $\nabla f(x)$ is the n-dimensional column vector of first-order partial derivatives of f evaluated at x. This condition may be phrased as a problem in matrix algebra: under what conditions on $A \equiv \nabla^2 f(x)$ and $b \equiv \nabla f(x)$ does there exist a scalar r such that $A - rbb^T$ is negative semidefinite? We answer this question in great detail in Section 8.2, providing several alternative and equivalent answers using bordered Hessian matrices, restricted eigenvalue criteria, and various determinantal conditions. This material may be of independent interest from the viewpoint of matrix algebra. (It should be mentioned that these matrix criteria assume that $n \geq 2$.) Given that an r exists that makes $A - rbb^T$ negative semidefinite, we derive three equivalent methods for computing the minimal r that will do the job; some of this material has not been published before. The two final necessary conditions for transconcavity of f in the twice-differentiable case, Conditions C and D, involve the boundedness of the minimal $r(x)$ such that $\nabla^2 f(x) - r(x)\nabla f(x)\nabla f(x)^T$ is negative semidefinite over x belonging to C. In the case $n = 1$, we require that $\sup_x \{f''(x)/[f'(x)]^2 : f'(x) \neq 0, x \in C\}$ be finite. This last condition is not implied by Condition A or B. Theorem 8.18 shows that Conditions A–D are necessary and sufficient for transconcavity.

Definition 8.2 defines a special class of transconcave or G-concave functions, namely, the class of r-concave functions. A function f defined over a convex set C is *r-concave* iff f is G-concave with $G(t) \equiv -e^{rt}$. This special class of transconcave functions plays an important role in Section 8.2.

The first two sections of Chapter 8 deal with the class of transconcave or G-concave functions. These are functions that can be transformed into concave functions by means of a monotonic transformation G of the images of the original function f. The last section of Chapter 8 deals with a more complicated class of functions, the class of (h, ϕ)-concave functions. A function f is *(h, ϕ)-concave* iff for every x^1 and x^2 belonging to C and scalar λ between 0 and 1, we have $\phi[f\{h^{-1}[(1-\lambda)h(x^1) + \lambda h(x^2)]\}] \geq (1-\lambda)\phi[f(x^1)] + \lambda\phi[f(x^2)]$, where ϕ is an increasing, continuous function of one variable and $h(x) \equiv \{h_1(x), \ldots, h_n(x)\}$ is a vector-valued function of n variables $x \equiv (x_1, \ldots, x_n)$ that is continuous and one-to-one over C. We also assume that the image set $h(C)$ is convex. Theorem 8.30 shows that f is (h, ϕ)-concave over the set C iff $\hat{f}(y) \equiv \phi[f\{h^{-1}(y)\}]$ is a concave function over the set $h(C)$, which we assume convex. Thus, f is (h, ϕ) concave if we can find a one-to-one and continuous function h such that $f\{h^{-1}(y)\}$ is a quasiconcave function and an increasing continuous function of one variable that further transforms $f\{h^{-1}\}$ into the concave function $\phi[f\{h^{-1}\}]$. It can be verified that a function f defined over a convex set C is (i) concave iff it is (h, ϕ)-concave with $h(x) \equiv x$ and $\phi(t) \equiv t$, (ii) G-concave iff it is (h, ϕ)-concave with $h(x) \equiv x$ and $\phi(t) \equiv G(t)$, and (iii) r-concave iff it is (h, ϕ)-concave with $h(x) \equiv x$ and $\phi(t) \equiv -\exp(-rt)$. Thus, the class of (h, ϕ)-concave functions contains our earlier classes of transconcave functions. Another interesting class of (h, ϕ)-concave functions is the class of (log, log)-concave functions. In this case, $h(x) \equiv [h_1(x_1, \ldots, x_n), \ldots, h_n(x_1, \ldots, x_n)] \equiv [\log x_1, \ldots, \log x_n]$ and $\phi(t) \equiv \log t$. Functions that are (log, log)-concave appear in a special branch of nonlinear programming called geometric programming.

Theorem 8.31 in Section 8.3 shows that differentiable (h, ϕ)-concave functions f have the same property that concave (and pseudoconcave) functions had; i.e., if there exists an x^* such that $\nabla f(x^*) = 0$, then x^* is the global maximizer for f. For a characterization of (h, ϕ)-concave functions in the twice-differentiable case, as well as some applications of (h, ϕ)-concavity to statistical decision making, see Ben-Tal (1977).

Chapter 9 provides an introduction to another two classes of generalized concave functions: the class of *F-concave functions* due to Ben-Tal and Ben-Israel (1976), and the class of *arcwise connected functions* due to Ortega and Rheinboldt (1970) and Avriel and Zang (1980).

In Chapter 2, it is shown that a differentiable concave function defined over an open convex set has the property that the first-order Taylor series approximation to the function around any point lies above the graph of the function, and moreover, this supporting hyperplane property can serve to characterize differentiable concave functions. Theorem 9.1 shows that this supporting hyperplane property can serve to characterize concave functions even in the nondifferentiable case. A natural generalization of concavity is obtained by considering functions whose graphs are supported from above by functions that are not necessarily linear. These support functions are drawn from a prespecified class of functions denoted by the set F. If F is defined to be the class of affine functions, then the class of F-concave functions reduces to the class of concave functions. In Section 9.1, we present several interesting examples of F-concave functions for various definitions of the supporting class of functions F.

Section 9.2 discusses various families of arcwise connected functions. In the definitions of the various kinds of concave and generalized concave functions, the behavior of the function along the straight line segment joining any two points in the domain of definition was restricted in some way. We define *quasiconnected*, *semistrictly quasiconnected*, *strictly quasiconnected*, *pseudoconnected*, and *strictly pseudoconnected* functions in a manner analogous to the definitions of quasiconcavity, semistrict quasiconcavity, strict quasiconcavity, pseudoconcavity, and strict pseudoconcavity, respectively, by replacing the straight line segments that occur in the latter definitions by *continuous arcs*. Moreover, the old convex domain of definition set is replaced by an *arcwise connected set*, and convex upper level sets are replaced by arcwise connected upper level sets. The extremal properties of the new classes of "connected" functions turn out to be virtually identical to the extremal properties of the corresponding class of quasiconcave functions. Since the "connected" classes of functions are much more general, the reader may well wonder why we devote so much space to the various classes of generalized concave functions and so little space to the "connected" functions. An answer is that it is much easier to characterize concave and generalized concave functions. For example, in the twice differentiable case, concave functions may be characterized by the local property of having a negative semidefinite Hessian matrix at each point of the domain of definition. A comparable local characterization for the various classes of "connected" functions does not appear to exist. Furthermore, there are no composition rules for "connected" functions that are analogous to the composition rules developed for generalized concave functions in Chapter 5. However, on the positive side, the reader will see that it is not very much more difficult to develop the theory of "connected" functions than it is to develop the properties of the corresponding classes of quasiconcave functions.

We conclude by noting that more detailed references to the literature on generalized concavity will appear at the ends of the respective chapters and in the Supplementary Bibliography.

References

ARMACOST, R.L., and FIACCO, A.V. (1974), Computational experience in sensitivity analysis for nonlinear programming, *Math. Programming* **6**, 301-326.

AVRIEL, M., and ZANG, I. (1980), Generalized arcwise connected functions and characterizations of local-global minimum properties, *J. Opt. Theory Appl.* **32**, 407-425.

BEN-TAL, A. (1977), On generalized means and generalized convexity, *J. Opt. Theory Appl.* **21**, 1-13.

BEN-TAL, A., and BEN-ISRAEL, A. (1976), A generalization of convex functions via support properties, *J. Australian Math. Soc.* **21**, 341-361.

DEBREU, G. (1959), *Theory of Value: An Axiomatic Analysis of Economic Equilibrium*, John Wiley, New York.

DIEWERT, W.E. (1973), On a theorem of Negishi, *Metroeconomica* **25**, 119-135.

DIEWERT, W.E. (1984), Sensitivity analysis in economics, *Comput. Oper. Res.* **11**, 141-156.

DIEWERT, W. E., AVRIEL, M., and ZANG, I. (1981), Nine kinds of quasiconcavity and concavity, *J. Economic Theory* **25**, 397-420.

KAKUTANI, S. (1941), A generalization of Brower's fixed point theorem, *Duke Mathematical Journal* **8**, 457-459.

KANNAI, Y. (1977), Concavifiability and the construction of concave utility functions, *J. Math. Econ.* **4**, 1-56.

KANNAI, Y. (1981), Concave utility functions—Existence, constructions and cardinality, in *Generalized Concavity in Optimization and Economics*, Edited by S. Schaible and W.T. Ziemba, Academic Press, New York, pp. 543-611.

KARLIN, S. (1959), *Mathematical Methods and Theory in Games, Programming and Economics*, Vol. 1, Addison-Wesley, Reading, Massachusetts.

KARUSH, W. (1939), Minima of functions of several variables with inequalities as side conditions, M.Sc. dissertation, Department of Mathematics, University of Chicago.

KUHN, H.W., and TUCKER, A.W. (1951), Nonlinear programming, in *Proceedings of the Second Berkeley Symposium on Mathematical Statistics and Probability*, Edited by J. Neyman, University of California Press, Berkeley, pp. 481-492.

NEGISHI, T. (1960), Welfare economics and existence of an equilibrium for a competitive economy, *Metroeconomica* **12**, 92-97.

ORTEGA, J.M., and RHEINBOLDT, W.C. (1970), *Interactive Solution of Nonlinear Equations in Several Variables*, Academic Press, New York.

SAMUELSON, P.A. (1947), *Foundation of Economic Analysis*, Harvard University Press, Cambridge, Massachusetts.

SCARF, H. (1967), The approximation of fixed points of a continuous mapping, *SIAM J. Appl. Math.* **15**, 1328-1343.

SCHAIBLE, S. (1971), Beiträge zur quasikonvexen Programmierung, Doctoral Dissertation, Köln.

SCHAIBLE, S. (1972), Quasiconvex optimization in general real linear spaces, *Z. Oper. Res.* **16**, 205-213.

SCHAIBLE, S. (1978), *Analyse und Anwendungen von Quotienten Programmen*, Hain-Verlag, Meisenheim.

SCHAIBLE, S. (1981a), Quasiconvex, pseudoconvex and strictly pseudoconvex quadratic functions, *J. Opt. Theory Appl.* **35**, 303-338.

SCHAIBLE, S. (1981b), Generalized convexity of quadratic functions, in *Generalized Concavity in Optimization and Economics*, Edited by S. Schaible and W.T. Ziemba, Academic Press, New York, pp. 183-197.

SCHAIBLE, S. (1981c), A survey of fractional programming, in *Generalized Concavity in Optimization and Economics*, Edited by S. Schaible and W.T. Ziemba, Academic Press, New York, pp. 417-440.

UZAWA, H. (1958), The Kuhn-Tucker theorem in concave programming, in *Studies in Linear and Nonlinear Programming*, Edited by K. Arrow, L. Hurwicz, and H. Uzawa, Stanford University Press, Stanford, California, pp. 32-37.

VON NEUMANN, J. (1945), A model of general economic equilibrium, *Rev. Econ. Stud.* **13**, 1-9.

2

Concavity

In this chapter we present some basic results on concave functions that will be generalized in subsequent chapters. Since the discussion of generalized concavity in this book is oriented toward applications in economics and optimization, only selected properties of concave functions pertinent to the above applications are mentioned here. The origins of the concepts of convex sets and concave and convex functions can be traced back to the turn of the century; see Hölder (1889), Jensen (1906), and Minkowski (1910, 1911).

2.1. Basic Definitions

The domain of a concave function is always a convex set. We therefore begin our formal discussion of concavity with the following definition.

Definition 2.1. A subset C of the n-dimensional real Euclidean space R^n is a convex set if for every $x^1 \in C$, $x^2 \in C$ and $0 \leq \lambda \leq 1$ we have $[\lambda x^1 + (1-\lambda)x^2] \in C$.

The study of convex sets is an important topic of mathematics in itself. We, however, restrict our discussion of convex sets to those properties that are relevant to concave functions and their generalizations. Readers interested in convex sets may find ample material on this subject in Eggleston (1958), Fenchel (1951), Lay (1982), Rockafellar (1970), or Valentine (1964).

Unless mentioned otherwise, we shall always refer to functions that are real-valued. This is somewhat restrictive, since in recent works in optimization one can find a treatment of functions that can also take on infinite values; we nevertheless sacrifice generality for the sake of simplicity. We then have the following definition.

Definition 2.2. A function f defined on the convex set $C \subset R^n$ is called concave if for every $x^1 \in C$, $x^2 \in C$, and $0 \leq \lambda \leq 1$ we have

$$f(\lambda x^1 + (1-\lambda)x^2) \geq \lambda f(x^1) + (1-\lambda)f(x^2). \tag{2.1}$$

It is called strictly concave if the inequality in (2.1) is strict for $x^1 \neq x^2$ and $0 < \lambda < 1$. If f is (strictly) concave then $g \equiv -f$ is a (strictly) convex function.

2.2. Single-Variable Concave Functions

In this section we present some elementary results on concave functions of a single variable. These results are shown here separately in order to assist the reader in an intuitive understanding of the more advanced topics that will follow. This is also the historical path along which the subject of concavity has been developed.

The definition of concave functions given in inequality (2.1) can be easily illustrated in the case where f is a function of a single variable. The left-hand side of the inequality in (2.1) is the expression for the values of $f(x)$ between the points x^1 and x^2, whereas the right-hand side is the values of the affine function $F(\lambda) \equiv \lambda f(x^1) + (1-\lambda)f(x^2)$ for $0 \leq \lambda \leq 1$; see Figure 2.1. Equivalently, if A, B, and C are any three points on the graph of f

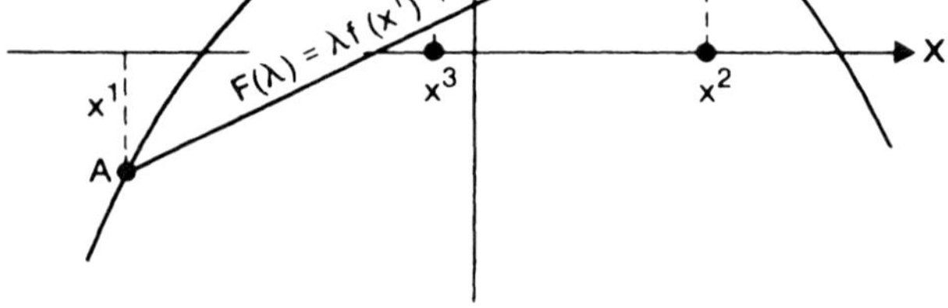

Figure 2.1. The "graph above the chord" property of concave functions.

such that B is between A and C, then B is on or above the line segment (chord) AC. Letting $A = f(x^1)$, $B = f(x^3)$, and $C = f(x^2)$, the reader can easily verify the following relationship for a concave function that holds for $x^1 \le x^3 \le x^2$:

$$\frac{f(x^3) - f(x^1)}{x^3 - x^1} \ge \frac{f(x^2) - f(x^1)}{x^2 - x^1} \ge \frac{f(x^2) - f(x^3)}{x^2 - x^3}. \tag{2.2}$$

Example 2.1. Here are some simple examples of concave functions:

1. $f(x) = \log x$, defined on the open convex set $C = \{x: x > 0\}$.
2. $f(x) = \sin x$, defined on the closed convex set $C = \{x: 0 \le x \le \pi\}$.

These two functions are actually strictly concave.

3. $f(x) = ax + b$, defined on the whole real line R.

This is an affine function that is both concave and convex (but, clearly, neither strictly concave nor strictly convex). Note that the three functions above are continuous on their domain of definition.

These functions are illustrated in Figure 2.2.

Example 2.2. Let $C = \{x: x \ge 0\}$ and let $f(0) = 0$, $f(x) = k_1 + k_2(x)^{0.5}$ for $x > 0$, where k_1, k_2 are given positive numbers. This function may represent the cost of manufacturing some commodity, where k_1 is the *fixed charge* or *setup cost*, which is incurred only when the commodity is actually being manufactured. This function is strictly concave, having a discontinuity at a boundary point of its domain.

In general, concave functions are continuous in the interior of their domain but may have discontinuities at boundary points. In particular, concave functions are continuous on an open interval, a result we show next. A function f is said to be Lipschitz continuous (Bartle, 1976) on an interval $[a, b]$ if for any two points $x \in [a, b]$, $y \in [a, b]$ there exists a constant K such that

$$|f(x) - f(y)| \le K|x - y|. \tag{2.3}$$

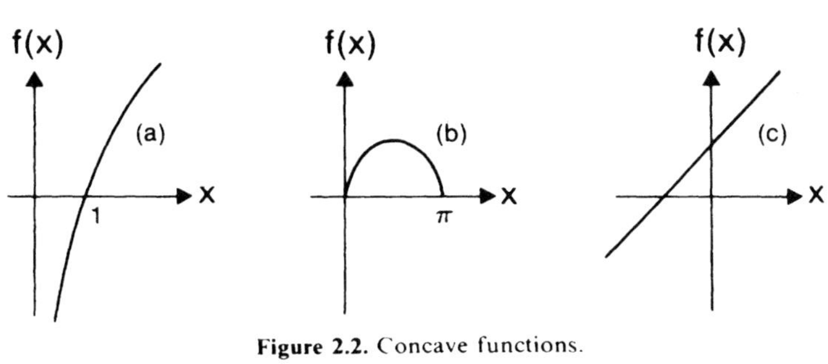

Figure 2.2. Concave functions.

It is easy to see that a Lipschitz continuous function is continuous.

Proposition 2.1 (Roberts and Varberg, 1973). Let f be a concave function on the convex set $C \subset R$. Then f is Lipschitz continuous on every closed interval contained in the interior of C, and consequently, f is continuous on the interior of C.

Proof. First, let $[a, b] \subset C$ be a closed interval and let $m = \min\{f(a), f(b)\}$. Then, for any $x = \lambda a + (1 - \lambda)b$, $0 \le \lambda \le 1$ we have

$$f(x) \ge \lambda f(a) + (1 - \lambda)f(b) \ge \lambda m + (1 - \lambda)m = m. \qquad (2.4)$$

Thus f is bounded from below by m on $[a, b]$. Similarly, let $(a - b)/2 \le t \le (b - a)/2$. Then, by the concavity of f

$$f\left(\frac{a+b}{2}\right) \ge \tfrac{1}{2}f\left(\frac{a+b}{2} + t\right) + \tfrac{1}{2}f\left(\frac{a+b}{2} - t\right) \qquad (2.5)$$

or

$$f\left(\frac{a+b}{2} + t\right) \le 2f\left(\frac{a+b}{2}\right) - f\left(\frac{a+b}{2} - t\right). \qquad (2.6)$$

But, since $\left(\frac{a+b}{2} - t\right) \in [a, b]$, we have

$$-f\left(\frac{a+b}{2} - t\right) \le -m \qquad (2.7)$$

and

$$f\left(\frac{a+b}{2} + t\right) \le 2f\left(\frac{a+b}{2}\right) - m = M. \qquad (2.8)$$

Thus f is bounded from above on $[a, b]$ by M. Now let $[a, b]$ be in the interior of C. Choose an $\varepsilon > 0$ such that $[a - \varepsilon, b + \varepsilon] \subset C$ and let $\bar{m}$ and $\bar{M}$ be the lower and upper bounds for f on $[a - \varepsilon, b + \varepsilon]$. We assume that

f is nonconstant on $[a-\varepsilon, b+\varepsilon]$, hence $\bar{m} < \bar{M}$. Let $x \in [a, b]$, $y \in [a, b]$, $x \neq y$, and define

$$z = y - \frac{\varepsilon}{|y-x|}(x-y), \qquad \lambda = \frac{|x-y|}{\varepsilon + |x-y|}. \tag{2.9}$$

Then, $z \in [a-\varepsilon, b+\varepsilon]$, $y = \lambda z + (1-\lambda)x$, and

$$f(y) \geq \lambda f(z) + (1-\lambda)f(x) = \lambda[f(z) - f(x)] + f(x). \tag{2.10}$$

Hence,

$$f(x) - f(y) \leq \lambda[f(x) - f(z)] \leq \lambda(\bar{M} - \bar{m}) < \frac{|x-y|}{\varepsilon}(\bar{M} - \bar{m}) \tag{2.11}$$

and, since x and y are arbitrary points in $[a, b]$, we conclude that

$$|f(x) - f(y)| \leq K|x-y|, \tag{2.12}$$

where $K = (\bar{M} - \bar{m})/\varepsilon$. Thus f is Lipschitz continuous on any closed interval contained in the interior of C, and, consequently, f is continuous on the interior of C. □

Turning now to the differential properties of concave functions of a single variable, it is convenient to start with considering the left and right derivatives of a function f defined on $C \subset R$ at a point $x^0 \in C$. The left and right derivatives are, respectively, defined as

$$f'_-(x^0) = \lim_{t \to 0^-} \frac{f(x^0 + t) - f(x^0)}{t} \tag{2.13}$$

$$f'_+(x^0) = \lim_{t \to 0^+} \frac{f(x^0 + t) - f(x^0)}{t} \tag{2.14}$$

where $t \to 0^-$ and $t \to 0^+$ mean that t approaches 0 through negative and positive numbers, respectively. If $f'_-(x^0) = f'_+(x^0)$, the common value is called $f'(x^0)$, the derivative of f at x^0, and given by

$$f'(x^0) = \lim_{t \to 0} \frac{f(x^0 + t) - f(x^0)}{t}. \tag{2.15}$$

Example 2.3. Let $f(x) = x$ for $0 \le x \leqq 1$ and $f(x) = \frac{1}{2}x + \frac{1}{2}$ for $1 \le x \le 3$. This may be a cost function for manufacturing certain goods with economies of scale. It is a piecewise linear concave function that does not have a derivative at $x = 1$. However, it has a left and right derivative there; $f'_-(1) = 1$, $f'_+(1) = \frac{1}{2}$.

If f is concave on C then by the inequalities in (2.2), Fenchel (1951) showed that the left and right derivatives of f exist at every point in the interior of C. For more on one-sided derivatives the reader is referred to Avriel (1976) and Roberts and Varberg (1973).

The last two examples of concave functions are shown in Figure 2.3.

In the next theorems we examine some properties of differentiable concave functions and present characterizations using first and second derivatives.

Definition 2.3. Let f be a function defined on the open set $U \subset R$ and let $x^0 \in U$. Then f is said to be differentiable at x^0 if there exists a real number $\tau(x^0)$ such that for all $x \in R$ satisfying $x^0 + x \in U$

$$f(x^0 + x) = f(x^0) + \tau(x^0)x + \alpha(x^0, x)|x| \tag{2.16}$$

where α is a real function such that $\lim_{x\to 0} \alpha(x^0, x) = 0$. Moreover, f is said to be differentiable on U if it is differentiable at every $x^0 \in U$.

It can be shown (see, for example, Bartle, 1976), that if f is differentiable at x^0, then f is continuous at x^0 and $\tau(x^0)$ is equal to $f'(x^0)$, the derivative of f as defined above.

Theorem 2.2. Let f be a differentiable function on the open convex set $C \subset R$. It is concave if and only if for every $x^0 \in C$, $x \in C$ we have

$$f(x) \le f(x^0) + f'(x^0)(x - x^0). \tag{2.17}$$

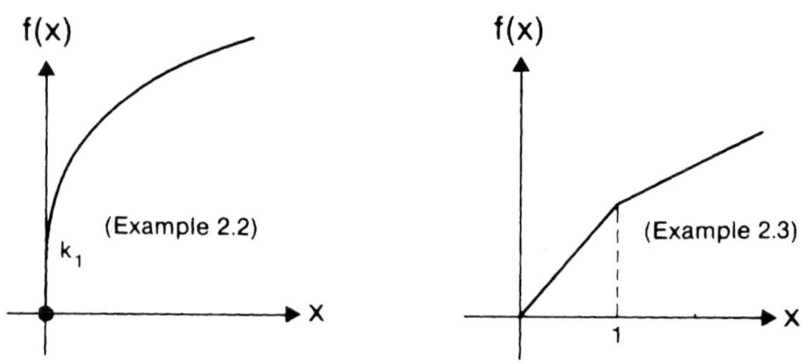

Figure 2.3. Additional concave functions.

It is strictly concave if and only if the inequality in (2.17) is strict for $x \neq x^0$.

Proof. Suppose that f is concave. Without loss of generality assume that $x < x^0$. Let t be a sufficiently small positive number such that $x^0 + t \in C$. Then, by the definition of concave functions and by letting $x^0 = \lambda x + (1-\lambda)(x^0 + t)$ we get

$$f(x^0) \geq \frac{t}{x^0 + t - x} f(x) + \frac{x^0 - x}{x^0 + t - x} f(x^0 + t). \tag{2.18}$$

Subtracting $f(x)$ from both sides of (2.18) and dividing by $x^0 - x$ yields

$$\frac{f(x^0) - f(x)}{x^0 - x} \geq \frac{f(x^0 + t) - f(x)}{x^0 + t - x} \geq \frac{f(x^0 + t) - f(x^0)}{t}. \tag{2.19}$$

Letting $t \to 0$ we obtain

$$\frac{f(x^0) - f(x)}{x^0 - x} \geq f'(x^0) \tag{2.20}$$

yielding (2.17) by a few simple algebraic manipulations.

Conversely, let x^1 and x^2 be any two points in C and let $x^3 = \lambda x^1 + (1-\lambda)x^2$ for some $0 < \lambda < 1$. Then, by (2.17)

$$f(x^1) \leq f(x^3) + (1-\lambda) f'(x^3)(x^1 - x^2) \tag{2.21}$$

$$f(x^2) \leq f(x^3) + \lambda f'(x^3)(x^2 - x^1). \tag{2.22}$$

Multiplying (2.21) and (2.22) by λ and $(1-\lambda)$, respectively, and adding up, we get

$$f(x^3) = f(\lambda x^1 + (1-\lambda)x^2) \geq \lambda f(x^1) + (1-\lambda) f(x^2) \tag{2.23}$$

for $0 \leq \lambda \leq 1$ and f is concave. The strictly concave case can be proven in a similar way. □

Inequality (2.17) also has a simple interpretation. The right-hand side of (2.17) is the linear approximation (Taylor expansion) of f around the point x^0. The inequality then says that the linear Taylor expansion of a concave function never underestimates the function values of f at any other point or, equivalently, that f lies on or below the affine function tangent to f at every point $x^0 \in C$; see Figure 2.4. We can also characterize differentiable concave functions by the monotonic behavior of their derivatives.

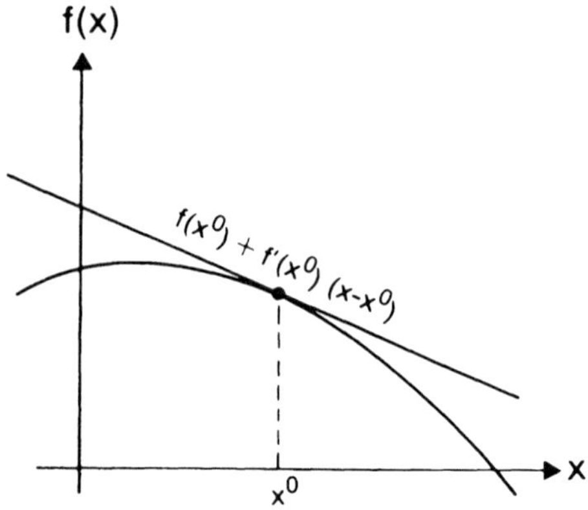

Figure 2.4. The "graph below the tangent" property of concave functions.

Theorem 2.3 (Fenchel, 1951). Let f be a differentiable function on the open convex set $C \subset R$. It is concave (strictly concave) if and only if f' is a nonincreasing (decreasing) function.

Proof. Suppose f is (strictly) concave and let x^1, x^2 be two points in C such that $x^1 < x^2$. Then, by (2.17)

$$\frac{f(x^2) - f(x^1)}{x^2 - x^1} \geq f'(x^2) \tag{2.24}$$

and

$$f'(x^1) \geq \frac{f(x^2) - f(x^1)}{x^2 - x^1}. \tag{2.25}$$

Hence

$$f'(x^1) \geq f'(x^2) \tag{2.26}$$

and f' is nonincreasing. The inequalities in (2.24)–(2.26) are strict if f is strictly concave. Conversely, let $x^1 \in C$, $x^2 \in C$, $x^1 < x^2$ and $x^3 = \lambda x^1 + (1 - \lambda)x^2$ for some $0 < \lambda < 1$.

By the Mean Value Theorem (see Bartle, 1976),

$$f(x^2) = f(x^3) + \lambda f'(\bar{x})(x^2 - x^1), \qquad x^3 < \bar{x} < x^2 \tag{2.27}$$

$$f(x^3) = f(x^1) + (1 - \lambda)f'(\bar{\bar{x}})(x^2 - x^1), \qquad x^1 < \bar{\bar{x}} < x^3. \tag{2.28}$$

If f' is nonincreasing, we have

$$f(x^3) \geq f(x^1) + (1-\lambda)f'(\bar{x})(x^2 - x^1) \tag{2.29}$$

and if f' is decreasing, the inequality in (2.29) is strict. Multiplying (2.27) and (2.29) by $(\lambda - 1)$ and λ, respectively, and adding up, we obtain

$$\lambda f(x^3) + (\lambda - 1)f(x^2) \geq \lambda f(x^1) + (\lambda - 1)f(x^3) \tag{2.30}$$

or

$$f(x^3) = f(\lambda x^1 + (1-\lambda)x^2) \geq \lambda f(x^1) + (1-\lambda)f(x^2). \tag{2.31}$$

Thus, if f' is nonincreasing (decreasing), then f is concave (strictly concave). □

Note that instead of saying that f' is nonincreasing, we could also say that for every two points $x^1 \in C$, $x^2 \in C$

$$[f'(x^2) - f'(x^1)](x^2 - x^1) \leq 0 \tag{2.32}$$

and f' is decreasing if and only if the inequality in (2.32) is strict. This last inequality is the one that can be easily extended to the multidimensional case if an analog of monotonicity of derivatives is sought.

Let us turn now our attention to the characterization of concave functions in terms of their second derivative f''. We have

Theorem 2.4 (Fenchel, 1951). Let f be a function on the open convex set $C \subset R$. Suppose that f'' exists on C. Then f is a concave function if and only if $f''(x) \leq 0$ for every $x \in C$. If $f''(x) < 0$ for every $x \in C$, then f is strictly concave.

Proof. By Theorem 2.3, f is concave if and only if f' is a nonincreasing function on C, that is, $f''(x) \leq 0$ for every $x \in C$. If $f''(x) < 0$ for every $x \in C$, then f' is decreasing on C, hence f is strictly concave. □

The last statement in Theorem 2.4 cannot be reversed. In fact, one can easily find examples of strictly concave functions whose second derivative vanishes at some points. Take, for example, $f(x) = -(x)^4$ on $C = R$, a strictly concave function [since $f'(x) = -4(x)^3$ is decreasing]. The second derivative is $f''(x) = -12(x)^2$ and, of course, $f''(0) = 0$.

2.3. Concave Functions of Several Variables

Most of the results of this section are straightforward extensions of the one-dimensional case. Before we present them, however, let us view concave functions from a new angle.

We have seen that concave functions and convex sets are related: the domain of a concave function is a convex set. Another relationship between concave functions and convex sets is given below. Let f be a function defined on the convex set $C \subset R^n$. The set

$$H(f) = \{(x, \alpha): x \in C, \alpha \in R, f(x) \geq \alpha\} \tag{2.33}$$

in R^{n+1} is called the *hypograph* of f. It is the set of all points lying on or below the graph of f. Similarly, the *epigraph* of f is the set

$$P(f) = \{(x, \alpha): x \in C, \alpha \in R, f(x) \leq \alpha\}. \tag{2.34}$$

The hypograph of a concave function is illustrated in Figure 2.5. We have then the following proposition.

Proposition 2.5 (Fenchel, 1951). Let f be a function defined on a convex set $C \subset R^n$. Then f is concave if and only if its hypograph $H(f)$ is a convex set. Similarly, f is convex if and only if its epigraph $P(f)$ if a convex set.

Proof. We prove the proposition for concave functions only. Suppose that f is concave and (x^1, α^1), (x^2, α^2) the two points in $H(f)$. Then for every $0 \leq \lambda \leq 1$

$$f(\lambda x^2 + (1 - \lambda)x^2) \geq \lambda f(x^1) + (1 - \lambda)f(x^2) \geq \lambda\alpha^1 + (1 - \lambda)\alpha^2. \tag{2.35}$$

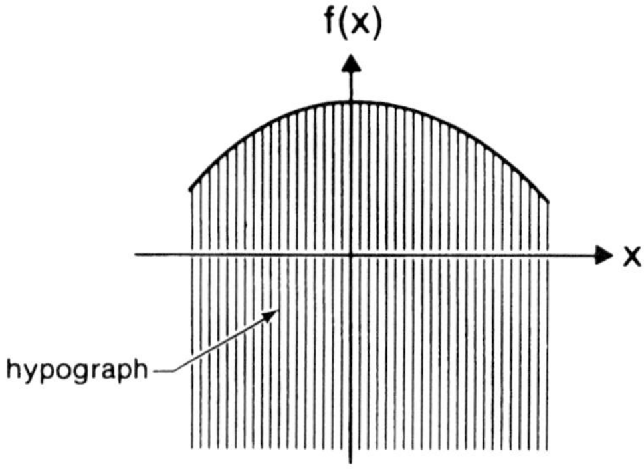

Figure 2.5. Hypograph of a concave function.

Thus the point

$$(x, \alpha) = (\lambda x^1 + (1 - \lambda)x^2, \lambda\alpha^1 + (1 - \lambda)\alpha^2) \tag{2.36}$$

is also in $H(f)$. Conversely, if $H(f)$ is a convex set, then for every two points $(x^1, \alpha^1) \in H(f)$, $(x^2, \alpha^2) \in H(f)$ and $0 \le \lambda \le 1$ we have $(\lambda x^1 + (1 - \lambda)x^2, \lambda\alpha^1 + (1 - \lambda)\alpha^2) \in H(f)$. Letting $\alpha^1 = f(x^1)$, $\alpha^2 = f(x^2)$ we obtain

$$f(\lambda x' + (1 - \lambda)x^2) \ge \lambda f(x^1) + (1 - \lambda)f(x^2), \tag{2.37}$$

and f is concave. □

For any function f on C and any $\alpha \in R$ the set $U(f, \alpha)$ defined by

$$U(f, \alpha) = \{x: x \in C, f(x) \ge \alpha\} \tag{2.38}$$

is called the *upper level set* of f. Similarly, the set

$$L(f, \alpha) = \{x: x \in C, f(x) \le \alpha\} \tag{2.39}$$

is called the *lower level set* of f. Note that the upper and lower level sets of f are, respectively, projections of the hypographs and epigraphs of f on C. We then obtain from Proposition 2.5 the following corollary.

Corollary 2.6 (Fenchel, 1951). Let f be a concave (convex) function on $C \subset R^n$. Then its upper (lower) level sets are convex sets for every real number α.

The notion of upper level sets is illustrated for the concave function $f(x) = \log x$ in Figure 2.6.

Note that the implications of the last corollary cannot be reversed—that is, convexity of the upper- or lower-level sets does not imply concavity or convexity of the function. In fact, we shall define later some families of functions, more general than concave functions, by the convexity of their level sets.

Upper level sets are a special case of more general sets, in terms of which we can characterize concave functions. For a given $\xi \in R^n$ and $\alpha \in R$ define the *generalized upper level set* of a real-valued function f on C by

$$\mathrm{GU}(f, \xi, \alpha) = \{x: x \in C, f(x) \ge \xi^T x + \alpha\} \tag{2.40}$$

where T denotes transpose. Note that $U(f, \alpha)$ as given by (2.38) is $\mathrm{GU}(f, 0, \alpha)$. A generalized upper level set for the function $f(x) = \log x$ is shown in Figure 2.7.

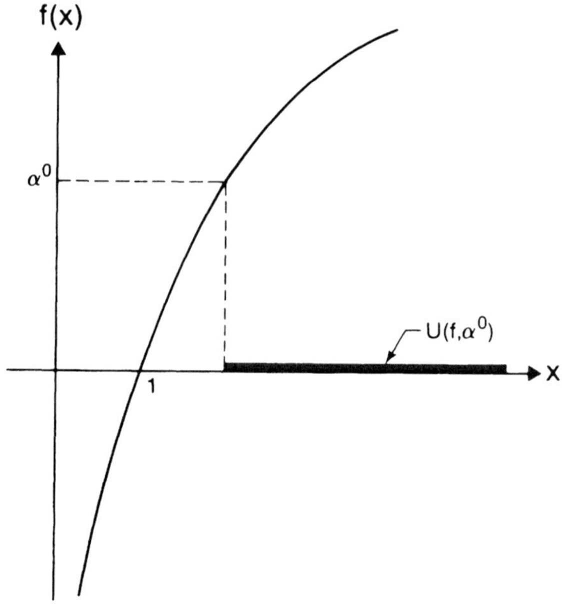

Figure 2.6. Upper-level set.

Proposition 2.7. The function f, defined on the convex set $C \subset R^n$, is concave if and only if $\mathrm{GU}(f, \xi, \alpha)$ is a convex set for every $\xi \in R^n$ and $\alpha \in R$.

Proof. Suppose that f is concave and let $x^1 \in \mathrm{GU}(f, \xi^0, \alpha^0)$, $x^2 \in \mathrm{GU}(f, \xi^0, \alpha^0)$ for some ξ^0, α^0. Then

$$f(x^1) \geq (\xi^0)^T x^1 + \alpha^0 \tag{2.41}$$

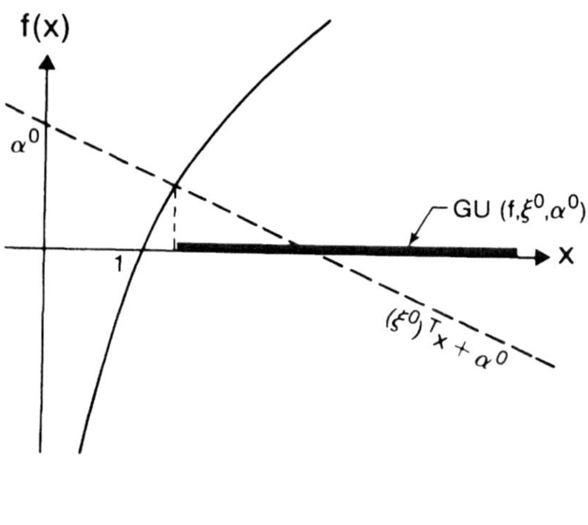

Figure 2.7. A generalized upper-level set.

$$f(x^2) \geq (\xi^0)^T x^2 + \alpha^0 \tag{2.42}$$

and

$$f(\lambda x^1 + (1-\lambda)x^2) \geq \lambda f(x^1) + (1-\lambda)f(x^2). \tag{2.43}$$

It follows from (2.41)-(2.43) that

$$f(\lambda x^1 + (1-\lambda)x^2) \geq \lambda[(\xi^0)^T x^1 + \alpha^0] + (1-\lambda)[(\xi^0)^T x^2 + \alpha^0] \tag{2.44}$$

$$= (\xi^0)^T[\lambda x^1 + (1-\lambda)x^2] + \alpha^0. \tag{2.45}$$

Hence $\mathrm{GU}(f, \xi^0, \alpha^0)$ is a convex set.

Conversely, let $\mathrm{GU}(f, \xi, \alpha)$ be convex for every $\xi \in R^n$ and $\alpha \in R$. Let x^1 and x^2 be two distinct points of C, the convex domain of f. Without loss of generality assume that $x_1^1 \neq x_1^2$. Suppose that f is not concave, that is, for some $0 < \bar{\lambda} < 1$, and $\bar{x} = \bar{\lambda}x^1 + (1-\bar{\lambda})x^2$ we have

$$f(\bar{x}) < \bar{\lambda}f(x^1) + (1-\bar{\lambda})f(x^2). \tag{2.46}$$

Then, for α^0, ξ^0 such that

$$\alpha^0 = \frac{x_1^1 f(x^2) - x_1^2 f(x^1)}{x_1^1 - x_1^2} \tag{2.47}$$

$$\xi_1^0 = \frac{f(x^1) - f(x^2)}{x_1^1 - x_1^2}, \qquad \xi_j^0 = 0, \qquad j = 2, \ldots, n \tag{2.48}$$

we have

$$(\xi^0)^T\bar{x} + \alpha^0 = (\xi^0)^T[\bar{\lambda}x^1 + (1-\bar{\lambda})x^2] + \alpha^0 \tag{2.49}$$

$$= \xi_1^0[\bar{\lambda}(x_1^1 - x_1^2) + x_1^2] + \alpha^0 \tag{2.50}$$

$$= \bar{\lambda}[f(x^1) - f(x^2)] + \frac{f(x^1) - f(x^2)}{x_1^1 - x_1^2}x_1^2 + \frac{x_1^1 f(x^2) - x_1^2 f(x^1)}{x_1^1 - x_1^2} \tag{2.51}$$

$$= \bar{\lambda}[f(x^1) - f(x^2)] + f(x^2) \tag{2.52}$$

$$> f(\bar{x}) \tag{2.53}$$

where (2.53) follows from (2.46). Hence $\bar{x} \notin \mathrm{GU}(f, \xi^0, \alpha^0)$. But

$$(\xi^0)^T x^1 + \alpha^0 = f(x^1) \tag{2.54}$$

$$(\xi^0)^T x^2 + \alpha^0 = f(x^2) \tag{2.55}$$

and $x^1 \in \mathrm{GU}(f, \xi^0, \alpha^0)$, $x^2 \in \mathrm{GU}(f, \xi^0, \alpha^0)$, contradicting that $\mathrm{GU}(f, \xi^0, \alpha^0)$ is a convex set. □

We have already seen in Example 2.2 that a concave function may have discontinuities at boundary points of its domain. It is easy to see that the hypograph of such a function is not a closed convex set.

Definition 2.4. A function f defined on the convex set $C \subset R^n$ is called a closed concave function if its hypograph $H(f)$ is a closed convex subset of R^{n+1}.

It follows that if f is a closed concave function then its upper-level sets $U(f, \alpha)$ are closed convex subsets of R^n for every real α.

Next we derive and prove a result stating that concave functions are continuous on open sets. The line of proof follows that of Roberts and Varberg (1973). First we need the following lemma.

Lemma 2.8. Let f be a concave function on the open convex set $C \subset R^n$. If f is bounded from below in a neighborhood of one point $x^0 \in C$, then it is locally bounded—that is, each $x \in C$ has a neighborhood on which f is bounded.

Proof. We first show that if f is bounded from below in a neighborhood of x^0, it is also bounded from above in the same neighborhood. Suppose that f is bounded from below by a number m in $N_\varepsilon(x^0) = \{z: z \in C, \|z - x^0\| < \varepsilon\}$. We can express every $z \in N_\varepsilon(x^0)$ as $z = x^0 + \theta y$, where $y \in R^n$ is a vector such that $\|y\| = 1$ and θ is a sufficiently small positive number. Then

$$x^0 = \tfrac{1}{2}(x^0 + \theta y) + \tfrac{1}{2}(x^0 - \theta y) \tag{2.56}$$

and

$$f(x^0) \geq \tfrac{1}{2} f(x^0 + \theta y) + \tfrac{1}{2} f(x^0 - \theta y), \tag{2.57}$$

$$2f(x^0) - f(x^0 - \theta y) \geq f(x^0 + \theta y). \tag{2.58}$$

By the hypothesis $f(x^0 - \theta y) \geq m$, hence

$$2f(x^0) - m \geq f(x^0 + \theta y) = f(z) \tag{2.59}$$

and f is bounded from above for every $z \in N_\varepsilon(x^0)$.

Let $x \in C$, $x \neq x^0$. Then $x = x^0 + \alpha y$, where again $y \in R^n$, $\|y\| = 1$ and α is a positive number. Choose $\rho > \alpha$ such that $u = x^0 + \rho y \in C$ and let $\lambda = \alpha/\rho$. Then,

$$N_\delta(x) = \{v\colon v \in C, v = (1-\lambda)z + \lambda u, z \in N_\varepsilon(x^0)\} \tag{2.60}$$

is a neighborhood of x with radius $\delta = (1-\lambda)\varepsilon$. Also, for $v \in N_\delta(x)$

$$f(v) \geq (1-\lambda)f(z) + \lambda f(u) \geq (1-\lambda)m + \lambda f(u). \tag{2.61}$$

That is, f is bounded from below on $N_\delta(x)$, and by the first part of the proof, f is also bounded from above on $N_\delta(x)$. □

A function f defined on an open set $U \subset R^n$ is said to be *locally Lipschitz continuous* if at each $x^0 \in U$ there is a neighborhood $N_\varepsilon(x^0)$ and a constant $K(x^0)$ such that if $x^1 \in U$, $x^2 \in U$, then

$$|f(x^1) - f(x^2)| \leq K(x^0)\|x^1 - x^2\|. \tag{2.62}$$

Note that locally Lipschitz continuity implies continuity. Now we have the following proposition.

Proposition 2.9 (Fenchel, 1951). Let f be a concave function on the open convex set $C \subset R^n$. If f is bounded from below in a neighborhood of one point of C, then f is locally Lipschitz continuous in C, hence continuous on C.

Proof. By Lemma 2.8, for every $x^0 \in C$ we can find a neighborhood $N_{2\varepsilon}(x^0)$ on which f is bounded, say by M, that is, $x \in N_{2\varepsilon}(x^0)$ implies $|f(x)| \leq M$. We now show that (2.62) holds on $N_\varepsilon(x^0)$. Suppose, to the contrary, that we can find points $x^1 \in N_\varepsilon(x^0)$, $x^2 \in N_\varepsilon(x^0)$ such that

$$\frac{f(x^1) - f(x^2)}{\|x^1 - x^2\|} > \frac{2M}{\varepsilon}. \tag{2.63}$$

Let now $x^3 \in N_{2\varepsilon}(x^0)$ be the point such that $x^2 = \lambda x^3 + (1-\lambda)x^1$ and $\|x^3 - x^2\| = \varepsilon$. Restricting f to the line segment $[x^2 - x^3]$ it follows from (2.2) that

$$\frac{f(x^2) - f(x^3)}{\|x^3 - x^2\|} \geq \frac{f(x^1) - f(x^2)}{\|x^1 - x^2\|} > \frac{2M}{\varepsilon} \tag{2.64}$$

implying that $f(x^2) - f(x^3) > 2M$ and hence $|f(x^3)| > M$, a contradiction, and f must be locally Lipschitz continuous at every $x^0 \in C$, hence continuous on C. □

The last result can be generalized to show that concave functions are continuous on the relative interior of their domain. To state this more general result we need a few preliminaries.

Let A be a subset of R^n such that for every two points $x^1 \in A$, $x^2 \in A$ and for every real number α, also $\alpha x^1 + (1 - \alpha)x^2 \in A$. Such a set is called an *affine set.* Clearly, affine sets are convex. Examples of affine sets in R^n are single points, lines, hyperplanes, and the whole space R^n. For a convex set $C \subset R^n$ the intersection of all affine sets containing C is called the *affine hull* of C. The *relative interior* of C, denoted by ri C, is the interior of C relative to its affine hull. The interior of C (relative to R^n) may be empty although ri C is nonempty as in Example 2.4 below. Generally this is the case if the affine hull is a proper subset of R^n. If the affine hull of C is R^n, then the interior of C is equal to the relative interior of C.

Example 2.4. Let C be the unit (two-dimensional) circle in three-dimensional space R^3—that is, let

$$C = \{x: x \in R^3, (x_1)^2 + (x_2)^2 \leq 1, x_3 = 0\}. \tag{2.65}$$

This set has no interior (relative to R^3) since one cannot find an open sphere in R^3 contained in C. The affine hull of C is the (x_1, x_2) plane and

$$\text{ri } C = \{x: x \in R^3, (x_1)^2 + (x_2)^2 < 1, x_3 = 0\}, \tag{2.66}$$

that is, the interior of C relative to its affine hull.

We can state now the following proposition.

Proposition 2.10. If f is a concave function defined on the convex set $C \subset R^n$ then it is continuous on ri C. In particular, if C is an open convex set then f is continuous on C.

Proof of this theorem requires additional background material and will be omitted here. Readers interested in the proof are referred to Fenchel (1951) or Rockafellar (1970).

An important property of concave functions useful in proving several results is that a function of n variables is concave if and only if it is concave on every line segment included in its domain. Formally we have the following.

Proposition 2.11. The function f defined on the convex set $C \subset R^n$ is concave if and only if for every $x^1 \in C$, $x^2 \in C$ the single-variable function F, defined by

$$F(\lambda) \equiv f(\lambda x^1 + (1-\lambda)x^2) \tag{2.67}$$

is concave for every $0 \le \lambda \le 1$.

The proof is straightforward and is left for the reader. Proposition 2.11 enables us to extend several results holding for concave functions of a single variable to the n-variable case.

Let us turn now to differentiable concave functions. First we need to define differentiable functions on subsets of R^n.

Definition 2.5. Let f be a function defined on the open set $U \subset R^n$ and let $x^0 \in U$. Then f is said to be differentiable at x^0 if there exists a vector $T(x^0)$ such that for all $x \in R^n$ satisfying $x^0 + x \in U$ we have

$$f(x^0 + x) = f(x^0) + x^T T(x^0) + \alpha(x^0, x)\|x\| \tag{2.68}$$

where α is a real function such that $\lim_{x\to 0} \alpha(x^0, x) = 0$. Moreover, f is said to be differentiable on U if it is differentiable at every $x^0 \in U$.

As in the single-variable case, if f is differentiable at x^0 then f is continuous there and $T(x^0)$ is equal to $\nabla f(x^0)$, the *gradient vector* of f at x^0 (see, e.g., Bartle, 1976), where

$$\nabla f(x^0) = \begin{bmatrix} \partial f(x^0)/\partial x_1 \\ \vdots \\ \partial f(x^0)/\partial x_n \end{bmatrix}. \tag{2.69}$$

The next result is a direct extension of Theorem 2.2.

Theorem 2.12 (Avriel, 1976; Mangasarian, 1969). Let f be a differentiable function on the open convex set $C \subset R^n$. It is concave if and only if for every $x^0 \in C$, $x \in C$ we have

$$f(x) \le f(x^0) + (x - x^0)^T \nabla f(x^0). \tag{2.70}$$

It is strictly concave if and only if the inequality in (2.70) is strict for $x \ne x^0$.

Proof. Let f be concave on C and let $x^0 \in C$, $x \in C$, such that $x^0 \ne x$. It follows that for $0 < \lambda \le 1$

$$f(x^0 + \lambda(x - x^0)) = f(\lambda x + (1-\lambda)x^0) \ge \lambda f(x) + (1-\lambda)f(x^0) \tag{2.71}$$

or

$$f(x) - f(x^0) \le (1/\lambda)[f(x^0 + \lambda(x - x^0)) - f(x^0)]. \tag{2.72}$$

Substituting (2.68) into the last inequality we obtain

$$f(x) - f(x^0) \le (x - x^0)^T \nabla f(x^0) + \alpha(x^0, \lambda(x - x^0))\|x - x^0\|. \tag{2.73}$$

Since $\alpha(x^0, \lambda(x - x^0))$ approaches zero as λ approaches zero we obtain (2.70) from (2.73). Conversely, let $x^1 \in C$, $x^2 \in C$ and let $0 \le \lambda \le 1$. Then

$$f(\lambda x^1 + (1 - \lambda)x^2) + (1 - \lambda)(x^1 - x^2)^T \nabla f(\lambda x^1 + (1 - \lambda)x^2) \ge f(x^1) \tag{2.74}$$

and

$$f(\lambda x^1 + (1 - \lambda)x^2) + \lambda(x^2 - x^1)^T \nabla f(\lambda x^1 + (1 - \lambda)x^2) \ge f(x^2). \tag{2.75}$$

Multiplying (2.74) and (2.75) by λ and $(1 - \lambda)$, respectively, and adding up we get

$$f(\lambda x^1 + (1 - \lambda)x^2) \ge \lambda f(x^1) + (1 - \lambda)f(x^2) \tag{2.76}$$

as asserted.

To prove the result for the strictly concave case, suppose that f is strictly concave and $x^0 \in C, x \in C$, such that $x \ne x^0$. Then, since f is concave, by (2.71)–(2.73) we have that (2.70) holds.

We now show that equality cannot hold in (2.70). Suppose, to the contrary, that

$$f(x) = f(x^0) + (x - x^0)^T \nabla f(x^0). \tag{2.77}$$

Then, for $0 < \lambda < 1$,

$$\begin{aligned} f(\lambda x^0 + (1 - \lambda)x) &> \lambda f(x^0) + (1 - \lambda)f(x) \\ &= f(x^0) + (1 - \lambda)(x - x^0)^T \nabla f(x^0), \end{aligned} \tag{2.78}$$

where the inequality follows from the strict concavity of f and the equality from (2.77). Now let $x^2 = \lambda x^0 + (1 - \lambda)x$. Then, since $x^2 \in C$ and f is concave we obtain (2.70) with x^2 replacing x—that is,

$$f(\lambda x^0 + (1 - \lambda)x) \le f(x^0) + (1 - \lambda)(x - x^0)^T \nabla f(x^0) \tag{2.79}$$

holds. However, (2.79) contradicts (2.78), hence (2.77) cannot hold. The proof of the converse statement for the strictly concave case is similar to the concave case and will be omitted here. □

Theorem 2.3 can be generalized to n-dimensional space along the lines of the remark preceding (2.32), as can be seen in the following theorem.

Theorem 2.13 (Fenchel, 1951; Mangasarian, 1969). Let f be a differentiable function on the open convex set $C \subset R^n$. It is concave if and only if for every two points $x^1 \in C$, $x^2 \in C$

$$(x^2 - x^1)^T[\nabla f(x^2) - \nabla f(x^1)] \leq 0. \tag{2.80}$$

It is strictly concave if and only if the inequality in (2.80) is strict for $x^1 \neq x^2$.

Proof. Suppose that f is concave on C and let $x^1 \in C$, $x^2 \in C$. From Theorem 2.12 we get

$$f(x^2) - f(x^1) - (x^2 - x^1)^T \nabla f(x^1) \leq 0, \tag{2.81}$$

$$f(x^1) - f(x^2) - (x^1 - x^2)^T \nabla f(x^2) \leq 0. \tag{2.82}$$

Adding the last two inequalities yields (2.80). Conversely, suppose that (2.80) holds, and let $x^1 \in C$, $x^2 \in C$. By the Mean Value Theorem we have

$$f(x^2) - f(x^1) = (x^2 - x^1)^T \nabla f(x^1 + \lambda(x^2 - x^1)) \tag{2.83}$$

for some $0 < \lambda < 1$. But by (2.80)

$$\lambda(x^2 - x^1)^T[\nabla f(x^1 + \lambda(x^2 - x^1)) - \nabla f(x^1)] \leq 0 \tag{2.84}$$

and

$$(x^2 - x^1)^T \nabla f(x^1 + \lambda(x^2 - x^1)) \leq (x^2 - x^1)^T \nabla f(x^1). \tag{2.85}$$

From (2.83) and (2.85) we obtain

$$f(x^2) \leq f(x^1) + (x^2 - x^1)^T \nabla f(x^1) \tag{2.86}$$

and it follows from Theorem 2.11 that f is concave. The proof of the strictly concave case is similar. □

Let us generalize now the characterization of concave functions in terms of second derivatives, as given in Theorem 2.4. First we need the following definition.

Definition 2.6. The $n \times n$ symmetric matrix B of real numbers is said to be negative semidefinite if $y^T B y \le 0$ for every $y \in R^n$. It is negative definite if $y^T B y < 0$ for every $y \neq 0$. Similarly, B is said to be positive (semi) definite if the sense of the above inequalities is reversed.

Let f be a function defined on $C \subset R^n$, whose second partial derivatives exist at x. The $n \times n$ matrix of second partial derivatives $\nabla^2 f(x) \equiv H(x)$, evaluated at a point $x \in C$ is called the *Hessian matrix* of f, that is

$$\nabla^2 f(x) \equiv H(x) = \begin{bmatrix} \partial^2 f(x)/\partial x_1\, \partial x_1 & \cdots & \partial^2 f(x)/\partial x_1\, \partial x_n \\ \vdots & & \vdots \\ \partial^2 f(x)/\partial x_n\, \partial x_1 & \cdots & \partial^2 f(x)/\partial x_n\, \partial x_n \end{bmatrix} \tag{2.87}$$

We have then the n-dimensional generalization of Theorem 2.4:

Theorem 2.14 (Fenchel, 1951). Let f be a twice-differentiable function on an open convex set $C \subset R^n$. Then f is concave if and only if its Hessian matrix is negative semidefinite for every $x \in C$. That is, for every $x \in C$ and $y \in R^n$ we have

$$y^T H(x) y \le 0. \tag{2.88}$$

If $H(x)$ is negative definite for every $x \in C$, then f is strictly concave.

The reader can prove this theorem directly by using Theorem 2.12, or by using Proposition 2.11 and Theorem 2.4.

The last statement in the theorem cannot be reversed, since there exist strictly concave functions whose Hessians are not negative definite, as we have already seen for the single-variable case.

Characterizations of concave functions are usually given in terms of the behavior of the function at every point of its domain. For example, if one would want to use the characterizations of the last two theorems, an infinite number of inequalities would have to be checked. A notable exception is the case of quadratic functions. Here the Hessian is a fixed matrix whose (semi) definiteness can be checked in a finite number of arithmetic operations. It is well known from matrix algebra that negative definiteness of a symmetric matrix B is equivalent to its eigenvalues being negative numbers. Similarly, negative semidefiniteness is equivalent to nonpositive eigenvalues, see Noble (1969).

Now we state a few results on the closedness of concave functions under functional operations. Additional results will be derived in Chapters 5–7. First we have the following.

Proposition 2.15 (Mangasarian, 1969). Let f be a concave function and let λ be a nonnegative number. Then $F(x) = \lambda f(x)$ is also a concave function. Let f_1 and f_2 be concave functions. Then $F(x) = f_1(x) + f_2(x)$ is also concave.

The proof is straightforward. It follows from the above proposition that every nonnegative linear combination of concave functions is also concave.

Denote by ϕ the mapping from $C \subset R^n$ into $D \subset R^m$, defined by

$$\phi(x) = (\phi_1(x), \ldots, \phi_m(x)). \tag{2.89}$$

Proposition 2.16 (Mangasarian, 1969). Let $\phi_1, \ldots, \phi_m$ be concave (strictly concave) functions on $C \subset R^n$ and let f be a componentwise nondecreasing (increasing) and concave function on $D \subset R^m$. Supose that D contains the range of $\phi_1, \ldots, \phi_m$. Then the composite function $f\phi(x) = f(\phi_1(x), \ldots, \phi_m(x))$ is concave (strictly concave) on C.

Proof. We prove only the strictly concave case. Since $\phi_1, \ldots, \phi_m$ are strictly concave, for every $x^1 \in C$, $x^2 \in C$, $x^1 \neq x^2$

$$\phi(\lambda x^1 + (1-\lambda)x^2) > \lambda\phi(x^1) + (1-\lambda)\phi(x^2), \tag{2.90}$$

thus

$$f\phi(\lambda x^1 + (1-\lambda)x^2) > f(\lambda\phi(x^1) + (1-\lambda)\phi(x^2)). \tag{2.91}$$

By the concavity of f

$$f(\lambda\phi(x^1) + (1-\lambda)\phi(x^2)) \geq \lambda f\phi(x^1) + (1-\lambda)f\phi(x^2). \tag{2.92}$$

Hence

$$f\phi(\lambda x^1 + (1-\lambda)x^2) > \lambda f\phi(x^1) + (1-\lambda)f\phi(x^2). \tag{2.93}$$

□

Proposition 2.17. Let f be a concave function defined on a convex set $D \subset R^m$ and let A be a given $m \times n$ matrix. Let C be a convex set in R^n such that $Ax \in D$ for every $x \in C$. Then $f(Ax)$ is a concave function on C.

Proof. The reader can easily verify that for every $x^1 \in C$, $x^2 \in C$ and $0 \leq \lambda \leq 1$ we have

$$f(\lambda Ax^1 + (1-\lambda)Ax^2) \geq \lambda f(Ax^1) + (1-\lambda)f(Ax^2). \tag{2.94}$$

□

Proposition 2.18 (Fenchel, 1951). Let f_i, $i \in I$ be a finite or infinite collection of concave functions, defined on a convex set $C \subset R^n$. For every $x \in C$ let

$$f(x) = \inf\{f_i(x), i \in I\}. \tag{2.95}$$

Then f is a concave function on C.

The proof is straightforward and left for the reader. Note that if f_i are convex, then $f(x) = \sup\{f_i(x), i \in I\}$ is also convex.

We conclude this section by introducing the class of strongly concave functions, a subclass of strictly concave functions. Strongly concave functions have applications in optimization and economics. In optimization, proofs of convergence of certain maximization algorithms have been presented for twice differentiable strongly concave functions that are characterized by having the eigenvalues of their Hessian bounded away from zero; see Avriel (1976). Economic applications of strongly concave functions can be found in Diewert (1981), Ginsberg (1973), Jorgenson and Lau (1974), and Lau (1978), and will also be discussed in Chapter 4.

Definition 2.7. A function f defined on the convex set $C \subset R^n$ is said to be strongly concave if it can be written as

$$f(x) = \phi(x) - \tfrac{1}{2}\alpha x^T x \tag{2.96}$$

where ϕ is a concave function defined on C and α is a positive number. Similarly, g is said to be strongly convex if it can be written as

$$g(x) = \psi(x) + \tfrac{1}{2}\alpha x^T x \tag{2.97}$$

where ψ is a convex function defined on C and α is a positive number.

Since f is the sum of a concave function and a strictly concave quadratic function, every strongly concave function is also strictly concave. A strongly convex function is similarly defined as the sum of a convex function and a strictly convex quadratic function. An example of a strictly concave function that is not strongly concave is again the function $f(x) = -(x)^4$ on R. Let $f(x) = \phi(x) - \frac{1}{2}\alpha(x)^2$, where ϕ is some concave function defined on R, and $\alpha > 0$. We can see that at the origin $f''(0) = 0 = \phi''(0) - \alpha$, or $\phi''(0) = \alpha > 0$, contradicting that ϕ is concave.

We can now characterize strongly concave functions.

Proposition 2.19 (Diewert, Avriel, and Zang, 1981; Poljak, 1966; Rockafellar, 1976; Vial, 1982, 1983). Let f be a function defined on the convex set $C \subset R^n$. Then f is strongly concave if and only if there exists a positive number α such that for every $x^1 \in C$, $x^2 \in C$, and $0 \leq \lambda \leq 1$ we have

$$f(\lambda x^1 + (1-\lambda)x^2) \geq \lambda f(x^1) + (1-\lambda)f(x^2) + \tfrac{1}{2}\lambda(1-\lambda)\alpha\|x^2 - x^1\|^2. \tag{2.98}$$

If f is differentiable on the open convex set $C \subset R^n$, then it is strongly concave if and only if there exists an $\alpha > 0$ such that for every $x^1 \in C$, $x^2 \in C$ we have

$$f(x^2) - f(x^1) + (x^2 - x^1)^T \nabla f(x^1) \leq -\tfrac{1}{2}\alpha\|x^2 - x^1\|^2. \tag{2.99}$$

If f is twice continuously differentiable on the open convex set $C \subset R^n$, then it is strongly concave if and only if there exists an $\alpha > 0$ such that for every $x \in C$

$$\nabla^2 f(x) + \alpha I \tag{2.100}$$

is negative semidefinite, where I is the identity matrix.

Proof of this proposition follows closely the proofs of results presented earlier for concave and strictly concave functions. It follows from this proposition that the eigenvalues of the Hessian matrix of a twice differentiable strongly concave function are uniformly bounded away from zero on C by the negative number $\beta = -\alpha$.

2.4. Extrema of Concave Functions

The importance of concave functions from the optimization point of view lies in some properties of concave functions with regard to their extrema.

Theorem 2.20 (Fenchel, 1951). Let f be a concave function defined on a convex set $C \subset R^n$. Then every local maximum of f at $x^* \in C$ is a global maximum of f over all C.

Proof. If $x^* \in C$ is a local maximum of f then there exists a positive number ε such that

$$f(x^*) \geq f(x) \tag{2.101}$$

for all $x \in C$ satisfying $\|x - x^*\| < \varepsilon$. Let y be any point in C. It follows that $(1-\lambda)x^* + \lambda y \in C$ for all $0 \leq \lambda \leq 1$, and for sufficiently small $\lambda^0 > 0$

$$\|(1-\lambda^0)x^* + \lambda^0 y - x^*\| < \varepsilon. \tag{2.102}$$

Thus,

$$f(x^*) \geq f((1-\lambda^0)x^* + \lambda^0 y). \tag{2.103}$$

From the concavity of f it follows that

$$f((1-\lambda^0)x^* + \lambda^0 y) \geq (1-\lambda^0)f(x^*) + \lambda^0 f(y). \tag{2.104}$$

Combining the last two inequalities and dividing the result by λ^0, we obtain

$$f(x^*) \geq f(y). \tag{2.105}$$

Hence x^* is the global maximum of f over C. □

Clearly, the maximum of a concave function can be attained at more than one point. The next result characterizes these points.

Theorem 2.21 (Fenchel, 1951). The set of points at which a concave function f attains its maximum over C is a convex set.

Proof. Let α^* be the value of f at a maximizing point. Then, by Corollary 2.6, the upper level set

$$U(f, \alpha^*) = \{x : x \in C, f(x) \geq \alpha^*\} \tag{2.106}$$

is convex. □

For strictly concave functions we have the following theorem.

Theorem 2.22 (Fenchel, 1951). Let f be a strictly concave function, defined on the convex set $C \subset R^n$. If f attains its maximum at $x^* \in C$, this maximizing point is unique.

The proof of this result is left for the reader. Note that the assumption of strict concavity on all C can be weakened to strict concavity in a neighborhood of x^*.

Do concave functions always have a maximum? The answer to this question is, of course, negative, as can be seen in the following example.

Example 2.5. (a) Let $C = \{x: x \in R, x \geq 1\}$ and let $f(x) = \log x$. This concave function has no maximum since x is not bounded from above. By changing C to a compact (closed and bounded) set, however, this function will always have a maximum.

(b) A compact domain for a concave function is not always sufficient for having a maximum as demonstrated by the case of $C = \{x: x \in R, 1 \leq x \leq 2\}$ and

$$f(x) = \begin{cases} x, & 1 \leq x < 2 \\ 1, & x = 2. \end{cases} \tag{2.107}$$

It is well known that if a differentiable function f attains its maximum at an interior point x^* of its domain, then $\nabla f(x^*) = 0$; see Avriel (1976). The converse statement generally does not hold. For concave functions, however, we have the following theorem.

Theorem 2.23. Let f be a differentiable concave (strictly concave) function on the convex set C. If

$$\nabla f(x^*) = 0 \tag{2.108}$$

at a point $x^* \in C$, then f attains its maximum (unique maximum) at x^*.

Proof. It follows from the assumptions on f that for every $x \in C$

$$f(x) \leq f(x^*) + (x - x^*)^T \nabla f(x^*) \tag{2.109}$$

and the inequality above is strict if f is strictly concave. If $\nabla f(x^*) = 0$, then for every $x \in C$

$$f(x) \leq f(x^*) \tag{2.110}$$

and the inequality is strict if f is strictly concave. Thus x^* is a maximum (unique maximum) of f. □

Let us turn our attention to minima of concave functions. The first result is as follows.

Theorem 2.24 (Fenchel, 1951). If f is a concave function defined on the convex set C and attains its global minimum over C at an interior point x^0 of C, then f is constant on C.

Proof. Suppose that f is not constant on C. Let $y \in C$ such that $f(y) > f(x^0)$. Since x^0 is an interior point, we can find a point $z \in C$ and a number $0 < \lambda < 1$ such that

$$x^0 = \lambda y + (1 - \lambda)z. \tag{2.111}$$

By concavity we have

$$f(x^0) \geq \lambda f(y) + (1 - \lambda)f(z) > \lambda f(x^0) + (1 - \lambda)f(x^0) = f(x^0), \tag{2.112}$$

a contradiction. □

Definition 2.8. A point x^0 belonging to a convex set $C \subset R^n$ is called an extreme point of C if there are no points $x^1 \in C$, $x^2 \in C$, $x^1 \neq x^2$, such that $x^0 = \lambda x^1 + (1 - \lambda)x^2$ with $0 < \lambda < 1$.

Now we can state a result on minima of concave functions over compact sets.

Proposition 2.25. If f is a continuous concave function on the compact convex set $C \subset R^n$, then f attains a global minimum at an extreme point of C.

Proof. A complete proof of this proposition would require certain results on convex sets that are not stated here, and, therefore, we present here only an outline of the proof.

Since f is continuous and C is compact, the minimum of f on C is attained at some point $x^* \in C$. It can be shown that a compact convex set $C \subset R^n$ is equal to the *convex hull* of its extreme points—that is, the intersection of all convex sets containing the extreme points of C; see, for example, Roberts and Varberg (1973). If S is any subset of R^n, the convex hull of S can be also obtained by taking the union of all *convex combinations* of elements of S. If $x^1, \ldots, x^m$ are points in R^n, the convex combination of these points is given by $\lambda_1 x^1 + \cdots + \lambda_m x^m$, where $\lambda_1 \geq 0, \ldots, \lambda_m \geq 0$ and $\lambda_1 + \cdots + \lambda_m = 1$. It can be shown that if $C \subset R^n$ is a convex set, $x^1, \ldots, x^m$ are elements of C, and if f is a concave function defined on C, then

$$f(\lambda_1 x^1 + \cdots + \lambda_m x^m) \geq \lambda_1 f(x^1) + \cdots + \lambda_m f(x^m) \tag{2.113}$$

where

$$\lambda_1 \geq 0, \ldots, \lambda_m \geq 0, \qquad \sum_{i=1}^{m} \lambda_i = 1. \tag{2.114}$$

Now, the classical Carathéodory's theorem states that every element of the convex hull of S can be expressed as the convex combination of m elements of S, where $m \leq n + 1$. For more on Carathéodory's theorem the reader is referred to Roberts and Varberg (1973), or Rockafellar (1970). In our case we can, therefore, write

$$x^* = \lambda_1 v^1 + \cdots + \lambda_m v^m \tag{2.115}$$

where $v^1, \ldots, v^m$ are extreme points of C. Then, from (2.113) it follows that

$$f(x^*) \geq \lambda_1 f(v^1) + \cdots + \lambda_m f(v^m) \geq \min[f(v^1), \ldots, f(v^m)]. \tag{2.116}$$

But x^* is a global minimum, so f must attain the value $f(x^*)$ at some extreme point $v^i \in C$. □

2.5. Concave Mathematical Programs

We continue here the discussion of extrema of concave functions by considering the following mathematical program:

$$\text{CP} \qquad \max f(x) \tag{2.117}$$

subject to the constraints

$$g_i(x) \leq 0, \qquad i = 1, \ldots, m, \tag{2.118}$$

$$h_j(x) = 0, \qquad j = 1, \ldots, p, \tag{2.119}$$

where f, the g_i, and the h_j are, respectively, concave, convex, and affine functions on the nonempty open convex set $C \subset R^n$, containing the set of points satisfying (2.118)–(2.119). It is assumed that the vectors of coefficients a^j of the affine functions $h_j(x) = (a^j)^T x - b_j$ are linearly independent. Program CP is called a *concave program*.

The set of points $x \in C$ satisfying (2.118) and (2.119) is called the *feasible set* of CP and, by Corollary 2.6, it is a convex set. Thus CP is the problem of finding the maximum of a concave function over a convex set. By Theorem 2.20 every local maximum of f at a point x^* belonging to the feasible set is also a global maximum of f over the whole feasible set. It is also called a *solution* of CP. If the feasible set is nonempty, CP is called a *consistent program.* If there exists a point $\hat{x} \in C$ such that

$$g_i(\hat{x}) < 0, \qquad i = 1, \ldots, m, \tag{2.120}$$

$$h_j(\hat{x}) = 0, \qquad j = 1, \ldots, p, \tag{2.121}$$

then CP is called *strongly consistent.* The strong consistency condition is also known as *Slater's constraint qualifications*; see Slater (1950). A general result on the consistency of a system of equations and inequalities is given in the following proposition.

Proposition 2.26 (Fan, Glicksberg, and Hoffman, 1957; Mangasarian, 1969). Let $\phi_1, \ldots, \phi_m$, and $\phi_{m+1}, \ldots, \phi_p$ be convex and affine functions, respectively, on the nonempty convex set $C \subset R^n$. If the system

$$\phi_i(x) < 0, \qquad i = 1, \ldots, m \tag{2.122}$$

$$\phi_i(x) = 0, \qquad i = m+1, \ldots, p \tag{2.123}$$

has no solution $x \in C$ then there exist numbers $\alpha_1, \ldots, \alpha_p$, not all zero, such that $\alpha_i \geq 0$ for $i = 1, \ldots, m$ and for every $x \in C$

$$\sum_{i=1}^{p} \alpha_i \phi_i(x) \geq 0. \tag{2.124}$$

Proof. Again, we present an outline of the proof. Let

$$G(x) = \{y: y \in R^p, y_i > \phi_i(x), i = 1, \ldots, m, y_i = \phi_i(x), i = m+1, \ldots, p\} \tag{2.125}$$

and let

$$G = \bigcup_{x \in C} G(x). \tag{2.126}$$

By the assumptions, G does not contain the origin. This set is, however, a nonempty convex set, for let $y^1 \in G$, $y^2 \in G$, then

$$\lambda y_i^1 + (1-\lambda)y_i^2 > \lambda\phi_i(x^1) + (1-\lambda)\phi_i(x^2) \geq \phi_i(\lambda x^1 + (1-\lambda)x^2),$$
$$i = 1, \ldots, m \qquad (2.127)$$

$$\lambda y_i^1 + (1-\lambda)y_i^2 = \lambda\phi_i(x^1) + (1-\lambda)\phi_i(x^2) = \phi_i(\lambda x^1 + (1-\lambda)x^2),$$
$$i = m+1, \ldots, p. \qquad (2.128)$$

Since G is a nonempty convex set not containing the origin, it is well known from separation theorems for convex sets that there exists a hyperplane separating G and the origin. That is, there exists a set of points in R^p, given by

$$H = \{z: z \in R^p, \alpha^T z = 0, \alpha \in R^p, \alpha \neq 0\} \qquad (2.129)$$

and called a *hyperplane* with the property that for every $y \in G$ we have $\alpha^T y \geq 0$. Such a hyperplane is called a *separating hyperplane*. Various theorems on separating hyperplanes for convex sets can be found in the literature; see, for example, Mangasarian (1969), or Avriel (1976).

Since each y_i, $i = 1, \ldots, m$ can be as large as desired, we must have $\alpha_i \geq 0$ for $i = 1, \ldots, m$. Now let $\varepsilon > 0$ and for every $x \in C$ let

$$\hat{y}_i = \phi_i(x) + \varepsilon, \qquad i = 1, \ldots, m, \qquad (2.130)$$

$$\hat{y}_i = \phi_i(x), \qquad i = m+1, \ldots, p. \qquad (2.131)$$

Then $\hat{y} \in G(x) \subset G$, and

$$\sum_{i=1}^{p} \alpha_i \hat{y}_i = \sum_{i=1}^{p} \alpha_i \phi_i(x) + \varepsilon \sum_{i=1}^{m} \alpha_i \geq 0 \qquad (2.132)$$

or

$$\sum_{i=1}^{p} \alpha_i \phi_i(x) \geq -\varepsilon \sum_{i=1}^{m} \alpha_i. \qquad (2.133)$$

Since ε is an arbitrary positive number and $\alpha_i \geq 0$, $i = 1, \ldots, m$, it follows that (2.124) holds. □

The above proposition clearly applies to concave programs that are not strongly consistent.

We shall characterize the optimal solution of concave programs in terms of properties of the *Lagrangian function* associated with CP and defined by

$$L(x, \lambda, \mu) = f(x) - \sum_{i=1}^{m} \lambda_i g_i(x) - \sum_{j=1}^{p} \mu_j h_j(x). \tag{2.134}$$

Note that for every fixed λ and μ such that $\lambda \geq 0$, the Lagrangian is a concave function of x on C, and for every fixed $x \in C$, it is a convex (affine) function of λ and μ. Such functions exhibit some interesting saddlepoint properties as we shall see below.

First we need the following definition.

Definition 2.9. Let Φ be a real-valued function of the two vector variables $x \in D \subset R^n$ and $y \in E \subset R^m$. A point (x^0, y^0) with $x^0 \in D$ and $y^0 \in D$ is said to be a saddlepoint of Φ if

$$\Phi(x^0, y) \geq \Phi(x^0, y^0) \geq \Phi(x, y^0) \tag{2.135}$$

for every $x \in D$ and $y \in E$.

Saddlepoints are closely related to minimax (inf-sup) and maximin (sup-inf) points of a function in two vector variables:

Proposition 2.27. Let Φ be a function of the two vector variables $x \in D \subset R^n$ and $y \in E \subset R^m$. Then

$$\inf_{y \in E} \sup_{x \in D} \Phi(x, y) \geq \sup_{x \in D} \inf_{y \in E} \Phi(x, y). \tag{2.136}$$

If Φ has a saddlepoint at (x^0, y^0) then

$$\inf_{y \in E} \sup_{x \in D} \Phi(x, y) = \Phi(x^0, y^0) = \sup_{x \in D} \inf_{y \in E} \Phi(x, y). \tag{2.137}$$

Proof. For every $y \in E$ we have

$$\sup_{x \in D} \Phi(x, y) \geq \Phi(x, y). \tag{2.138}$$

Similarly, for every $x \in D$

$$\Phi(x, y) \geq \inf_{y \in E} \Phi(x, y). \tag{2.139}$$

Hence, for every $x \in D$, and $y \in E$

$$\sup_{x \in D} \Phi(x, y) \geq \inf_{y \in E} \Phi(x, y) \tag{2.140}$$

and consequently (2.136) holds. Suppose now that (x^0, y^0) is a saddlepoint of Φ. Then

$$\sup_{x \in D} \Phi(x, y^0) \leq \Phi(x^0, y^0) \leq \inf_{y \in E} \Phi(x^0, y). \tag{2.141}$$

Also,

$$\inf_{y \in E} \sup_{x \in D} \Phi(x, y) \leq \sup_{x \in D} \Phi(x, y^0) \tag{2.142}$$

$$\inf_{y \in E} \Phi(x^0, y) \leq \sup_{x \in D} \inf_{y \in E} \Phi(x, y). \tag{2.143}$$

Combining (2.141), (2.142), and (2.143) we obtain

$$\inf_{y \in E} \sup_{x \in D} \Phi(x, y) \leq \Phi(x^0, y^0) \leq \sup_{x \in D} \inf_{y \in E} \Phi(x, y). \tag{2.144}$$

Comparing (2.144) with (2.136) we conclude that (2.137) must hold. □

Associated with the concave program CP there is a so-called *saddlepoint problem*:

SP Find a point $(x^0, (\lambda^0, \mu^0))$, where $x^0 \in C \subset R^n$, $\lambda^0 \in R^m$, $\lambda^0 \geq 0$, $\mu^0 \in R^p$ such that $(x^0, (\lambda^0, \mu^0))$ is a saddlepoint of the Lagrangian function L.

The relationship between the optimal solution of a concave program CP and the solution of the associated saddlepoint problem SP is summarized in the following theorem.

Theorem 2.28 (Karlin, 1959; Uzawa, 1958). If $(x^*, (\lambda^*, \mu^*))$ is a solution of SP, then x^* is a solution of CP. If CP is strongly consistent and x^* is a solution of CP, then there exists λ^*, μ^* such that $(x^*, (\lambda^*, \mu^*))$ is a solution of SP. In both cases $f(x^*) = L(x^*, \lambda^*, \mu^*)$.

Proof. Suppose that $(x^*, (\lambda^*, \mu^*))$ is a solution of SP. Then for all $x \in R^n$, $\lambda \in R^m$, $\lambda \geq 0$, and $\mu \in R^p$

$$\begin{aligned} f(x^*) - \sum_{i=1}^{m} \lambda_i g_i(x^*) - \sum_{j=1}^{p} \mu_j h_j(x^*) &\geq f(x^*) - \sum_{i=1}^{m} \lambda_i^* g_i(x^*) - \sum_{j=1}^{p} \mu_j^* h_j(x^*) \\ &\geq f(x) - \sum_{i=1}^{m} \lambda_i^* g_i(x) - \sum_{j=1}^{p} \mu_j^* h_j(x). \end{aligned} \tag{2.145}$$

From the first inequality we have for all $\lambda \geq 0$ and μ

$$\sum_{i=1}^{m} (\lambda_i^* - \lambda_i) g_i(x^*) + \sum_{j=1}^{p} (\mu_j^* - \mu_j) h_j(x^*) \geq 0. \tag{2.146}$$

Suppose that $h_k(x^*) \neq 0$. Choose $\lambda_i = \lambda_i^*$ for $i = 1, \ldots, m$ and $\mu_j = \mu_j^*$ for $j = 1, \ldots, p$, $j \neq k$. By an appropriate choice of μ_k we get a contradiction to (2.146). Hence $h_j(x^*) = 0$ for $j = 1, \ldots, p$ and for all $\lambda \geq 0$ we have

$$\sum_{i=1}^{m} (\lambda_i^* - \lambda_i) g_i(x^*) \geq 0. \tag{2.147}$$

If $g_k(x^*) > 0$ for some k, let $\lambda_k = \lambda_k^* + 1$ and $\lambda_i = \lambda_i^*$ for $i = 1, \ldots, m$, $i \neq k$, contradicting (2.147). Hence $g_i(x^*) \leq 0$ for $i = 1, \ldots, m$ and x^* is feasible for CP. Next let $\lambda = 0$. Then

$$0 \geq -\sum_{i=1}^{m} \lambda_i^* g_i(x^*). \tag{2.148}$$

Since $\lambda_i^* \geq 0$ and $g_i(x^*) \leq 0$ for $i = 1, \ldots, m$ we obtain $\lambda_i^* g_i(x^*) = 0$ for all i. Hence $f(x^*) = L(x^*, \lambda^*, \mu^*)$. Moreover, the second inequality of (2.145) means

$$f(x^*) \geq f(x) - \sum_{i=1}^{m} \lambda_i^* g_i(x) - \sum_{j=1}^{p} \mu_j^* h_j(x). \tag{2.149}$$

If x is feasible for CP, then $g_i(x) \leq 0$ and $h_j(x) = 0$. Thus

$$f(x^*) \geq f(x) \tag{2.150}$$

and x^* is a solution of CP.

Suppose now that x^* is a solution of CP. Then the system of inequalities

$$f(x^*) - f(x) < 0 \tag{2.151}$$

$$g_i(x) \leq 0, \qquad i = 1, \ldots, m \tag{2.152}$$

$$h_j(x) = 0, \qquad j = 1, \ldots, p \tag{2.153}$$

has no solution $x \in C$. It follows from Proposition 2.26 that there exist numbers $\alpha_0 \geq 0$, $\alpha_1 \geq 0, \ldots, \alpha_m \geq 0$, $\beta_1, \ldots, \beta_p$, not all zero, such that

$$\alpha_0(f(x^*) - f(x)) + \sum_{i=1}^{m} \alpha_i g_i(x) + \sum_{j=1}^{p} \beta_j h_j(x) \geq 0 \tag{2.154}$$

for every $x \in C$. Suppose that $\alpha_0 > 0$. Then dividing (2.154) by α_0 and rearranging we obtain

$$f(x^*) \geq L(x, \lambda^*, \mu^*) \tag{2.155}$$

where $\lambda_i^* = \alpha_i/\alpha_0$ and $\mu_j^* = \beta_j/\alpha_0$. Since $g_i(x^*) \leq 0$ and $h_j(x^*) = 0$ we get

$$L(x^*, \lambda, \mu) = f(x^*) - \sum_{i=1}^{m} \lambda_i g_i(x^*) - \sum_{j=1}^{p} \mu_j h_j(x^*) \geq f(x^*) \tag{2.156}$$

for all $\lambda \geq 0$ and μ. Letting $x = x^*$ in (2.155), and, $\lambda = \lambda^*$ and $\mu = \mu^*$ in (2.156), we conclude that $f(x^*) = L(x^*, \lambda^*, \mu^*)$. Substituting $L(x^*, \lambda^*, \mu^*)$ instead of $f(x^*)$ in the left- and right-hand sides of the inequalities in (2.155) and (2.156), respectively, we obtain that $(x^*, (\lambda^*, \mu^*))$ is a saddlepoint of the Lagrangian, as asserted.

Suppose now that $\alpha_0 = 0$. Then

$$\sum_{i=1}^{m} \alpha_i g_i(x) + \sum_{j=1}^{p} \beta_j h_j(x) \geq 0 \tag{2.157}$$

for all $x \in C$. Since CP is strongly consistent there exists an $\hat{x} \in C$ such that (2.120) and (2.121) hold. Hence $\alpha_1 = \alpha_2 = \cdots = \alpha_m = 0$. Then $\beta \neq 0$ and for all $x \in C$

$$\sum_{j=1}^{p} \beta_j \left(\sum_{k=1}^{n} a_{jk} x_k - b_j \right) \geq 0. \tag{2.158}$$

Observe now that the point x given by

$$x_k = \hat{x}_k - \varepsilon \left(\sum_{j=1}^{p} \beta_j a_{jk} \right), \qquad k = 1, \ldots, n \tag{2.159}$$

is also in C for a sufficiently small positive ε. However, the reader can verify that for such an x

$$\sum_{j=1}^{p} \beta_j \left(\sum_{k=1}^{n} a_{jk} x_k - b_j \right) < 0 \tag{2.160}$$

contradicting (2.158). Consequently α_0 cannot vanish. □

For concave programs consisting of functions that are not necessarily differentiable, Theorem 2.28 states a set of necessary and sufficient conditions. In a more general case, the relationship between a mathematical program (such as CP, but without the concavity and convexity assumptions) and the associated SP is only one-sided: a solution of SP is optimal for the mathematical program, but not conversely. These relationships were first established by Karush (1939) and Kuhn and Tucker (1951).

Let us turn now to concave programs in which the functions involved are differentiable. For these programs, necessary and sufficient optimality conditions involving the Lagrangian function were derived by Karush, and Kuhn and Tucker. The necessary conditions are given in the following theorem.

Theorem 2.29. Suppose that the functions appearing in CP are differentiable and that CP is strongly consistent. If x^* is a solution of CP, then there exist vectors λ^* and μ^* such that

$$\nabla_x L(x^*, \lambda^*, \mu^*) \equiv \nabla f(x^*) - \sum_{i=1}^{m} \lambda_i^* \nabla g_i(x^*) - \sum_{j=1}^{p} \mu_j^* \nabla h_j(x^*) = 0, \tag{2.161}$$

$$\lambda_i^* g_i(x^*) = 0, \qquad i = 1, \ldots, m, \tag{2.162}$$

$$\lambda^* \geq 0. \tag{2.163}$$

Proof. We have seen in Theorem 2.28 that under hypotheses weaker than those of the present theorem there exist vectors λ^* and μ^* such that $L(x, \lambda, \mu)$ has a saddlepoint at $(x^*, (\lambda^*, \mu^*))$ with $\lambda^* \geq 0$, hence (2.163) holds. By the definition of saddlepoints, the Lagrangian $L(x, \lambda^*, \mu^*)$ has an unconstrained maximum at x^*. Since L is differentiable, the gradient of $L(x, \lambda^*, \mu^*)$, evaluated at x^*, must vanish, and (2.161) holds. We have also seen that $f(x^*) = L(x^*, \lambda^*, \mu^*)$. We know that $h_1(x^*) = \cdots = h_p(x^*) = 0$ and, consequently

$$\sum_{i=1}^{m} \lambda_i^* g_i(x^*) = 0. \tag{2.164}$$

But $\lambda_i^* \geq 0$ and $g_i(x^*) \leq 0$, hence $\lambda_i^* g_i(x^*) \leq 0$ for all i. It follows that (2.162) must hold. □

The above theorem also holds for more general mathematical programs without the concavity, convexity, and affinity assumptions, provided that the strong consistency assumption is replaced by some conditions that Kuhn and Tucker (1951) termed "constraint qualifications." The reader is referred to Mangasarian (1969), where an extensive review of constraint qualifications is presented, and to Avriel (1976), Bazaraa and Shetty (1979), and Peterson (1973), for a more recent discussion. Optimality conditions without constraint qualifications can be found in Ben-Israel, Ben-Tal, and Zlobec (1981).

Sufficient optimality conditions for concave programs with differentiable functions are given in the following theorem.

Theorem 2.30. If there exists an x^* feasible for CP and λ^*, μ^* such that

$$\nabla_x L(x^*, \lambda^*, \mu^*) = \nabla f(x^*) - \sum_{i=1}^{m} \lambda_i^* \nabla g_i(x^*) - \sum_{j=1}^{p} \mu_j^* \nabla h_j(x^*) = 0, \tag{2.165}$$

$$\lambda_i^* g_i(x^*) = 0, \qquad i = 1, \ldots, m, \tag{2.166}$$

$$\lambda^* \geq 0, \tag{2.167}$$

then x^* is a solution of CP.

Proof. Let x be feasible for CP. Then, by (2.167)

$$f(x) \leq f(x) - \sum_{i=1}^{m} \lambda_i^* g_i(x) - \sum_{j=1}^{p} \mu_j^* h_j(x). \tag{2.168}$$

By the concavity, convexity, and affinity assumptions we obtain from (2.168)

$$f(x) \leq f(x^*) + (x - x^*)^T \nabla f(x^*) - \sum_{i=1}^{m} \lambda_i^* g_i(x^*) - \sum_{i=1}^{m} \lambda_i^* (x - x^*)^T \nabla g_i(x^*)$$
$$- \sum_{j=1}^{p} \mu_j^* h_j(x^*) - \sum_{j=1}^{p} \mu_j^* (x - x^*)^T \nabla h_j(x^*). \tag{2.169}$$

Rearranging and noting that x^* is feasible for CP yields

$$f(x) \leq f(x^*) - \sum_{i=1}^{m} \lambda_i^* g_i(x^*) + (x - x^*)^T \left[\nabla f(x^*) - \sum_{i=1}^{m} \lambda_i^* \nabla g_i(x^*) \right.$$
$$\left. - \sum_{j=1}^{p} \mu_j^* \nabla h_j(x^*) \right] \tag{2.170}$$

and by (2.165)(-(2.167) we obtain for every feasible x

$$f(x) \le f(x^*) \tag{2.171}$$

as asserted. □

These sufficient conditions, involving only the first derivatives of the functions appearing in a concave program, do not hold in the case of programs with arbitrary functions. We shall, however, see in the following chapters thay they do hold if the concave, convex, and affine functions of CP are replaced by certain generalized concave and convex functions.

We conclude this section by a few brief remarks concerning duality in concave programs. Consider the following special case of CP in which there are no equality constraints:

PCP
$$\max f(x) \tag{2.172}$$

subject to

$$g_i(x) \le 0, \qquad i = 1, \ldots, m \tag{2.173}$$

where f and $g_1, \ldots, g_m$ are, respectively, concave and convex functions on the convex set $C \subset R^n$. PCP is called the *primal concave program.* Note that the Lagrangian function associated with PCP is given by

$$L(x, \lambda) = f(x) - \sum_{i=1}^{m} \lambda_i g_i(x) \tag{2.174}$$

where $x \in C$ and $\lambda \in R^m$, $\lambda \ge 0$. Corresponding to PCP we define the *dual concave program*

DCP
$$\min_{\lambda} \{\phi(\lambda)\} = \min_{\lambda} \{\sup_{x \in C} L(x, \lambda)\} \tag{2.175}$$

subject to

$$\lambda \ge 0. \tag{2.176}$$

The function ϕ is the supremum of a family of affine functions [since for every $x \in C$ the function $L(x, \lambda)$ is affine in λ], and by Proposition 2.18 it is convex. Hence DCP is the problem of minimizing a convex function subject to linear inequality (nonnegativity) constraints. Note that ϕ is not necessarily finite for every $\lambda \geq 0$, and if we denote by V the set of points $\lambda \geq 0$ for which ϕ is finite, then V can, in general, be empty. Formally, $\phi(\lambda) = +\infty$ for $\lambda \notin V$. We then have the following duality theorem, proof of which follows closely Roberts and Varberg (1973).

Theorem 2.31. If $x \in C$ is feasible for PCP and $\lambda \geq 0$ then

$$f(x) \leq \phi(\lambda). \tag{2.177}$$

Suppose that PCP is strongly consistent and has an optimal solution x^*. Then DCP is also consistent and has an optimal solution λ^*. Moreover,

$$f(x^*) = \phi(\lambda^*) \tag{2.178}$$

and (x^*, λ^*) is a saddlepoint of $L(x, \lambda)$ as given by (2.174).

Proof. For $\lambda \notin V$ (2.177) trivially holds. Now let $\lambda \in V$ and notice that ϕ is defined as the supremum of the Lagrangian function that is,

$$\phi(\lambda) = \sup_{x \in C} L(x, \lambda). \tag{2.179}$$

Hence, for all feasible $x \in C$

$$f(x) \leq f(x) - \sum_{i=1}^{m} \lambda_i g_i(x) \leq \phi(\lambda) \tag{2.180}$$

where the left-hand inequality follows from the feasibility of x and the nonnegativity of λ. Thus the inequality in (2.177) holds for every feasible x and $\lambda \geq 0$.

Now suppose that PCP is strongly consistent and has an optimal solution x^*. Then, by Theorem 2.28 there exists a $\lambda^* \geq 0$ such that (x^*, λ^*) is a saddlepoint of the Lagrangian $L(x, \lambda)$ as given by (2.174)—that is,

$$L(x^*, \lambda) \geq L(x^*, \lambda^*) \geq L(x, \lambda^*) \tag{2.181}$$

and

$$L(x^*, \lambda^*) = f(x^*). \tag{2.182}$$

But

$$L(x^*, \lambda^*) = \sup_{x \in C} L(x, \lambda^*) = \phi(\lambda^*) \tag{2.183}$$

is finite, consequently $\lambda^* \in V$ and DCP is consistent. Moreover, (2.182)–(2.183) imply (2.178) and by (2.177)

$$\phi(\lambda) \geq f(x^*) = \phi(\lambda^*) \tag{2.184}$$

must hold for all $\lambda \geq 0$. Hence λ^* is optimal for DCP. □

For a discussion of duality in nonlinear programming and for more general duality results see Avriel (1976), Fenchel (1951), and Rockafellar (1970).

Example 2.6. Consider the following primal concave program

$$\max f(x) = 4x - (x)^2 \tag{2.185}$$

subject to

$$g(x) = x - \tfrac{1}{2} \leq 0. \tag{2.186}$$

To find the dual program we first define the Lagrangian

$$L(x, \lambda) = 4x - (x)^2 - \lambda(x - 1/2). \tag{2.187}$$

Hence,

$$\phi(\lambda) = \sup_{x \in C} [4x - (x)^2 - \lambda(x - 1/2)] \tag{2.188}$$

where $C = R^1$. The function appearing in the brackets in (2.188) is differentiable and concave in x. We can, therefore easily calculate that the maximizing point in C is given by

$$x^* = 2 - \tfrac{1}{2}\lambda \tag{2.189}$$

and

$$\phi(\lambda) = \tfrac{1}{4}(\lambda)^2 - \tfrac{3}{2}\lambda + 4. \tag{2.190}$$

Clearly, ϕ is finite for all λ and, consequently, $V = \{\lambda : \lambda \geq 0\}$. Hence, in the dual program $\phi(\lambda)$ is minimized subject to $\lambda \geq 0$. The reader can easily verify that the optimal solution of the primal program is $x^* = \frac{1}{2}$ and $f(x^*) = \frac{7}{4}$. Similarly, the optimal dual solution is $\lambda^* = 3$ and $\phi(\lambda^*) = \frac{7}{4}$. Note that for all $x \leq \frac{1}{2}$ and $\lambda \geq 0$

$$\phi(\lambda) = \tfrac{1}{4}(\lambda)^2 - \tfrac{3}{2}\lambda + 4 \geq \tfrac{7}{4} \geq 4x - (x)^2 = f(x). \tag{2.191}$$

By substituting the appropriate quantities into (2.174) it is easy to verify that the saddlepoint relation also holds for the above (x^*, λ^*).

References

AVRIEL, M. (1976), *Nonlinear Programming: Analysis and Methods*, Prentice-Hall, Englewood Cliffs, New Jersey.

BARTLE, R. G. (1976), *The Elements of Real Analysis*, 2nd ed., Wiley, New York.

BAZARAA, M. S., and SHETTY, C. M. (1979), *Nonlinear Programming, Theory and Algorithms*, Wiley, New York.

BEN-ISRAEL, A., BEN-TAL, A., and ZLOBEC, S. (1981), *Optimality in Nonlinear Programming: A Feasible Directions Approach*, Wiley, New York.

DIEWERT, W. E. (1981), Generalized concavity and economics, in *Generalized Concavity in Optimization and Economics*, Edited by S. Schaible and W. T. Ziemba, Academic Press, New York.

DIEWERT, W. E., AVRIEL, M., and ZANG, I. (1981), Nine kinds of quasiconcavity and concavity, *J. Econ. Theory* **25**, 397–420.

EGGLESTON, H. G. (1958), *Convexity*, Cambridge University Press, Cambridge, England.

EAN, K. I., GLICKSBERG, I., and HOFFMAN, A. J. (1957), Systems of inequalities involving convex functions, *Proc. Am. Math. Soc.* **8**, 617–622.

FENCHEL, W. (1951), Convex cones, sets and functions, Mimeographed lecture notes, Princeton University, Princeton, New Jersey.

GINSBERG, W. (1973), Concavity and quasiconcavity in economics, *J. Econ. Theory* **6**, 596–605.

HÖLDER, O. (1889), Über einen Mittelwertsatz, *Nachr. Ges. Wiss. Goettingen* 38–47.

JENSEN, J. L. W. V. (1906), Sur les Fonctions Convexes et les Inégalities Entre les Valeurs Moyennes, *Acta Math.* **30**, 175–193.

JORGENSON, D. W., and LAU, L. J. (1974), Duality and differentiability in production, *J. Econ. Theory* **9**, 23–42.

KARLIN, S. (1959), *Mathematical Methods and Theory in Games, Programming and Economics*, Vol. 1, Addison-Wesley, Reading, Massachusetts.

KARUSH, W. (1939), Minima of functions of several variables with inequalities as side conditions, M.S. dissertation, Department of Mathematics, University of Chicago.

KUHN, H. W., and TUCKER, A. W. (1951), Nonlinear programming, in *Proceedings of the Second Berkeley Symposium on Mathematical Statistics and Probability*, Edited by J. Neymann, University of California Press, Berkeley, California.

LAU, L. J. (1978), Applications of profit functions, in *Production Economics: A Dual Approach to Theory and Applications*, Vol. 1, Edited by M. Fuss and D. McFadden, North-Holland, Amsterdam.

LAY, S. R. (1982), *Convex Sets and their Applications*, Wiley, New York.

MANGASARIAN, O. L. (1969), *Nonlinear Programming*, McGraw-Hill, New York.

MINKOWSKI, H. (1910), *Geometrie der Zahlen*, Teubner, Leipzig.

MINKOWSKI, H. (1911), *Theorie der Konvexen Korper, Inbesondere Begründung ihres Oberflächenbegriffs*, Gesammelte Abhandlungen II, Teubner, Leipzig.

NOBLE, B. (1969), *Applied Linear Algebra*, Prentice-Hall, Englewood Cliffs, New Jersey.

PETERSON, D. W. (1973), A review of constraint qualifications in finite-dimension spaces, *SIAM Rev.* **15**, 639–654.

POLJAK, B. T. (1966), Existence theorems and convergence of minimizing sequences in extremum problems with restrictions, *Dokl. Akad. Nauk C.C.P.* **166**, 287–290 (English transl. *Sov. Math.* **7**, 72–75).

ROBERTS, A. W., and VARBERG, D. E. (1973), *Convex Functions*, Academic Press, New York.

ROCKAFELLAR, R. T. (1970), *Convex Analysis*, Princeton University Press, Princeton, New Jersey.

ROCKAFELLAR, R. T. (1976), Saddle points of Hamiltonian systems in convex Lagrange problems having a nonzero discount rate, *J. Econ. Theory* **12**, 71–113.

SLATER, M. (1950), Lagrange multipliers revisited: A contribution to nonlinear programming, Cowles Commission Discussion Paper, Math. 403.

UZAWA, H. (1958), The Kuhn–Tucker theorem in concave programming, in *Studies in Linear and Non-Linear Programming*, Edited by K. H. Arrow, L. Hurwicz, and H. Uzawa, Stanford University Press, Stanford, California.

VALENTINE, F. A. (1964), *Convex Sets*, McGraw-Hill, New York.

VIAL, J.-PH. (1982), Strong convexity of sets and functions, *J. Math. Econ.* **9**, 187–207.

VIAL, J.-PH. (1983), Strong and weak convexity of sets and functions, *Math. Oper. Res.* **8**, 231–259.

3

Generalized Concavity

We begin generalizing concave functions by recalling from the preceding chapter that the upper-level sets of a concave function are convex sets. As will become apparent below, concavity of a function is a sufficient condition for this property but not a necessary one. We have also seen in Chapter 1 that the fundamental property of utility functions in the theory of consumer demand is that indifference curves define convex sets. In other words, if the consumer has a utility function f defined on some domain $C \subset R^n$ that quantifies his preference ordering, then the set of all those commodity bundles $x \in C$ that yield a satisfaction level $f(x)$ of at least value α is a convex set. We have, therefore, that the upper-level sets of f are convex sets. One can readily construct examples of nonconcave functions exhibiting this property, and, in fact, we define a family of functions by the convexity of their upper-level sets. Such functions are called quasiconcave functions. They are generalized concave functions since it is easy to show that every concave function is quasiconcave, but not conversely. Quasiconcave functions are the most widely used generalized concave functions in economics. They are also important in optimization, since many properties of concave functions can be extended to quasiconcave functions as well; see for example, Greenberg and Pierskalla (1973).

3.1. Quasiconcave Functions—Definitions and General Properties

There are a number of equivalent definitions and characterizations of quasiconcave functions. We shall formally define them by the convexity of their upper-level sets:

Definition 3.1. Let f be defined on the convex set $C \subset R^n$. It is said to be quasiconcave if its upper-level sets

$$U(f, \alpha) = \{x: x \in C, f(x) \geq \alpha\} \tag{3.1}$$

are convex sets for every real α. Similarly, f is said to be quasiconvex if its lower-level sets

$$L(f, \alpha) = \{x: x \in C, f(x) \leq \alpha\} \tag{3.2}$$

are convex sets for every real α.

Functions defined this way have been first mentioned in de Finetti (1949). Properties of quasiconcave functions and the relationship between them and concave functions were subsequently studied by Fenchel (1951).

An example of a quasiconcave function of two variables is shown in Figure 3.1.

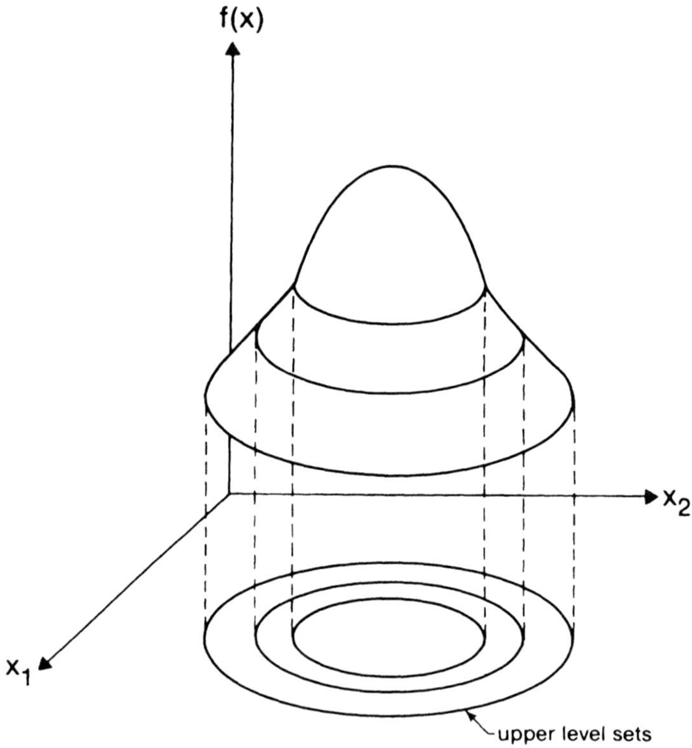

Figure 3.1. A quasiconcave function.

Let us present now a few characterizations and properties of quasiconcave functions. Equivalent results for quasiconvex functions can be easily obtained by recognizing that f is quasiconvex if and only if $-f$ is quasiconcave.

First we have the following theorem.

Theorem 3.1 (Fenchel, 1951). Let f be defined on the convex set $C \subset R^n$. It is a quasiconcave function if and only if

$$f(\lambda x^1 + (1-\lambda)x^2) \geq \min [f(x^1), f(x^2)] \tag{3.3}$$

for every $x^1 \in C$, $x^2 \in C$, and $0 \leq \lambda \leq 1$.

Proof. Suppose that f is quasiconcave. Then $U(f, \alpha)$ is convex for every α. Let $x^1 \in C$, $x^2 \in C$, and let $\bar{\alpha} = \min [f(x^1), f(x^2)]$. From the convexity of the set $U(f, \bar{\alpha})$ it follows that

$$f(\lambda x^1 + (1-\lambda)x^2) \geq \bar{\alpha} = \min [f(x^1), f(x^2)] \tag{3.4}$$

for every $0 \leq \lambda \leq 1$. Conversely, let $U(f, \alpha)$ be any upper-level set of f and let $x^1 \in U(f, \alpha)$, $x^2 \in U(f, \alpha)$. Then

$$f(x^1) \geq \alpha \quad \text{and} \quad f(x^2) \geq \alpha \tag{3.5}$$

and by (3.3) it follows that

$$f(\lambda x^1 + (1-\lambda)x^2) \geq \alpha \tag{3.6}$$

for every $0 \leq \lambda \leq 1$. Hence $U(f, \alpha)$ is a convex set and f is quasiconcave. □

It is clear that a concave function is also quasiconcave, since

$$f(\lambda x^1 + (1-\lambda)x^2) \geq \lambda f(x^1) + (1-\lambda)f(x^2) \geq \min [f(x^1), f(x^2)] \tag{3.7}$$

where the first inequality follows from the concavity of f.

An interesting question concerning concave and quasiconcave functions can be raised in view of the following result:

Proposition 3.2 (Fenchel, 1951). Let ϕ be a quasiconcave function defined on $C \subset R^n$ and let f be a nondecreasing function on $D \subset R$, containing the range of ϕ. Then the composite function $f\phi(x)$ is also quasiconcave.

Proof. Let ϕ be quasiconcave on C. Then for every $x^1 \in C$, $x^2 \in C$, and $0 \le \lambda \le 1$ we have

$$\phi(\lambda x^1 + (1-\lambda)x^2) \ge \min[\phi(x^1), \phi(x^2)]. \tag{3.8}$$

If f is nondecreasing, then

$$f\phi(\lambda x^1 + (1-\lambda)x^2) \ge f(\min[\phi(x^1), \phi(x^2)]) = \min[f\phi(x^1), f\phi(x^2)]. \tag{3.9}$$

Hence $f\phi$ is also quasiconcave. □

Various results on the quasiconcavity of composite functions were developed in Schaible (1971), where they were used to characterize products, ratios, and quadratic functions that are quasiconcave. These as well as more recent results will be discussed in detail in Chapters 5 and 6.

Proposition 3.2 cannot be strengthened to concave functions—that is, if ϕ is concave, then $f\phi$ is not necessarily concave too. An example can be $\phi(x) = x$ on $C = R^1$, and $f(y) = (y)^3$ on R^1. Then ϕ is concave, f is nondecreasing, but $f\phi(x) = (x)^3$ is not a concave function.

Note that in Proposition 2.16 f was restricted to be nondecreasing and concave. Under such conditions the composite function $f\phi$ preserves concavity.

It follows from the proposition that every nondecreasing transformation of a concave function is quasiconcave. Is it possible then that every quasiconcave function is a nondecreasing transformation of some concave function and consequently can be "concavified"—that is, transformed back into a concave function by an (inverse) nondecreasing function? The answer, in general, is negative. It was given by Fenchel (1951, 1956) and more recently by Crouzeix (1977), and Kannai (1977). We shall, however, see in Chapter 8 that for certain subfamilies of quasiconcave functions such transformation can be carried out; see also Zang (1981).

Although the upper-level sets of both concave and quasiconcave functions are convex, quasiconcave functions can considerably differ from concave functions as illustrated by Figure 3.2. We see, for instance, that in contrast to concave functions, quasiconcave functions can be discontinuous in the interior of their domain, and not every local maximum is a global one. Local maxima, however, that are not global cannot be strict maxima, as we show below.

Proposition 3.3. Let f be a quasiconcave function defined on the convex set $C \subset R^n$. If $x^* \in C$ is a strict local maximum of f, then x^* is also a strict global maximum of f on C. The set of points at which f attains its global maximum over C is a convex set.

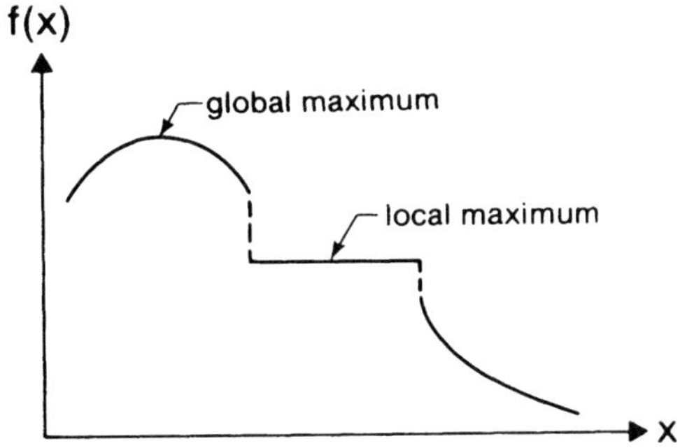

Figure 3.2. A quasiconcave function of one variable.

Proof. Suppose that $x^* \in C$ is a strict local maximum—that is, there is a $\delta > 0$ such that for every $x \neq x^*$ in the set $C \cap N_\delta(x^*)$, where

$$N_\delta(x^*) = \{x: x \in R^n, \|x - x^*\| < \delta\} \tag{3.10}$$

we have

$$f(x^*) > f(x). \tag{3.11}$$

If x^* is not a strict global maximum of f then there exists an $\bar{x} \in C$, $\bar{x} \neq x^*$ such that

$$f(\bar{x}) \geq f(x^*). \tag{3.12}$$

By the quasiconcavity of f

$$f(\lambda\bar{x} + (1 - \lambda)x^*) \geq f(x^*) \tag{3.13}$$

for all $0 \leq \lambda \leq 1$. But for sufficiently small λ it follows that $x = (\lambda\bar{x} + (1 - \lambda)x^*) \in C \cap N_\delta(x^*)$, contradicting (3.11). The convexity of the global maximizing set follows from the convexity of the upper-level sets of quasiconcave functions. □

We may mention here that if a local maximum x^0 of a quasiconcave function f is not a global maximum, then f is constant on the intersection of some neighborhood of x^0 and the line segment between x^0 and any global maximum x^*; see Ponstein (1967).

In order to better acquaint the reader with quasiconcave functions we turn now to some properties of single-variable quasiconcave functions—that is, functions defined on an interval of the real line. First let us introduce the concept of *line segment minimum property* (Diewert, Avriel, and Zang (1981), which will be used in the characterization of quasiconcave functions.

Definition 3.2. Let f be defined on the set $S \subset R^n$. Then f is said to have the line segment minimum property if f attains its minimum over every closed line segment $[x^1, x^2] \subset S$ at some point $x \in [x^1, x^2]$.

Note that no continuity assumptions appear in this definition. Clearly, if f is a lower semicontinuous (or continuous) function over S (see Definition 3.4 below), then it must have the line segment minimum property. Now we can state and prove the following proposition

Proposition 3.4 (Diewert, Avriel, and Zang, 1981). A quasiconcave function defined on the convex set $C \subset R^n$ has the line segment minimum property.

Proof. Let $[x^1, x^2] \subset C$ be a closed line segment. Then every point in $[x^1, x^2]$ can be written as $x = \lambda x^1 + (1 - \lambda)x^2$, $0 \leq \lambda \leq 1$, and by (3.3)

$$f(\lambda x^1 + (1 - \lambda)x^2) \geq \min [f(x^1), f(x^2)]. \tag{3.14}$$

Thus, the minimum of f over $[x^1, x^2]$ is attained at x^1 or x^2. □

The next concept, concerning functions of a single variable, was also introduced by Diewert, Avriel, and Zang (1981).

Definition 3.3. Let f be defined on the open interval $C \subset R$. Then f is said to attain a semistrict local minimum at a point $x^0 \in C$ if there exist two points $x^1 \in C$, $x^2 \in C$, $x^1 < x^0 < x^2$ such that

$$f(x^0) \leq f(\lambda x^1 + (1 - \lambda)x^2) \tag{3.15}$$

for every $0 \leq \lambda \leq 1$ and

$$f(x^0) < \min [f(x^1), f(x^2)]. \tag{3.16}$$

Note that if f attains a semistrict local minimum at x^0, then even if f is constant around x^0, the function eventually increases on both sides of x^0. Strict local minima are, clearly, semistrict local minima. Also note that whereas a local minimum in general can be a global maximum, this cannot happen in the case of a semistrict local minimum. An example of a semistrict local minimum is given in Figure 3.3.

For quasiconcave functions we have now the following proposition.

Proposition 3.5 (Diewert, Avriel, and Zang, 1981). Let f be defined on the open interval $C \subset R$. If f is quasiconcave, then there is no point $x^0 \in C$ that is a semistrict local minimum of f.

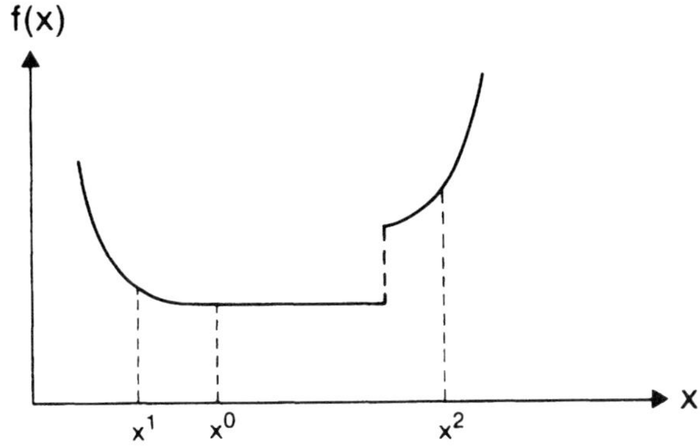

Figure 3.3. Semistrict local minimum.

Proof. If f is quasiconcave, then for every $x^1 \in C$, $x^2 \in C$, and $0 \leq \lambda \leq 1$

$$f(\lambda x^1 + (1-\lambda)x^2) \geq \min [f(x^1), f(x^2)] \tag{3.17}$$

and, clearly, for any two points of C no intermediate point can become a semistrict local minimum of f. □

The converse of this proposition is generally not true, as can be seen from the following example.

Example 3.1. Let

$$f(x) = \begin{cases} (x)^2 & \text{if } x < 0, \\ (x)^2 + \frac{3}{4} & \text{if } x \geq 0. \end{cases} \tag{3.18}$$

This function is not quasiconcave since the upper-level set of f at $\alpha = 1$, given by

$$U(f, 1) = \{x: -1 \geq x, x \geq 1/2\} \tag{3.19}$$

is not convex. However, the reader can easily verify that this function has no semistrict local minimum. Also, it can be shown that this function does not have the line segment minimum property.

Next, we present a characterization of quasiconcave functions of n variables.

Theorem 3.6 (Diewert, Avriel, and Zang, 1981). A function f defined on the open convex set $C \subset R^n$ is quasiconcave if and only if (i) it has the line segment minimum property, and (ii) for every $x \in C$ and $v \in R^n$ such that $v^T v = 1$, and $\bar{t} \in R$, $\bar{t} > 0$ satisfying $x + \bar{t}v \in C$, the function $F(t) = f(x + tv)$ does not attain a semistrict local minimum at any $t \in (0, \bar{t})$.

Proof. If f is quasiconcave then (i) follows from Proposition 3.4. Since quasiconcavity of f on C implies quasiconcavity of f on line segments in C (see Proposition 3.9 below), (ii) follows from Proposition 3.5. Conversely, let $x^1 \in C$, $x^2 \in C$, $x^1 \neq x^2$, $v = (x^2 - x^1)/\|x^2 - x^1\|$, and let $\bar{t} = \|x^2 - x^1\|$. Since f has the line segment minimum property, the function $F(t)$ attains its minimum over $[0, t]$ at some point $t^0 \in [0, \bar{t}]$. If F has no semistrict local minimum point on $(0, t)$, then for every $0 \leq t \leq \bar{t}$

$$F(\bar{t}) \geq F(t^0) \geq \min [F(0), F(1)] \tag{3.20}$$

where the right-hand inequality is actually an equation. Thus, for every $0 \leq \lambda \leq 1$

$$f(\lambda x^1 + (1 - \lambda)x^2) \geq \min [f(x^1), f(x^2)] \tag{3.21}$$

and f is quasiconcave. □

As mentioned earlier, if the function under consideration satisfies certain continuity conditions, then it will also satisfy the line segment minimum property.

First we need the following definition.

Definition 3.4. Let f be defined on $C \subset R^n$. It is said to be lower (upper) semicontinuous on C if its lower (upper) level sets $L(f, \alpha)$ $(U(f, \alpha))$ are closed relative to C for every $\alpha \in R$. It is continuous on C if it is both lower and upper semicontinuous.

Equivalently, it can be shown that f is lower semicontinuous on C if and only if for every $x^0 \in C$ and every sequence $\{x^k\} \subset C$ converging to x^0

$$\limsup_{k \to \infty} f(x^k) \geq f(\lim_{k \to \infty} x^k) = f(x^0) \tag{3.22}$$

holds. For an upper semicontinuous function the sense of the above inequality is reversed. Moreover, it is well known from mathematical analysis that a lower semicontinuous function attains its minimum over every nonempty compact subset of its domain; see Bartle (1976) or Berge (1963). The following proposition is then a special case of this result.

Proposition 3.7. Let f be a lower semicontinuous function defined on $S \subset R^n$. Then f has the line segment minimum property.

It follows that a lower semicontinuous function defined on an open convex set $C \subset R^n$ is quasiconcave if and only if there are no semistrict local minima in any open line segment contained in C. Another characterization of quasiconcavity is in terms of unimodal functions. We first look at the single-variable case.

Definition 3.5. Let f be defined on the interval $C \subset R$. It is said to be unimodal on C if there exists an $x^* \in C$ at which f attains its maximum on C and for any $x^1 \in C$, $x^2 \in C$, $x^1 < x^2$

$$x^2 \leq x^* \quad \text{implies that} \quad f(x^1) \leq f(x^2) \tag{3.23}$$

and

$$x^* \leq x^1 \quad \text{implies that} \quad f(x^2) \leq f(x^1). \tag{3.24}$$

The above definition simply means that for a single-variable unimodal function the point at which the maximum is attained divides C, the domain of f, into two subintervals (one of them possibly empty) such that in the subinterval to the left of x^* the function is nondecreasing, and in the subinterval to the right of x^* the function is nonincreasing. Examples of unimodal functions are shown in Figure 3.4. We then have the following proposition.

Proposition 3.8 (Martos, 1975). Let f be defined on the interval $C \subset R$ and suppose that it attains its maximum at a point $x^* \in C$. Then f is quasiconcave if and only if it is unimodal on C.

Proof. Suppose that f is quasiconcave and let $x^1 \in C$, $x^2 \in C$, $x^1 < x^2$. Then

$$x^1 < x^2 \leq x^* \quad \text{implies that} \quad f(x^2) \geq \min\,[f(x^1), f(x^*)] = f(x^1) \tag{3.25}$$

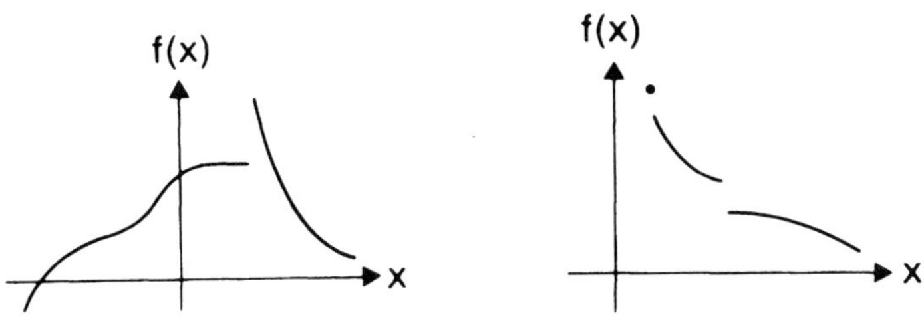

Figure 3.4. Unimodal functions.

and

$$x^* \leq x^1 < x^2 \quad \text{implies that} \quad f(x^1) \geq \min[f(x^*), f(x^2)] = f(x^2) \tag{3.26}$$

and f is unimodal. Conversely, suppose that f is unimodal and let again $x^1 \in C, x^2 \in C, x^1 < x^2$, and $x^0 = \lambda x^1 + (1-\lambda)x^2$ for some $0 < \lambda < 1$. Then

$$x^1 < x^0 \leq x^* \quad \text{implies that} \quad f(x^0) \geq f(x^1) \geq \min[f(x^1), f(x^2)] \tag{3.27}$$

and

$$x^* \leq x^0 < x^2 \quad \text{implies that} \quad f(x^0) \geq f(x^2) \geq \min[f(x^1), f(x^2)] \tag{3.28}$$

which shows that f is quasiconcave. □

Note that if C is a nonempty compact interval and f is upper semicontinuous then it attains a maximum over C. Hence, under such conditions quasiconcavity and unimodality are completely equivalent. A more elaborate version of the above theorem can be found in Martos (1975). The notion of unimodality has been generalized by Newman (1965) and Wilde (1964) to n dimensions:

Definition 3.6. Let f be defined on the convex set $C \subset R^n$. It is said to be unimodal on C if for every closed interval $[x^1, x^2] \subset C$ on which f attains a maximum at a point x^* we have $f(x^1) \leq f(x) \leq f(x^*)$, where $x = \lambda x^1 + (1-\lambda)x^*$, $0 < \lambda < 1$.

Using this definition it can be shown that for upper semicontinuous functions quasiconcavity and unimodality are equivalent. The reader is referred to Elkin (1968) for a proof of this result.

We present now a result analogous to Proposition 2.11 in the preceding chapter.

Proposition 3.9. The function f defined on the convex set $C \subset R^n$ is quasiconcave if and only if for every $x^1 \in C$, $x^2 \in C$ the single-variable function F, defined by

$$F(\lambda) \equiv f(\lambda x^1 + (1-\lambda)x^2) \tag{3.29}$$

is quasiconcave on the convex set $J = \{\lambda : \lambda \in R, 0 \leq \lambda \leq 1\}$.

Proof. Let $x^1 \in C$, $x^2 \in C$ and suppose that f is quasiconcave. Let λ^0, λ^1, λ^2 be such that $0 \leq \lambda^1 < \lambda^0 < \lambda^2 \leq 1$ and define

$$y^i = \lambda^i x^1 + (1-\lambda^i)x^2, \qquad i = 0, 1, 2. \tag{3.30}$$

Then

$$f(y^0) \geq \min [f(y^1), f(y^2)], \tag{3.31}$$

implying that

$$F(\lambda^0) \geq \min [F(\lambda^1), F(\lambda^2)] \tag{3.32}$$

and F is quasiconcave. Conversely, let $x^1 \in C$, $x^2 \in C$ and suppose that F is quasiconcave on J. Then

$$F(\lambda) \geq \min [F(0), F(1)] \tag{3.33}$$

and

$$f(\lambda x^1 + (1-\lambda)x^2) \geq \min [f(x^1), f(x^2)] \tag{3.34}$$

as asserted. □

Thus f is a quasiconcave function on its convex domain $C \subset R^n$ if and only if it is quasiconcave on every line segment contained in C.

Let us mention here a family of functions that can be viewed as a generalization of single-variable monotonic functions on the one hand, and of affine (linear) functions, on the other hand.

A function is said to be *quasimonotonic* if it is both quasiconcave and quasiconvex. For such a function, clearly, both its upper- and lower-level sets are convex, and for every x^1, x^2 in the convex domain of f and for every $0 \leq \lambda \leq 1$

$$\max [f(x^1), f(x^2)] \geq f(\lambda x^1 + (1-\lambda)x^2) \geq \min [f(x^1), f(x^2)]. \tag{3.35}$$

For functions of a single variable the notions of monotonicity (increasing or decreasing functions) and quasimonotonicity are equivalent.

It is well known (see, for example, Fenchel, 1951), that a function is both concave and convex if and only if it is affine. Quasimonotonic functions are generalizations of affine functions in the case that concave and convex functions are replaced by quasiconcave and quasiconvex ones. For further details on quasimonotonic functions see Martos (1975), Stoer and Witzgall (1970), and Thompson and Parke (1973).

An interesting result concerning quasimonotonic functions in terms of level surfaces or, using economic terminology, indifference surfaces (iso-utility surfaces, or isoproduct surfaces), is given next.

Proposition 3.10. Let f be a continuous function defined on the convex set $C \subset R^n$. The level surface defined by

$$Y(f, \alpha) = U(f, \alpha) \cap L(f, \alpha) = \{x: x \in C, f(x) = \alpha\} \tag{3.36}$$

is a convex set for all $\alpha \in R$, if and only if f is quasimonotonic.

Proof. If f is quasimonotonic, then it is both quasiconcave and quasiconvex. Hence both $U(f, \alpha)$ and $L(f, \alpha)$ are convex sets for all $\alpha \in R$. It follows that the intersection is also a convex set.

Conversely, suppose that $Y(f, \alpha)$ is convex for all α. We have to show that f is both quasiconcave and quasiconvex. Since the proof of quasiconvexity is completely analogous to the proof of quasiconcavity, we only present the latter case here. Let $x^1 \in C$, $x^2 \in C$ and let $x = \lambda x^1 + (1 - \lambda)x^2$, $0 < \lambda < 1$. Then we must show that

$$f(x) \geq \min [f(x^1), f(x^2)] \equiv \bar{\alpha}. \tag{3.37}$$

Suppose that $f(x) < \bar{\alpha}$ and without loss of generality assume that $\bar{\alpha} = f(x^1)$. Then $f(x) < f(x^1) \leq f(x^2)$. Since f is continuous on the compact line segment $[x^1, x^2]$, there exists an $\bar{x}$ such that $\bar{x} = \bar{\lambda}x + (1 - \bar{\lambda})x^2$, $0 \leq \bar{\lambda} < 1$ and $f(\bar{x}) = \bar{\alpha}$. Hence, by simple algebraic manipulations we obtain

$$x = \gamma\bar{x} + (1 - \gamma)x^1 \tag{3.38}$$

where

$$\gamma = \frac{(1 - \lambda)}{1 - \bar{\lambda} + (1 - \lambda)\bar{\lambda}} \tag{3.39}$$

and $0 < \gamma < 1$. Since $Y(f, \bar{\alpha})$ is convex and $\bar{x} \in Y(f, \bar{\alpha})$, $x^1 \in Y(f, \bar{\alpha})$ we must have $x \in Y(f, \bar{\alpha})$, contradicting that $f(x) < \bar{\alpha}$. □

We conclude this section by two observations due to Ponstein (1967) and Thompson and Parke (1973). First, for continuous functions Definition 3.1 can be slightly changed as follows: Let f be a continuous function defined on the convex set $C \subset R^n$. It is said to be quasiconcave if its *strict upper-level sets*

$$U^0(f, \alpha) = \{x: x \in C, f(x) > \alpha\} \tag{3.40}$$

are convex sets for every real α. The definition of continuous quasiconvex functions is analogous. Second, without loss of generality, the inequality in (3.3) is equivalent to saying that $f(x^2) \geq f(x^1)$ implies $f(\lambda x^1 + (1-\lambda)x^2) \geq f(x^1)$. Then, the characterization of quasiconcave functions as given in Theorem 3.1 can be modified as follows: Let f be a continuous function defined on a convex set $C \subset R^n$. It is quasiconcave if and only if for every $x^1 \in C$, $x^2 \in C$ such that $f(x^2) > f(x^1)$ and $0 \leq \lambda \leq 1$ we have $f(\lambda x^1 + (1-\lambda)x^2) \geq f(x^1)$.

3.2. Differentiable Quasiconcave Functions

We start this section by a characterization of once-differentiable quasiconcave functions. (See Definition 2.5 of differentiable functions.)

Theorem 3.11 (Arrow and Enthoven, 1961). Let f be differentiable on the open convex set $C \subset R^n$. Then f is quasiconcave if and only if for every $x^1 \in C$, $x^2 \in C$

$$f(x^1) \geq f(x^2) \quad \text{implies that} \quad (x^1 - x^2)^T \nabla f(x^2) \geq 0. \tag{3.41}$$

Proof. If f is quasiconcave then

$$f(x^1) \geq f(x^2) \quad \text{implies that} \quad f(\lambda x^1 + (1-\lambda)x^2) \geq f(x^2) \tag{3.42}$$

for every $0 < \lambda < 1$. Thus

$$\frac{f(x^2 + \lambda(x^1 - x^2)) - f(x^2)}{\lambda} \geq 0 \tag{3.43}$$

and by letting $\lambda \to 0^+$, that is, taking the limit as λ approaches 0 through positive numbers, we obtain

$$(x^1 - x^2)^T \nabla f(x^2) \geq 0. \tag{3.44}$$

Conversely, let $x^1 \in C$, $x^2 \in C$ and, without loss of generality, assume that $f(x^1) \geq f(x^2)$.

By Proposition 3.9 we have to show that the differentiable function

$$F(\lambda) = f(\lambda x^1 + (1-\lambda)x^2), \qquad 0 \leq \lambda \leq 1 \tag{3.45}$$

is quasiconcave—that is, $F(\lambda) \geq F(0)$ for all $0 \leq \lambda \leq 1$. Suppose that $F(\lambda) < F(0)$ for some λ. Since $F(1) \geq F(0)$, there must exist a $\lambda^0 > 0$ such that $F(0) > F(\lambda^0)$ and $F'(\lambda^0) > 0$. Then

$$F'(\lambda^0) = (x^1 - x^2)^T \nabla f(x^0) > 0, \tag{3.46}$$

where $x^0 = \lambda^0 x^1 + (1 - \lambda^0)x^2$. Since $f(x^2) > f(x^0)$ it follows from (3.41) that

$$(x^2 - x^0)^T \nabla f(x^0) \geq 0. \tag{3.47}$$

By rearranging and dividing both sides by λ^0 we get

$$(x^1 - x^2)^T \nabla f(x^0) \leq 0, \tag{3.48}$$

contradicting (3.46). □

The reader is urged to compare the above characterization of differentiable quasiconcave functions with the corresponding one given for concave functions in Theorem 2.12. Observe that inequality (2.70) implies (3.41).

We have already seen that a function is quasimonotonic if and only if it is both quasiconcave and quasiconvex. We can, therefore, state the following proposition.

Proposition 3.12. Let f be differentiable on the open convex set $C \subset R^n$. Then f is quasimonotonic if and only if for every $x^1 \in C$, $x^2 \in C$

$$f(x^1) \geq f(x^2) \quad \text{implies that} \quad (x^1 - x^2)^T \nabla f(x^2) \geq 0 \tag{3.49}$$

and

$$f(x^1) \leq f(x^2) \quad \text{implies that} \quad (x^1 - x^2)^T \nabla f(x^2) \leq 0. \tag{3.50}$$

From this proposition we get the following corollary, proof of which is left for the reader.

Corollary 3.13 (Martos, 1965). Let f be differentiable on the open convex set $C \subset R^n$. Then f is quasimonotonic if and only if for every $x^1 \in C$, $x^2 \in C$, and $x = \lambda x^1 + (1 - \lambda)x^2$, $0 \leq \lambda \leq 1$

$$f(x^1) \geq f(x^2) \quad \text{implies that} \quad (x^1 - x^2)^T \nabla f(x) \geq 0. \tag{3.51}$$

Another characterization of differentiable quasiconcave functions can be given in terms of semistrict local minima. We start with the single-variable case.

Proposition 3.14 (Diewert, Avriel, and Zang, 1981). Let f be a differentiable function defined on the open interval $C \subset R$. Then f is quasiconcave if and only if for every $x^0 \in C$ such that $f'(x^0) = 0$ the function f does not attain a semistrict local minimum at x^0.

The proof is a straightforward consequence of Theorem 3.6. In view of Proposition 3.9 it is also straightforward to generalize Proposition 3.14 to functions of n variables.

Theorem 3.15 (Diewert, Avriel, and Zang, 1981). Let f be a differentiable function defined on the open convex set $C \subset R^n$. Then f is quasiconcave if and only if for every $x^0 \in C$ and $v \in R^n$ such that $v^T v = 1$ and $v^T \nabla f(x^0) = 0$ the function $F(t) = f(x^0 + tv)$ does not attain a semistrict local minimum at $t = 0$.

The geometric interpretation of the condition in the above theorem is that the function f, restricted to a line passing through x^0 in a direction v, orthogonal to the gradient of f at x^0, does not have a semistrict local minimum at x^0. In particular, if $\nabla f(x^0) = 0$ at some point $x^0 \in C$, then x^0 is not a semistrict local minimum of f along any line passing through x^0.

For twice differentiable quasiconcave functions a direction v, orthogonal to ∇f has some interesting properties as we shall see below. First we present a necessary condition for twice continuously differentiable quasiconcave functions.

Proposition 3.16 (Arrow and Enthoven, 1961; Avriel, 1972). Let f be a twice continuously differentiable quasiconcave function on the open convex set $C \subset R^n$. If $x^0 \in C$, $v \in R^n$ and $v^T \nabla f(x^0) = 0$, then $v^T \nabla^2 f(x^0) v \leq 0$.

Proof. Suppose that f is quasiconcave, $v^T \nabla f(x^0) = 0$ and $v^T \nabla^2 f(x^0) v > 0$. By continuity, there exists an $\varepsilon > 0$ such that $\|\bar{x} - x^0\| < \varepsilon$ implies

$$v^T \nabla^2 f(\lambda \bar{x} + (1 - \lambda) x^0) v > 0 \tag{3.52}$$

for all $0 \leq \lambda \leq 1$. Let $\bar{x} = x^0 + \mu v$ with $0 < \mu < \varepsilon$. Then (3.52) can be written as

$$(\bar{x} - x^0)^T \nabla^2 f(\bar{x} + (1 - \lambda) x^0)(\bar{x} - x^0) > 0. \tag{3.53}$$

Hence, by Taylor's theorem,

$$f(\bar{x}) = f(x^0) + (\bar{x} - x^0)^T \nabla f(x^0) + \tfrac{1}{2}(\bar{x} - x^0)^T \nabla^2 f(\bar{\lambda}\bar{x} + (1 - \bar{\lambda})x^0)(\bar{x} - x^0) \tag{3.54}$$

holds for some $0 < \bar{\lambda} < 1$. Then, it follows from $v^T \nabla f(x^0) = 0$ and (3.52) that

$$f(\bar{x}) > f(x^0). \tag{3.55}$$

Letting $\hat{x} = x^0 + \mu(-v)$ we get, by similar arguments, that

$$f(\hat{x}) > f(x^0). \tag{3.56}$$

Since $x^0 = \frac{1}{2}(\bar{x} + \hat{x})$ it follows from (3.55) and (3.56) that f is not quasiconcave, hence $v^T \nabla^2 f(x^0) v$ must be nonpositive. □

We have immediately the following corollary.

Corollary 3.17. Let f be a twice continuously differentiable quasiconcave function on the open convex set $C \subset R^n$. If $x^0 \in C$ and $\nabla f(x^0) = 0$, then $z^T \nabla^2 f(x^0) z \leq 0$ for every $z \in R^n$.

Necessary and sufficient conditions for quasiconcavity in terms of second derivatives are somewhat more complicated than those presented in Chapter 2 for concave functions.

The first result in this direction follows from the last proposition.

Corollary 3.18 (Gerencsér, 1973). Let f be a twice continuously differentiable quasiconcave function on the open convex set $C \subset R^n$. Then $\nabla^2 f$, the Hessian of f, has at most one positive eigenvalue at every $x \in C$.

Proof. Suppose that $x^0 \in C$ and $\nabla^2 f(x^0)$ has two (or more) positive eigenvalues. Let $e^1, \ldots, e^n$ denote a set of n orthogonal eigenvectors of $\nabla^2 f(x^0)$ and, without loss of generality, assume that the eigenvalues associated with e^1 and e^2 are positive. Let T denote the subspace spanned by the set of vectors orthogonal to $\nabla f(x^0)$ and let E denote the subspace spanned by e^1 and e^2. Then $T \cap E$ must contain at least one nonzero vector v^0, given by $v^0 = \alpha_1 e^1 + \alpha_2 e^2$, where α_1, α_2 are not both zero. Clearly, $(v^0)^T \nabla f(x^0) = 0$. Computing $(v^0)^T \nabla^2 f(x^0) v^0$ and noting that e^1, e^2 are orthogonal eigenvectors with positive eigenvalues we find that the quadratic form is positive, contradicting Proposition 3.16. □

We have seen in Chapter 2 that the Hessian of a concave function cannot have positive eigenvalues. By the last corollary, we see that the generalization of concavity to quasiconcavity is, therefore, equivalent to allowing the existence of at most one positive eigenvalue in the Hessian.

In this connection we can mention the concept of negative subdefinite matrices, introduced by Martos (1969) and defined as follows:

Definition 3.7. The $n \times n$ symmetric matrix B of real numbers is said to be negative subdefinite if

$$y^T B y > 0 \quad \text{implies} \quad By \geq 0 \quad \text{or} \quad By \leq 0 \tag{3.57}$$

for every $y \in R^n$. If B is negative subdefinite, then $(-B)$ is called positive subdefinite.

In analogy with this definition, B is said to be negative semidefinite if

$$y^T B y \geq 0 \quad \text{implies} \quad By = 0. \tag{3.58}$$

It can be shown that this definition of a negative semidefinite matrix is equivalent to the one given in Definition 2.6. Furthermore, if B is negative semidefinite, then it is also negative subdefinite and the family of negative subdefinite matrices is precisely the family of $n \times n$ symmetric matrices that have at most one positive eigenvalue. It follows that the Hessian matrix of a twice continuously differentiable quasiconcave function is subdefinite.

Example 3.2. The necessary condition of Proposition 3.16 is not sufficient for quasiconcavity. Consider the function $f(x) = (x)^4$ on $C = \{x\colon x \in R, -1 < x < 1\}$. Then

$$\nabla f(x) = f'(x) = 4(x)^3 \tag{3.59}$$

and the only point $x^0 \in C$ for which a $v \neq 0$ can be found such that $v^T \nabla f(x^0) = 0$ is $x^0 = 0$. At this point $v^T \nabla^2 g(x^0) v = 0$ for every $v \in R$, hence the condition of Proposition 3.16 is satisfied, but f is not quasiconcave. It is interesting to note that f is quasiconcave on any other open interval not containing the origin (the point where the gradient vanishes).

We shall now see that the condition of Proposition 3.16 together with some additional conditions are sufficient for quasiconcavity. The first result in this direction is in Katzner (1970), where it was shown that a twice continuously differentiable function f defined on an open convex set $C \subset R^n$ is quasiconcave if it (i) has a gradient whose components are negative, and (ii) satisfies the condition

$$x^0 \in C, \qquad v \in R^n, \qquad v^T \nabla f(x^0) = 0 \quad \Rightarrow \quad v^T \nabla^2 f(x^0) v \leq 0. \tag{3.60}$$

This result was strengthened in Diewert, Avriel, and Zang (1977, 1981), where (i) was replaced by the condition that the gradient should not vanish on C. A yet stronger result is the following theorem.

Theorem 3.19 (Crouzeix, 1980). Let f be a twice differentiable function on the open convex set $C \subset R^n$. Suppose that (3.60) holds and also

$$x^0 \in C, \qquad \nabla f(x^0) = 0 \;\Rightarrow\; \nabla^2 f(x^0) \text{ is negative definite.} \tag{3.61}$$

Then f is quasiconcave.

Proof. Suppose that f is not quasiconcave, i.e., there exist $x^1 \in C$, $x^2 \in C$, $x^1 \neq x^2$, such that

$$M = \min_{0 \le \lambda \le 1} f(x^1 + \lambda(x^2 - x^1)) < \min [f(x^1), f(x^2)]. \tag{3.62}$$

Let $\bar{x} \in (x^1, x^2)$ be the global minimizer of f on $[x^1, x^2]$, which is the closest to x^1, and let $\bar{\lambda}$ be defined by $\bar{x} = x^1 + \bar{\lambda}(x^2 - x^1)$. Also let $h(\lambda) = f(\bar{x} + \lambda(x^2 - x^1))$. Then,

$$h(\lambda) > h(0) = M, \qquad -\bar{\lambda} \le \lambda < 0, \tag{3.63}$$

$$h(\lambda) \ge h(0) = M, \qquad 0 \le \lambda \le 1 - \bar{\lambda}. \tag{3.64}$$

Also notice that

$$h'(0) = (x^2 - x^1)^T \nabla f(\bar{x}) = 0. \tag{3.65}$$

Two cases are now possible:

1. $\nabla f(\bar{x}) = 0$ holds. Then, by (3.61)

$$h''(0) = (x^2 - x^1)^T \nabla^2 f(\bar{x})(x^2 - x^1) < 0. \tag{3.66}$$

Applying a second-order Taylor expansion to $h(\lambda)$ around $\lambda = 0$ we have

$$h(\lambda) = h(0) + \lambda h'(0) + \tfrac{1}{2}(\lambda)^2 h''(0) + (\lambda)^2 \alpha(\lambda), \tag{3.67}$$

where $\alpha(\lambda) \to 0$ as $\lambda \to 0$. From (3.66) it follows that $(\lambda)^2[\tfrac{1}{2}h''(0) + \alpha(\lambda)] < 0$ for sufficiently small λ, hence using (3.65) we obtain $h(\lambda) < h(0)$, contradicting (3.63)-(3.64).

2. $\nabla f(\bar{x}) \neq 0$ holds. Note that by adding a constant to f, and by applying a nonsingular linear transformation to the variables, we can obtain, without loss of generality, that $\bar{x} = 0$, $f(\bar{x}) = 0$, and

$$\nabla f(\bar{x}) = (0, \ldots, 0, 1)^T. \tag{3.68}$$

Now let $\hat{x} \in R^{n-1}$ be the vector $\hat{x} = (x_1, \ldots, x_{n-1})^T$, hence any $x \in R^n$ is given by $x = (\hat{x}^T, x_n)^T$. It follows from (3.65) and (3.68) that $(x^2 - x^1) = ((\hat{x}^2 - \hat{x}^1)^T, 0)^T$. Invoking the Implicit Function Theorem (Bartle, 1976), there exist a convex neighborhood $U \subset R^{n-1}$ of $\hat{0}$ and a twice-differentiable function g from U to R such that for every $\hat{x} \in U$ we have

$$f(\hat{x}, g(\hat{x})) = 0, \tag{3.69}$$

$$f_n(\hat{x}, g(\hat{x})) > 0, \tag{3.70}$$

and

$$\nabla g(x) = -\frac{1}{f_n(\hat{x}, g(\hat{x}))} \nabla_{\hat{x}} f(\hat{x}, g(x)) \tag{3.71}$$

where $\nabla_{\hat{x}} f \in R^{n-1}$ is the vector of first partial derivatives of f with respect to its first $n-1$ arguments, and f_n is the first partial derivative of f with respect to its nth argument. Note that (3.70) follows from (3.68) and that (3.71) is obtained from differentiating (3.69) with respect to $\hat{x}$. Also note that $\bar{x} = 0$ implies $g(\hat{0}) = 0$ and that by (3.68) and (3.71) we obtain $\nabla g(\hat{0}) = 0$. Further differentiation of (3.71) with respect to $\hat{x}$ yields

$$\begin{aligned}\nabla^2 g(\hat{x}) = & -\frac{1}{f_n(\hat{x}, g(\hat{x}))} \{\nabla^2_{\hat{x}\hat{x}} f(\hat{x}, g(\hat{x})) + \nabla g(\hat{x}) \nabla^2_{n\hat{x}} f(\hat{x}, g(\hat{x}))^T \\ & + \nabla^2_{\hat{x}n} f(\hat{x}, g(\hat{x})) \nabla g(\hat{x})^T + \nabla^2_{nn} f(\hat{x}, g(\hat{x})) \nabla g(\hat{x}) \nabla g(\hat{x})^T\}\end{aligned} \tag{3.72}$$

for every $\hat{x} \in U$, where the second derivatives in (3.72) partition the Hessian of f as follows:

$$\nabla^2 f = \begin{vmatrix} \nabla^2_{\hat{x}\hat{x}} f & \nabla^2_{\hat{x}n} f \\ \nabla^2_{n\hat{x}} f^T & \nabla^2_{nn} f \end{vmatrix}. \tag{3.73}$$

Consider now the vector $\bar{z} \in R^n$, given by

$$\bar{z} = (\hat{z}^T, \hat{z}^T \nabla g(\hat{x}))^T. \tag{3.74}$$

It follows from (3.71) that $\bar{z}^T\nabla f(\hat{x}, g(\hat{x})) = 0$, and by (3.60) we obtain

$$\bar{z}^T\nabla^2 f(\hat{x}, g(\hat{x})) \le 0. \tag{3.75}$$

Using the partition (3.73) we get from (3.72)

$$\hat{z}^T\nabla^2 g(\hat{x})\hat{z} = -\frac{\bar{z}^T\nabla^2 f(\hat{x}, g(\hat{x}))\bar{z}}{f_n} \ge 0 \tag{3.76}$$

for all $\hat{x} \in U$ and $\hat{z} \in R^{n-1}$, where the inequality in (3.76) follows from (3.75). Hence g is convex on u.

Letting $\hat{u} = \hat{x}^2 - \hat{x}^1$ we obtain from (3.63) in terms of the new coordinates that

$$f(\lambda\hat{u}, 0) > f(\hat{0}, 0) = 0 \tag{3.77}$$

implying, in view of (3.69), that $g(\lambda\hat{u}) \ne 0$ must hold for every $\lambda \in [-\bar{\lambda}, 0)$ such that $\lambda\hat{u} \in u$. Moreover, by Taylor's theorem

$$g(\lambda\hat{u}) = g(\hat{0}) + \lambda\hat{u}\nabla g(\hat{0}) + \tfrac{1}{2}(\lambda)^2\hat{u}^T\nabla^2 g(\hat{v})\hat{u} \tag{3.78}$$

where $\hat{v} \in (0, \lambda\hat{u})$. Since g is convex and the first two terms on the right-hand side of (3.78) vanish, it follows that $g(\lambda\hat{u}) \ge 0$ and in particular $g(\lambda\hat{u}) > 0$ for all $\lambda \in [-\lambda, 0) \cap \{\lambda: \lambda\hat{u} \in U\}$. Also, the continuity of ∇f and (3.70) imply the existence of a convex neighborhood V of $\hat{0}$, $V \subset U$ and a number $\delta > 0$ such that

$$(\hat{x}, x_n) \in V \times [-\delta, \delta] \quad \Rightarrow \quad f_n(\hat{x}, x_n) > 0. \tag{3.79}$$

The continuity of g implies the existence of a $\tilde{\lambda} \in [-\bar{\lambda}, 0)$ such that $\tilde{\lambda}\hat{u} \in V$ and $0 < g(\tilde{\lambda}\hat{u}) < \delta$.

Finally, we have from (3.77), (3.69), and (3.79), respectively,

$$f(\tilde{\lambda}\hat{u}, 0) > 0 \tag{3.80}$$

$$f(\tilde{\lambda}\hat{u}, g(\tilde{\lambda}\hat{u})) = 0 \tag{3.81}$$

and

$$(0, \ldots, 0, 1)^T\nabla f(\tilde{\lambda}\hat{u}, \lambda) = f_n(\tilde{\lambda}\hat{u}, x_n) > 0 \tag{3.82}$$

for every $x_n \in [0, g(\tilde{\lambda}\hat{u})]$. But (3.82) implies that f is increasing from $x_n = 0$ to $x_n = \tilde{\lambda}\hat{u}$ when the first argument is kept at $\tilde{\lambda}\hat{u}$, thereby contradicting (3.80) and (3.81); thus (3.62) cannot hold, and the proof is complete. □

From the above theorem and Proposition 3.16 we obtain the following result proved by Diewert, Avriel, and Zang (1977). A shorter proof can be found in Otani (1983).

Corollary 3.20. Let f be a twice continuously differentiable function on the open convex set $C \subset R^n$ and suppose that $\nabla f(x) \neq 0$ for every $x \in C$. Then f is quasiconcave if and only if

$$x \in C, \qquad v \in R^n, \qquad v^T \nabla f(x) = 0 \;\Rightarrow\; v^T \nabla^2 f(x) v \leq 0. \tag{3.83}$$

The following two results, appearing in Diewert, Avriel, and Zang (1981) and Avriel and Schaible (1978), complement Proposition 3.14 and Theorem 3.15 for twice-differentiable functions.

Proposition 3.21. Let f be a twice continuously differentiable function defined on the open interval $C \subset R$. Then f is quasiconcave if and only if for every $x^0 \in C$ such that $f'(x^0) = 0$ either $f''(x^0) < 0$ or $f''(x^0) = 0$ and f does not attain a semistrict local minimum at x^0.

For functions of n variables Proposition 3.21 generalizes to the following theorem.

Theorem 3.22. Let f be a twice continuously differentiable function defined on the open convex set $C \subset R^n$. Then f is quasiconcave if and only if for every $x^0 \in C$ and $v \in R^n$ such that $v^T v = 1$ and $v^T \nabla f(x^0) = 0$ either $v^T \nabla^2 f(x^0) v < 0$ or $v^T \nabla^2 f(x^0) v = 0$ and the function $F(t) = f(x^0 + tv)$ does not attain a semistrict local minimum at $t = 0$.

We conclude this section by mentioning some more necessary and sufficient conditions for quasiconcavity of twice-differentiable functions—this time in terms of "bordered determinants." Arrow and Enthoven (1961) were the first to derive these conditions for functions defined on the nonnegative orthant of R^n. Their conditions were subsequently extended by Ferland (1971) to functions defined on more general convex sets. Related results can also be found in Avriel and Schaible (1978), and in Crouzeix and Ferland (1981).

The kth-order leading submatrix $D_k(x)$ of the Hessian of a twice continuously differentiable function f at a point $x \in R^n$ is defined as

$$D_k(x) = \begin{vmatrix} 0 & \dfrac{\partial f}{\partial x_1} & \cdots & \dfrac{\partial f}{\partial x_k} \\ \dfrac{\partial f}{\partial x_1} & \dfrac{\partial^2 f}{\partial x_1\,\partial x_1} & \cdots & \dfrac{\partial^2 f}{\partial x_1\,\partial x_k} \\ \vdots & \vdots & & \vdots \\ \dfrac{\partial f}{\partial x_k} & \dfrac{\partial^2 f}{\partial x_k\,\partial x_1} & \cdots & \dfrac{\partial^2 f}{\partial x_k\,\partial x_k} \end{vmatrix}, \qquad k = 1, \ldots, n. \tag{3.84}$$

A subset of R^n is called *solid* if it has a nonempty interior (the nonnegative orthant of R^n is clearly a solid set). A necessary condition for f to be quasiconcave on a solid convex set $C \subset R^n$ is that

$$(-1)^k \det D_k(x) \geq 0, \qquad k = 1, \ldots, n \tag{3.85}$$

for every $x \in C$, where "det" denotes determinant. A sufficient condition for a twice continuously differentiable function f defined on an open set containing the convex set $C \subset R^n$ to be quasiconcave on C is that

$$(-1)^k \det D_k(x) > 0, \qquad k = 1, \ldots, n \tag{3.86}$$

for every $x \in C$.

3.3. Strict Quasiconcavity

Let us now strengthen quasiconcavity. We start by the characterization of quasiconcave functions as given in Theorem 3.1. We can use that characterization as a definition of quasiconcave functions: Let f be defined on the convex set $C \subset R^n$. It is said to be quasiconcave if

$$f(\lambda x^1 + (1 - \lambda)x^2) \geq \min\,[f(x^1), f(x^2)] \tag{3.87}$$

for every $x^1 \in C$, $x^2 \in C$, and $0 \leq \lambda \leq 1$. We then have the following definition.

Definition 3.8. Let f be defined on the convex set $C \subset R^n$. It is said to be strictly quasiconcave if

$$f(\lambda x^1 + (1 - \lambda)x^2) > \min [f(x^1), f(x^2)] \tag{3.88}$$

for every $x^1 \in C$, $x^2 \in C$, $x^1 \neq x^2$, and $0 < \lambda < 1$. If f is strictly quasiconcave, then $g \equiv -f$ is a strictly quasiconvex function.

There is considerable confusion in the literature concerning the terminology of various families of generalized concave functions. Strictly quasiconcave functions are called "strongly quasiconcave" by Avriel (1976), "unnamed" by Ponstein (1967), "X-concave" by Thompson and Parke (1973), and "strictly quasiconcave" by Elkin (1968), Ginsberg (1973), and Ortega and Rheinboldt (1970).

It is clear that every strictly quasiconcave function is quasiconcave. If so, what is the difference between the two families of functions? A function that is quasiconcave but not strictly quasiconcave is constant on some interval in its domain of definition; upper-level sets of upper semicontinuous (and continuous) strictly quasiconcave functions are "strictly convex" sets in the interior of their domain of definition (see Definition 3.10 later in this chapter); a strictly quasiconcave function attains its maximum at no more than one point, and so on. We shall return to all these properties later in this section. It should be mentioned that strict quasiconcavity is not a proper generalization of concavity, but only of strict concavity. A concave function that is "flat" over some region cannot be strictly quasiconcave, but every strictly concave function is strictly quasiconcave, as can be easily proven by using the definitions of the two families of functions.

As in the quasiconcave case we turn now to discuss line segment properties of strictly quasiconcave functions, beginning with the single-variable case.

Proposition 3.23. Let f be defined on the open interval $C \subset R$. If f is strictly quasiconcave, then there is no point $x^0 \in C$ that is a local minimum of f.

Proof. If f is strictly quasiconcave, then for every $x^1 \in C$, $x^2 \in C$, $x^1 \neq x^2$, and $0 < \lambda < 1$

$$f(\lambda x^1 + (1 - \lambda)x^2) > \min [f(x^1), f(x^2)]. \tag{3.89}$$

Thus, for any two distinct points of C there is no intermediate point that can be a local minimum of f. □

Next we have the following proposition.

Proposition 3.24 (Diewert, Avriel, and Zang, 1981). Let f have the line segment minimum property on the open convex set $C \subset R$. If there is no point $x^0 \in C$ that is a local minimum of f, then f is strictly quasiconcave.

Proof. Let $x^1 \in C$, $x^2 \in C$, $x^1 < x^2$. By the hypotheses, f attains its minimum over the closed interval $[x^1, x^2]$. If the set of minimizing points consists of one or both end points only, then

$$f(\lambda x^1 + (1-\lambda)x^2) > \min[f(x^1), f(x^2)]. \tag{3.90}$$

Suppose now that the set of minimizing points contains a point x^0 such that $x^1 < x^0 < x^2$. But this contradicts the hypothesis that there is no local minimum point in C. Thus our supposition is false, and f is strictly quasiconcave. □

Similar to the quasiconcave case we can conclude that for functions of a single variable having the line segment minimum property, the nonexistence of local minima characterizes strict quasiconcavity. One can easily derive an analogous result to Proposition 3.9, namely, that a function is strictly quasiconcave on the convex set $C \subset R^n$ if and only if it is strictly quasiconcave on every line segment contained in C. An analogous result to Theorem 3.6 is the following theorem.

Theorem 3.25 (Diewert, Avriel, and Zang, 1981). A function f defined on the open convex set $C \subset R^n$ is strictly quasiconcave if and only if (i) it has the line segment minimum property, and (ii) for every $x \in C$ and $v \in R^n$ such that $v^T v = 1$, and $\bar{t} \in R$, $\bar{t} > 0$ satisfying $x + \bar{t}v \in C$ the function $F(t) = f(x + tv)$ does not attain a local minimum at any $t \in (0, \bar{t})$.

The proof is very similar to that of Theorem 3.6 and is left for the reader.

Theorem 3.26. Let f be a differentiable function defined on the open convex set $C \subset R^n$. Then f is strictly quasiconcave if and only if for every $x^0 \in C$ and $v \in R^n$ such that $v^T v = 1$ and $v^T \nabla f(x^0) = 0$ the function $f(t) = f(x^0 + tv)$ does not attain a local minimum at $t = 0$. If f is twice continuously differentiable then a necessary and sufficient condition for strict quasiconcavity of f is that for every $x^0 \in C$ and $v \in R^n$ such that $v^T \nabla f(x^0) = 0$, and $v^T v = 1$, either $v^T \nabla^2 f(x^0) v < 0$ or $v^T \nabla^2 f(x^0) v = 0$ and $F(t) = f(x^0 + tv)$ does not attain a local minimum at $t = 0$.

Next we discuss the relationship between strict quasiconcavity and strict unimodality, defined by Newman (1965) and Wilde (1964). First we strengthen the concept of unimodality, introduced in Definition 3.6.

Definition 3.9. Let f be defined on the convex set $C \subset R^n$. It is said to be strictly unimodal on C if for every closed interval $[x^1, x^2] \subset C$ on which f attains a maximum at a point $x^* \neq x^1$ we have $f(x^1) < f(x) < f(x^*)$, where $x = \lambda x^1 + (1-\lambda)x^*$, $0 < \lambda < 1$.

We now have the following proposition.

Proposition 3.27 (Elkin, 1968). Let f be an upper semicontinuous function defined on the convex set $C \subset R^n$. Then f is strictly quasiconcave if and only if it is strictly unimodal.

Proof. Assume that f is strictly quasiconcave. Then it must be quasiconcave. Let $x^1 \in C$, $x^2 \in C$ and let x^* be a maximum of f on the closed interval $[x^1, x^2]$, such that $x^* \neq x^1$. Since f is quasiconcave, it is also unimodal and for every $x = \lambda x^1 + (1-\lambda)x^*$, $0 < \lambda < 1$ we have $f(x^1) \leq f(x) \leq f(x^*)$. If $f(x^1) = f(x)$ then f must be constant either on $[x^1, x]$ or on $[x, x^*]$. To see this, suppose that f is not constant on either interval. Then we can find an $\bar{x} \in [x^1, x]$ and an $\hat{x} \in [x, x^*]$ such that $f(\bar{x}) > f(x)$ and $f(\hat{x}) > f(x)$. Hence,

$$f(x) < \min\,[f(\bar{x}), f(\hat{x})] \tag{3.91}$$

and since x is in $[\bar{x}, \hat{x}]$, f cannot be quasiconcave. But a strictly quasiconcave function by definition cannot be constant on any interval, hence $f(x^1) < f(x)$. If $f(x) = f(x^*)$, then, by strict quasiconcavity, for every $x^0 = \lambda^0 x + (1-\lambda^0)x^*$, $0 < \lambda^0 < 1$, we must have $f(x^0) > f(x^*)$, contradicting that x^* is the maximum of f on $[x^1, x^2]$. Conversely, let f be strictly unimodal. Then it is unimodal and hence quasiconcave. Let $x^1 \in C$, $x^2 \in C$, $x^1 \neq x^2$. Let $x^0 = \lambda x^1 + (1-\lambda)x^2$, $0 < \lambda < 1$. By the quasiconcavity of f we have

$$f(x^0) \geq \min\,[f(x^1), f(x^2)]. \tag{3.92}$$

If equality holds in (3.92) then, by arguments similar to those in the first part of this proof, we could conclude that f is constant on some interval, contradicting strict unimodality. □

Next we turn our attention to the level sets of a strictly quasiconcave function. We would like to characterize this family of functions in terms of level sets, as we did in the definition of quasiconcave functions. It would be nice to give such a characterization in terms of strictly convex level sets, as defined below, but, unfortunately, such level sets are only necessary and not sufficient for strict quasiconcavity. First we need the following definition.

Definition 3.10. A convex set $C \subset R^n$ is said to be strictly convex if for any two points x^1, x^2 on its boundary, every point x given by $x = \lambda x^1 + (1-\lambda)x^2$, $0 < \lambda < 1$ is an interior point of C.

Alternatively, the set C is strictly convex if all its boundary points are extreme points (see Definition 2.8).

Proposition 3.28 (Elkin, 1968). Let f be a continuous function, defined on R^n. If f is strictly quasiconcave then its upper level sets are strictly convex.

Proof. If an upper level set of f is not convex then f is not quasiconcave, hence not strictly quasiconcave. If $U(f, \bar{\alpha})$ is convex but not strictly convex for some $\bar{\alpha} \in R$, then there exist points x^1, x^2 and $x = \lambda x^1 + (1 - \lambda)x^2$, for some $0 < \lambda < 1$, all three on the boundary of $U(f, \bar{\alpha})$. Since f is continuous on R^n, it follows that a point $\bar{x} \in U(f, \bar{\alpha})$ such that $f(\bar{x}) < \bar{\alpha}$ must be in the interior of $U(f, \bar{\alpha})$. Hence $f(x^1) = f(x^2) = \bar{\alpha}$, contradicting that f is strictly quasiconcave.

Although the last proposition was proven only for functions defined on the whole space R^n, the result still holds for functions defined on a proper convex subset $C \subset R^n$, provided that all topological notions (especially the notions of boundary and interior) are interpreted in the relative topology for C.

Example 3.3. As mentioned above, strict convexity of level sets is not sufficient for strict quasiconcavity. Consider the function f defined on R^n and given by

$$f(x) = \begin{cases} 0 & \text{if } \|x\| \leq 1, \\ 1 - \|x\|^2 & \text{if } \|x\| > 1. \end{cases} \tag{3.93}$$

This is a concave function, whose level sets are strictly convex, but f is not strictly quasiconcave, as can be asserted by, for example, Proposition 3.23, since the origin $x = 0$ is a local minimum of f (this function has, in fact, a "flat" maximum).

Motivated by this example, the reader can easily prove the following proposition.

Proposition 3.29. A strictly quasiconcave function f defined on the convex set $C \subset R^n$ attains its maximum on C at no more than one point.

An important property shared by concave and strictly quasiconcave functions is that every local maximum is a global one. This property, however, holds for more general families of functions as well, and we shall return to it in the next section.

3.4. Semistrict Quasiconcavity

In this section we shall be concerned with a family of functions that, subject to some continuity requirements, lies between the families of quasiconcave and strictly quasiconcave functions. Let us recall that we can define quasiconcave functions by the condition

$$f(x^2) \geq f(x^1) \quad \text{implies that} \quad f(\lambda x^1 + (1-\lambda)x^2) \geq f(x^1) \tag{3.94}$$

for every $0 \leq \lambda \leq 1$. Similarly, the condition for a strictly quasiconcave function can be written as

$$f(x^2) \geq f(x^1) \quad \text{implies that} \quad f(\lambda x^1 + (1-\lambda)x^2) > f(x^1) \tag{3.95}$$

for every $x^1 \neq x^2$, and $0 < \lambda < 1$. Next, we have a seemingly intermediate definition as follows:

Definition 3.11. Let f be defined on the convex set $C \subset R^n$. It is said to be semistrictly quasiconcave if

$$f(x^2) > f(x^1) \quad \text{implies that} \quad f(\lambda x^1 + (1-\lambda)x^2) > f(x^1) \tag{3.96}$$

where $x^1 \in C$, $x^2 \in C$, and $0 < \lambda < 1$. If f is semistrictly quasiconcave, then $g \equiv -f$ is a semistrictly quasiconvex function.

What is the relationship between the family of semistrictly quasiconcave functions and the first two families of generalized concave functions introduced earlier in this chapter? The following example is perhaps somewhat surprising in its implications.

Example 3.4. Consider the function f defined on $C = [-1, 1] \subset R$ and given by

$$f(x) = \begin{cases} 1 & \text{if } x \in C, \quad x \neq 0, \\ \frac{1}{2} & \text{if } x = 0. \end{cases} \tag{3.97}$$

This function is also illustrated in Figure 3.5. This function is semistrictly quasiconcave but neither quasiconcave nor strictly quasiconcave, since $U(f, 1)$ is not convex.

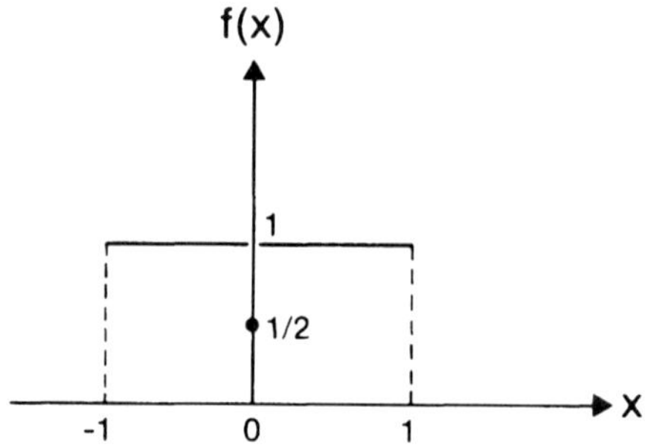

Figure 3.5. A semistrictly quasiconcave function.

The reason why the function appearing in the example is not quasiconcave lies in its (lack of) continuity properties. The family of semistrictly quasiconcave functions can fit nicely into the framework of concave and generalized concave functions if some continuity requirements are satisfied.

Proposition 3.30 (Karamardian, 1967). If f is an upper semicontinuous semistrictly quasiconcave function defined on the convex set $C \subset R^n$, then it is also quasiconcave.

Proof. We only have to show that for $x^1 \in C$, $x^2 \in C$ such that $f(x^1) = f(x^2)$ we have

$$f(\lambda x^1 + (1-\lambda)x^2) \geq f(x^1) \tag{3.98}$$

for every $0 < \lambda < 1$. Suppose now that (3.98) does not hold. Then there exists an x^0 such that

$$f(x^0) = f(\lambda^0 x^1 + (1-\lambda^0)x^2) < f(x^1) \tag{3.99}$$

for some $0 < \lambda^0 < 1$. By strict quasiconcavity, $f(x^1) > f(x^0)$ implies that

$$f(\bar{x}) > f(x^0) \tag{3.100}$$

for all $\bar{x} = \bar{\lambda}x^1 + (1-\bar{\lambda})x^0$, $0 < \bar{\lambda} < 1$. If for some $\bar{\lambda}$ $f(\bar{x}) > f(x^1) = f(x^2)$, then, since $x^0 \in [\bar{x}, x^2]$, we must have $f(x^0) > f(x^1)$, contradicting (3.99). If $f(\bar{x}) < f(x^1) = f(x^2)$, then (3.100) is contradicted. Hence, we must have $f(\bar{x}) = f(x^1) = f(x^2)$ for all $\bar{x}$, as defined above. By a similar argument we can show that $f(\hat{x}) = f(x^1) = f(x^2)$ for all $\hat{x} = \hat{\lambda}x^0 + (1-\hat{\lambda})x^2$, $0 < \hat{\lambda} < 1$.

Consider now the upper-level set $U(f, \alpha^1)$, where $\alpha^1 = f(x^1)$. We notice that $x^0 \notin U(f, \alpha^1)$ although x^0 is a cluster point of $U(f, \alpha^1)$, and $U(f, \alpha^1)$ is not closed in R^n. Since f is upper semicontinuous on C, it follows that $U(f, \alpha^1)$ is closed relative to C, i.e., it must contain all of its cluster points that are in C. Hence, $x^0 \in U(f, \alpha^1)$, a contradiction. □

The term "semistrictly quasiconcave" is due to Elkin (1968). Other terminology for semistrictly quasiconcave functions includes "strictly quasiconcave" by Avriel (1976), "functionally concave" by Hanson (1964), and "explicitly quasiconcave" by Martos (1975) if the function is also quasiconcave.

Comparing (3.95) with (3.96) we conclude that every strictly quasiconcave function is semistrictly quasiconcave. It is also easy to establish that concavity implies semistrict quasiconcavity: Let f be concave on $C \subset R^n$ and let $x^1 \in C$, $x^2 \in C$, such that $f(x^2) > f(x^1)$. Then

$$\lambda f(x^1) + (1 - \lambda) f(x^2) > f(x^1) \tag{3.101}$$

for every $0 < \lambda < 1$. Hence

$$f(\lambda x^1 + (1 - \lambda) x^2) \geq \lambda f(x^1) + (1 - \lambda) f(x^2) > f(x^1) \tag{3.102}$$

and f is semistrictly quasiconcave.

For continuous functions the intermediate position of semistrict quasiconcavity between strict quasiconcavity (nonexistence of local minima) and quasiconcavity (nonexistence of semistrict local minima) can be conveniently illustrated by a special type of (line segment) local minimum.

Definition 3.12. Let f be defined on the open interval $C \subset R$. Then f is said to attain a one-sided semistrict local minimum at a point $x^0 \in C$ if there exist two points $x^1 \in C$, $x^2 \in C$, $x^1 < x^0 < x^2$ such that

$$f(x^0) \leq f(\lambda x^1 + (1 - \lambda) x^2) \tag{3.103}$$

for every $0 \leq \lambda \leq 1$ and

$$f(x^0) < f(x^1) \quad \text{or} \quad f(x^0) < f(x^2). \tag{3.104}$$

It follows from this definition that if f attains a one-sided semistrict local minimum at x^0, then f attains a local minimum at x^0 and it must eventually increase on at least one side of x^0. Note that a semistrict local minimum is also a one-sided semistrict local minimum, so that the latter concept lies between the concepts of local minimum and semistrict local minimum.

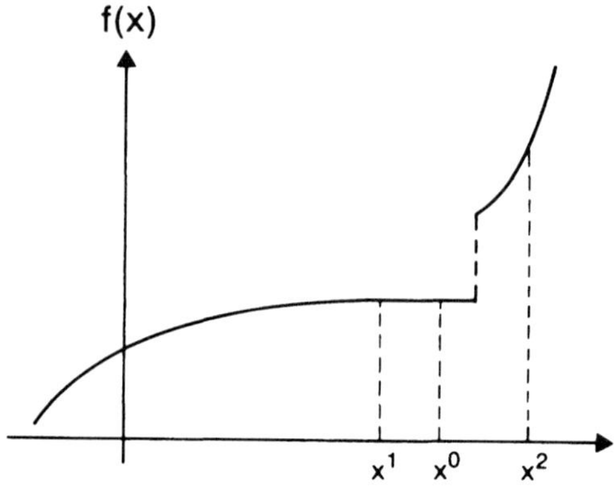

Figure 3.6. One-sided semistrict local minimum.

In Figure 3.6 we can see a function that attains a one-sided semistrict local minimum at x^0 that is, of course, a local minimum, but not a semistrict local minimum. We have then the following proposition.

Proposition 3.31 (Diewert, Avriel, and Zang, 1981). Let f be an upper semicontinuous function defined on the open interval $C \subset R$. If f is semistrictly quasiconcave, then there is no point $x^0 \in C$ that is a one-sided semistrict local minimum of f.

Proof. Suppose that $x^0 \in C$ is a one-sided semistrict local minimum of f. Without loss of generality we can assume that there exist points $x^1 \in C$, $x^1 < x^0$, $x^2 \in C$, $x^2 > x^0$, $f(x^2) > f(x^0)$ such that $f(x) \geq f(x^0)$ for all $x \in (x^1, x^2)$. Let $\{x^k\} \subset (x^1, x^0)$ be a sequence converging to x^0. Then, by the upper semicontinuity of f [see (3.22)] and since $f(x^k) \geq f(x^0)$ for all k we must have that $\{f(x^k)\}$ converges to $f(x^0)$. Hence, $x^k < x^0 < x^2$ and $f(x^0) \leq f(x^k) < f(x^2)$ for sufficiently large k, contradicting the semistrict quasiconcavity of f. □

Note that the upper semicontinuity requirement in the above proposition cannot be weakened as in the quasiconcave and strictly quasiconcave case. The semistrictly quasiconcave function appearing in Example 3.4 has a strict local minimum at $x = 0$ that is also a one-sided semistrict local minimum. This function is, of course, lower semicontinuous but not upper semicontinuous. We present now the converse result to the last proposition.

Proposition 3.32. Let f have the line segment minimum property on the open interval $C \subset R$. If there is no point $x^0 \in C$ that is a one-sided semistrict local minimum of f, then f is semistrictly quasiconcave.

Proof. If f is not semistrictly quasiconcave then there exist points $x^1 \in C$, $x^2 \in C$ such that $f(x^2) > f(x^1)$ and for some $0 < \bar{\lambda} < 1$

$$f(\bar{x}) = f(\bar{\lambda}x^1 + (1 - \bar{\lambda})x^2) \leq f(x^1) < f(x^2). \qquad (3.105)$$

It follows that f attains a one-sided semistrict local minimum over the closed line segment connecting x^1 and x^2 at some intermediate point x^0. □

Combining the last two results and the easily established fact that f is semistrictly quasiconcave on the convex set $C \subset R^n$ if and only if it is semistrictly quasiconcave on every line segment contained in C, we obtain the following theorem.

Theorem 3.33. An upper semicontinuous function f defined on the open convex set $C \subset R^n$ is semistrictly quasiconcave if and only if (i) it has the line segment minimum property and (ii) for every $x \in C$ and $v \in R^n$ such that $v^Tv = 1$, and $\bar{t} \in R$, $\bar{t} > 0$, satisfying $x + \bar{t}v \in C$, the function $F(t) = f(x + v)$ does not attain a one-sided semistrict local minimum at any $t \in (0, \bar{t})$.

For differentiable functions we have the following characterization of semistrictly quasiconcave functions:

Theorem 3.34 (Diewert, Avriel, and Zang, 1981). Let f be a differentiable function defined on the open convex set $C \subset R^n$. Then f is semistrictly quasiconcave if and only if for every $x^0 \in C$ and $v \in R^n$ such that $v^Tv = 1$ and $v^T\nabla f(x^0) = 0$ the function $F(t) = f(x^0 + tv)$ does not attain a one-sided semistrict local minimum at $t = 0$. If f is twice continuously differentiable then a necessary and sufficient condition for semistrict quasiconcavity of f is that for every $x^0 \in C$ and $v \in R^n$ such that $v^T\nabla f(x^0) = 0$, and $v^Tv = 1$ either $v^T\nabla^2 f(x^0)v < 0$ or $v^T\nabla^2 f(x^0)v = 0$ and $F(t)$ does not attain a one-sided semistrict local minimum at $t = 0$.

Let us turn our attention to the level sets and surfaces of semistrictly quasiconvex functions. Let us remind the reader that the upper-level set of a function f defined on $C \subset R^n$ is given by

$$U(f, \alpha) = \{x: x \in C, f(x) \geq \alpha\} \qquad (3.106)$$

and let

$$Y(f, \alpha) = \{x: x \in C, f(x) = \alpha\} \qquad (3.107)$$

be the level surface of f at α. Also let $B(f, \alpha)$ denote the boundary of $U(f, \alpha)$. In Figure 3.7 we illustrate these concepts for an upper semicontinuous quasiconcave function. Note that

$$B(f, \alpha) \subset Y(f, \alpha) \subset U(f, \alpha). \tag{3.108}$$

The level curves $Y(f, 2)$ are "thick," since the function has flat portions at function level 2. At this level the inclusions of (3.108) are all proper. At all other levels of this function $B(f, \alpha) = Y(f, \alpha)$, and for $\alpha = 3$ (the maximal value of f) $Y(f, \alpha) = U(f, \alpha)$. The following result is due to Elkin (1968).

Proposition 3.35. A continuous function f defined on R^n is semistrictly quasiconcave if and only if $U(f, \alpha)$ is convex and either $Y(f, \alpha) \subset B(f, \alpha)$ or $Y(f, \alpha) = U(f, \alpha)$ for all $\alpha \in R$.

Proof. Let f be semistrictly quasiconcave. Then by Proposition 3.30 it is also quasiconcave and $U(f, \alpha)$ is a convex set for all $\alpha \in R$. Suppose that $B(f, \alpha)$ is properly contained in $Y(f, \alpha)$. Then there exists a point $x^0 \in Y(f, \alpha^0)$ that is an interior point of $U(f, \alpha^0)$, and let x^1 be any other point of $U(f, \alpha^0)$. The line segment $[x^1, x^0]$ is properly included in the line

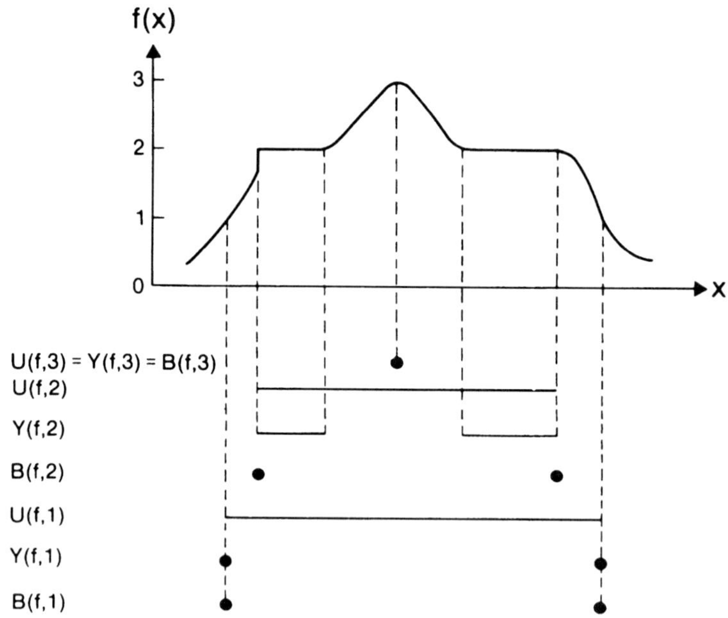

Figure 3.7. Upper-level sets and related concepts.

segment $[x^1, x^2] \subset U(f, \alpha^0)$ for some $x^2 \neq x^0$ and if $f(x^1) \neq f(x^2)$, say $f(x^2) > f(x^1)$, then by the semistrict quasiconcavity of f we have that

$$f(x^0) > f(x^1) \geq \alpha^0, \tag{3.109}$$

contradicting that $x^0 \in Y(f, \alpha^0)$, thus $f(x^2) = f(x^1)$. By the quasiconcavity of f we obtain

$$\alpha^0 = f(x^0) \geq f(x^1) \geq \alpha^0 \tag{3.110}$$

and $f(x^1) = \alpha^0$. Since x^0 is an arbitrary point of $U(f, \alpha^0)$ it follows that f is constant on $U(f, \alpha^0)$ and $Y(f, \alpha) = U(f, \alpha)$. Conversely, let $U(f, \alpha)$ be convex for all α. Hence f is quasiconcave and for all $x^1 \in R^n$, $x^2 \in R^n$, $0 < \lambda < 1$

$$f(x^2) > f(x^1) = \alpha^1 \quad \text{implies that} \quad f(\lambda x^1 + (1-\lambda)x^2) \geq \alpha^1. \tag{3.111}$$

It follows that $x^2 \in U(f, \alpha^1)$ and f is not constant on $U(f, \alpha^1)$, hence $Y(f, \alpha^1) \neq U(f, \alpha^1)$. Then $Y(f, \alpha^1) \subset B(f, \alpha^1)$. The point x^2 is in the interior of $U(f, \alpha^1)$ since, by continuity, there is a neighborhood $N_\varepsilon(x^2) \subset U(f, \alpha^1)$ such that for every x in this neighborhood $f(x) > \alpha^1$. Consider now the open interval (x^1, x^2). We want to show that every point in this interval is an interior point of $U(f, \alpha^1)$. Let $x = \lambda x^1 + (1-\lambda)x^2$, $0 < \lambda < 1$, be any point in the open interval (x^1, x^2). Then the neighborhood $N_{(1-\lambda)\varepsilon}(x)$ is in the convex hull of $N_\varepsilon(x^2) \cup x^1$ and thus in $U(f, \alpha^1)$. Hence the implication in (3.111) holds with strict inequality and f is semistrictly quasiconcave. □

This proposition shows that for continuous functions, defined on R^n, semistrict quasiconcavity is equivalent to assuming that the level surfaces $Y(f, \alpha)$ are in the boundary of the corresponding upper-level sets—that is, they are not "thick"—except possibly for that value of α at which f attains its global maximum on C. For strictly quasiconcave functions, discussed in the preceding section, we have the following corollary.

Corollary 3.36. Let f be a continuous strictly quasiconcave function f on R^n. Then its upper level sets $U(f, \alpha)$ are strictly convex and $Y(f, \alpha) \subset B(f, \alpha)$ for all $\alpha \in R$.

The main difference between semistrictly quasiconcave and strictly quasiconcave functions as shown in the last two results is that a semistrictly quasiconcave function can have a "flat" global maximum, whereas a strictly quasiconcave function can attain its global maximum at no more than one point. The latter result has been also proven separately in Proposition 3.29.

For both families of functions every local maximum is also global. Note again that the last proposition and corollary were proven only for functions defined on the whole space R^n. Using a relative topology these results can also be applied for functions defined on a proper convex subset of R^n.

Elkin (1968) showed that for continuous functions this local-global property can actually characterize semistrictly quasiconcave functions. A similar result was also proven by Martos (1969):

Theorem 3.37. Let f be a continuous quasiconcave function defined on R^n. Then f is semistrictly quasiconcave if and only if every local maximum $x^* \in C$ is also a global maximum of f on C.

Proof. The first part of the proof closely parallels the proof of Proposition 3.3 and will be omitted. For the second part, assume that f is quasiconcave and every local maximum of f is also a global maximum. By Proposition 3.35, it is sufficient to show that whenever $Y(f, \alpha)$ contains an interior point of $U(f, \alpha)$ then $Y(f, \alpha) = U(f, \alpha)$, that is, f is constant on $U(f, \alpha)$. If f is constant on R^n or $\alpha = \{\max f(x): x \in R^n\}$ the implication clearly follows. Assume, therefore, that f is not constant on R^n and let $x^0 \in Y(f, \alpha^0)$ be an interior point of $U(f, \alpha^0)$ for some α^0 such that the set

$$U^0(f, \alpha^0) = \{x: x \in R^n, f(x) > \alpha^0\} \tag{3.112}$$

is nonempty. Because of the continuity and quasiconcavity of f, $U^0(f, \alpha^0)$ is an open convex set, and $x^0 \notin U^0(f, \alpha^0)$. By the separation theorem for convex sets (see, for example, Avriel, 1976; Berge, 1963; or Mangasarian, 1969) there exists a hyperplane separating $U^0(f, \alpha^0)$ from x^0. Consequently, there exists a vector $c \in R^n$ and a real number a such that

$$c^T x < a \qquad \text{if } x \in U^0(f, \alpha^0), \tag{3.113}$$

$$c^T x = a \qquad \text{if } x = x^0. \tag{3.114}$$

Since x^0 is an interior point of $U(f, \alpha^0)$, we can find an $\hat{x}$, also in the interior of $U(f, \alpha^0)$, $\hat{x} \neq x^0$, such that $c^T\hat{x} > a$. By the continuity of f, there exists an $\varepsilon > 0$ such that $N_\varepsilon(\hat{x}) \subset U(f, \alpha^0)$ and $c^T x > a$ for every $x \in N_\varepsilon(\hat{x})$. Consequently, $N_\varepsilon(\hat{x}) \cap U^0(f, \alpha^0) = \varnothing$ and, therefore, $N_\varepsilon(\hat{x}) \subset Y(f, \alpha^0)$, hence $\hat{x}$ is a local maximum of f. By the hypothesis, $\hat{x}$ is also a global maximum of f, where, clearly, $Y(f, \alpha^0) = U(f, \alpha^0)$, and f is semistrictly quasiconcave. □

In Theorem 3.11 we presented a characterization of differentiable quasiconcave functions. This theorem cannot be strengthened to yield a corresponding straightforward characterization of semistrictly quasiconcave functions; See, however, some related results in Diewert (1981). It can be shown that if f is a differentiable function defined on the open convex set $C \subset R^n$, and for every $x^1 \in C$, $x^2 \in C$,

$$f(x^1) \geq f(x^2) \quad \text{implies that} \quad (x^1 - x^2)^T \nabla f(x^2) \geq 0, \tag{3.115}$$

then f is semistrictly quasiconcave. The converse of this result does not hold, as can be seen from the following example.

Example 3.5. Let f be defined on the interval $C = [0, 1] \subset R$ and given by

$$f(x) = (x)^2. \tag{3.116}$$

It is easy to see that for every $x^1 \in C$, $x^2 \in C$, and $0 < \lambda < 1$ such that $f(x^2) > f(x^1)$ we have $f(\lambda x^1 + (1 - \lambda)x^2) > f(x^1)$ and f is semistrictly quasiconcave. On the other hand, let $x^1 \in (0, 1]$ and let $x^2 = 0$. Then $\nabla f(x^2) = 0$ and (3.115) does not hold.

In the next section we shall use (3.115) as the defining relation for a new family of generalized concave functions.

3.5. Pseudoconcave Functions

The most commonly used definition of pseudoconcavity was introduced by Mangasarian (1965) for differentiable functions. A slightly different form was defined by Tuy (1964) under the name of "semiconcave function."

Definition 3.13. A differentiable function f defined on the open convex set $C \subset R^n$ is called pseudoconcave if for $x^1 \in C$ and $x^2 \in C$

$$f(x^1) > f(x^2) \quad \text{implies that} \quad (x^1 - x^2)^T \nabla f(x^2) > 0. \tag{3.117}$$

It is called strictly pseudoconcave if $x^1 \neq x^2$ and

$$f(x^1) \geq f(x^2) \quad \text{implies that} \quad (x^1 - x^2)^T \nabla f(x^2) > 0. \tag{3.118}$$

If f is (strictly) pseudoconcave, then $g \equiv -f$ is a (strictly) pseudoconvex function.

Pseudoconcave functions are intermediate between concave and semistrictly quasiconcave functions. Differentiable (strictly) concave functions are (strictly) pseudoconcave functions that, in turn, are semistrictly quasiconcave, hence quasiconcave. Ortega and Rheinboldt (1970) and Thompson and Parke (1973) define pseudoconcave functions without the differentiability assumption as follows:

Definition 3.14. Let f be defined on the open convex set $C \subset R^n$. It is called pseudoconcave if for $x^1 \in C$, $x^2 \in C$, and $0 < \lambda < 1$

$$f(x^1) > f(x^2) \quad \text{implies that} \quad f(\lambda x^1 + (1-\lambda)x^2) \geq f(x^2) + (1-\lambda)\lambda b(x^1, x^2) \tag{3.119}$$

where $b(x^1, x^2)$ is a positive number, depending, in general, on x^1 and x^2. It is called strictly pseudoconcave if $x^1 \neq x^2$ and

$$f(x^1) \geq f(x^2) \quad \text{implies that} \quad f(\lambda x^1 + (1-\lambda)x^2) \geq f(x^2) + (1-\lambda)\lambda b(x^1, x^2). \tag{3.120}$$

Diewert (1981) uses a related definition of pseudoconcave functions, using directional derivatives.

Let us now prove the following proposition.

Proposition 3.38. For differentiable functions Definitions 3.13 and 3.14 are equivalent.

Proof. Suppose that f satisfies Definition 3.14. If f is differentiable, then for any two points x^1, x^2 in the domain of f we have

$$f(x^1) = f(x^2) + (x^1 - x^2)^T \nabla f(x^2) + \alpha(x^1 - x^2)\|x^1 - x^2\|, \tag{3.121}$$

where $\alpha(x^1, x^2)$ is a function such that $\alpha(x^1, x^2) \to 0$ as $x^1 \to x^2$. Hence,

$$\lambda(x^1 - x^2)^T \nabla f(x^2) + \lambda\alpha(x^2 + \lambda(x^1 - x^2), x^2)\|x^1 - x^2\| \geq (1-\lambda)\lambda b(x^1, x^2) > 0 \tag{3.122}$$

for all $0 < \lambda < 1$. Dividing (3.122) by λ and taking the limit as $\lambda \to 0$ we get

$$(x^1 - x^2)^T \nabla f(x^2) \geq b(x^1, x^2) > 0 \tag{3.123}$$

and f satisfies Definition 3.13.

Conversely, suppose that $f(x^1) > f(x^2)$ and for every positive $b(x^1, x^2)$ there exists a $0 < \lambda < 1$ such that

$$\frac{f(x^2 + \lambda(x^1 - x^2)) - f(x^2)}{\lambda} < (1 - \lambda)b(x^1, x^2) < b(x^1, x^2). \quad (3.124)$$

In particular, there are $0 < \lambda^i < 1$ such that

$$\frac{f(x^2 + \lambda^i(x^1 - x^2)) - f(x^2)}{\lambda^i} < \frac{1}{i}, \qquad i = 1, 2, \ldots. \quad (3.125)$$

The sequence $\{\lambda^i\}$ has a convergent subsequence $\{\lambda^{i_k}\} \to \lambda^0$ such that $\{x^2 + \lambda^{i_k}(x^1 - x^2)\} \to x^0$. If $x^0 \neq x^2$, that is, $\lambda^0 \neq 0$, then $f(x^0) \leq f(x^2)$, and f is not semistrictly quasiconcave, hence it is not a pseudoconcave function satisfying Definition 3.13. If $x^0 = x^2$, then $\lambda^{i_k} \to 0$ and

$$0 \geq \lim_{i_k \to \infty} \frac{f(x^2 + \lambda^{i_k}(x^1 - x^2)) - f(x^2)}{\lambda^{i_k}} = (x^1 - x^2)^T \nabla f(x^2). \quad (3.126)$$

Consequently, f does not satisfy Definition 3.13. The proof of the strictly pseudoconcave case is identical. □

From Definition 3.13 we immediately obtain an important property of pseudoconcave functions.

Theorem 3.39 (Mangasarian, 1965). Let f be a differentiable (strictly) pseudoconcave function on the open convex set $C \subset R^n$. If $\nabla f(x^0) = 0$ for some $x^0 \in C$, then x^0 is a (strict) global maximum of f over C.

Proof. By (3.117), $\nabla f(x^0) = 0$ implies $f(x^0) \geq f(x)$ for all $x \in C$, and x^0 is a global maximum. Similarly, by (3.118), $\nabla f(x^0) = 0$ implies $f(x^0) > f(x)$ for all $x \in C$, and for a strictly pseudoconcave function the point x^0 is a strict global maximum. □

It may be interesting to note here that there is no known characterization of pseudoconcavity in terms of level sets. Upper-level sets of a pseudoconcave function are not necessarily strictly convex. However, it was shown by Ponstein (1967) that if f is differentiable, pseudoconcave, and also strictly quasiconcave (hence, its upper-level sets are strictly convex), then f is strictly pseudoconcave.

Example 3.6. Consider the function f defined on the convex set $C = \{x: x \in R^2, 0 < x_1 < 1, 0 < x_2 < 1\}$ and given by

$$f(x) = (x_1)^2 + x_1. \quad (3.127)$$

This function is pseudoconcave, since if $f(x^1) > f(x^2)$ then $x_1^1 > x_1^2$, and

$$(x^1 - x^2)^T \nabla f(x^2) = (x_1^1 - x_1^2)(2x_1^2 + 1) > 0. \tag{3.128}$$

It is clearly not concave (actually it is convex) and also not strictly pseudoconcave since if we take $x^1 = (\alpha, \beta)$ and $x^2 = (\alpha, \gamma)$, where α, β, γ are numbers in $(0, 1)$, $\alpha < \beta$, $\alpha < \gamma$, $\beta \neq \gamma$, then $f(x^1) = f(x^2)$ but $(x^1 - x^2)^T \nabla f(x^2)$ is not positive. This function is not strictly quasiconcave, as can be seen by substituting the same two points x^1 and x^2 into (3.88). The function f is, however, semistrictly quasiconcave and quasiconcave.

Let us turn now to line segment characterizations of differentiable (strictly) pseudoconcave functions, starting with functions on the real line.

Proposition 3.40 (Diewert, Avriel, and Zang, 1981). Let f be a continuously differentiable function defined on the open interval $C \subset R$. Then f is (strictly) pseudoconcave if and only if for every $x^0 \in C$ such that $f'(x^0) = 0$, x^0 is a (strict) local maximum of f.

Proof. If f is (strictly) pseudoconcave, then $f'(x^0) = 0$ clearly implies that x^0 is a (strict) local maximum of f. Conversely, let $x^1 \in C$, $x^2 \in C$, $x^1 \neq x^2$ be any two points such that $(x^1 - x^2)f'(x^2) \leq 0$ and suppose that $f(x^1) > f(x^2)$. If $(x^1 - x^2)f'(x^2) < 0$, then the function $g(\lambda) \equiv f(\lambda x^1 + (1 - \lambda)x^2)$ attains at least one local minimum over the open interval $(0, 1)$. Let λ^0 be the largest local minimizing point of g over $(0, 1)$. Then $g'(\lambda^0) = f'(\lambda^0 x^1 + (1 - \lambda^0)x^2) \equiv f'(x^0) = 0$, but x^0 is not a local maximum of f, contradicting the hypothesis. If $(x^1 - x^2)f'(x^2) = 0$ then $f'(x^2) = 0$, and by the hypothesis, x^2 is a local maximum of f. Since $f(x^1) > f(x^2)$ we can again conclude that the function $g(\lambda)$ defined above must attain at least one local minimum over the open interval $(0, 1)$, leading to a contradiction. Thus $f(x^1) \leq f(x^2)$ and f is pseudoconcave. The proof for strictly pseudoconcave functions is similar and will be omitted. □

One can easily establish the fact that a function defined on the convex set $C \subset R^n$ is (strictly) pseudoconcave if and only if it is (strictly) pseudoconcave over every line segment contained in C. Consequently we have the following results appearing in Diewert, Avriel, and Zang (1981).

Theorem 3.41. Let f be a continuously differentiable function defined on the convex set $C \subset R^n$. Then f is (strictly) pseudoconcave if and only if for every $x^0 \in C$ and $v \in R^n$ such that $v^T v = 1$ and $v^T \nabla f(x^0) = 0$ the function $F(t) = f(x^0 + tv)$ attains a (strict) local maximum at $t = 0$.

The geometric interpretation of the condition in this theorem is that the function f, restricted to a line passing through x^0 in a direction v, orthogonal to the gradient of f at x^0, has a local maximum at x^0.

For single-variable twice-differentiable functions the following proposition holds.

Proposition 3.42. Let f be a twice continuously differentiable function defined on the open interval $C \subset R$. Then f is (strictly) pseudoconcave if and only if for every $x^0 \in C$ such that $f'(x^0) = 0$ either $f''(x^0) < 0$ or $f''(x^0) = 0$ and x^0 is a (strict) local maximum of f.

The generalization of this proposition to the multidimensional case is straightforward; see Diewert, Avriel, and Zang (1981).

Theorem 3.43. Let f be a twice continuously differentiable function defined on the open convex set $C \subset R^n$. Then f is (strictly) pseudoconcave if and only if for every $x^0 \in C$ and $v \in R^n$ such that $v^T v = 1$ and $v^T \nabla f(x^0) = 0$ either $v^T \nabla^2 f(x^0) v < 0$ or $v^T \nabla^2 f(x^0) v = 0$ and the function $F(t) = f(x^0 + tv)$ attains a (strict) local maximum at $t = 0$.

A function that is both pseudoconcave and pseudoconvex is called *pseudomonotonic*, and it is also quasimonotonic. Pseudomonotonic functions were characterized by Thompson and Parke (1973). They will be mentioned in the next section in connection with optimality conditions for mathematical programs involving generalized concave functions.

We conclude this section by introducing the class of strongly pseudoconcave functions, a subclass of differentiable strictly pseudoconcave functions.

Definition 3.15 (Diewert, Avriel, and Zang, 1981). A differentiable function f defined on the open convex set $C \subset R^n$ is said to be strongly pseudoconcave if it is strictly pseudoconcave and for every $x^0 \in C$ and $v \in R^n$ such that $v^T v = 1$ and $v^T \nabla f(x^0) = 0$ there exist positive numbers ε and α such that $x^0 \pm \varepsilon v \in C$ and

$$F(t) = f(x^0 + tv) \leq F(0) - \tfrac{1}{2}\alpha(t)^2 \tag{3.129}$$

for $0 \leq t < \varepsilon$.

Strongly pseudoconcave functions have economic applications that will be discussed in Chapter 4. We shall also return to analyze such functions in Chapter 8, in connection with the problem of concavifying nonconcave functions. Strongly pseudoconcave functions were called "differentially strictly quasiconcave" by McFadden (1978). Ginsberg (1973) and Newman (1965) used the term "strongly quasiconcave" to describe a subclass of strongly pseudoconcave functions.

Next we define the concept of a strong local maximum on a line segment.

Definition 3.16 (Diewert, Avriel, and Zang, 1981). A function f defined on the open interval $C \subset R$ attains a strong local maximum at x^0 if there exist positive numbers ε and α such that $x^0 \pm \varepsilon \in C$ and

$$f(x) \leq f(x^0) - \tfrac{1}{2}\alpha(x - x^0)^2 \tag{3.130}$$

for $|x| \leq \varepsilon$.

We now have the following proposition.

Proposition 3.44 (Diewert, Avriel, and Zang, 1981). Let f be a differentiable function defined on the open convex set $C \subset R^n$. Then f is strongly pseudoconcave if and only if for every $x^0 \in C$ and $v \in R^n$ such that $v^T v = 1$ and $v^T \nabla f(x^0) = 0$ the function $F(t) = f(x^0 + tv)$ attains a strong local maximum at $t = 0$.

Finally we have the following proposition.

Proposition 3.45 (Diewert, Avriel, and Zang, 1981). Let f be a twice continuously differentiable function defined on the open convex set $C \subset R^n$. Then f is strongly pseudoconcave if and only if for every $x^0 \in C$ and $v \in R^n$ such that $v^T v = 1$ and $v^T \nabla f(x^0) = 0$ we have $v^T \nabla^2 f(x^0) v < 0$.

Proof of the last propositions follows closely the proofs of analoguous results presented earlier, and will be left for the reader.

3.6. Generalized Concave Mathematical Programs

In this section we discuss mathematical programs in which the objective and constraint functions are generalized concave. Such programs are called generalized concave mathematical programs. The books of Mangasarian (1969), Martos (1975), and Avriel (1976) were the first textbooks in which generalized concave mathematical programs were extensively studied. In

Chapter 2 we showed the relationship between concave mathematical programs and saddlepoints of the Lagrangian, presented necessary and sufficient conditions for optimality in a concave mathematical program, and briefly mentioned the basic concepts of duality in concave programming problems. Some of the results presented there are extended in this section.

The optimality conditions for concave programs that were given in Chapter 2 can be divided into two classes: Those that did not assume differentiability of the functions involved (saddlepoint conditions of the Lagrangian), and those that applied to differentiable functions (conditions on stationary points of the Lagrangian). Optimality in a concave program implies the existence of a saddlepoint of the corresponding Lagrangian. This result cannot be generalized if, for example, a concave objective function is replaced by a quasiconcave one.

Example 3.7. Consider the quasiconcave function defined on R

$$f(x) = \begin{cases} 0, & x > 0 \\ -1, & x \leq 0 \end{cases} \tag{3.131}$$

and the mathematical program

$$\max f(x) \tag{3.132}$$

subject to

$$x \leq 0. \tag{3.133}$$

The corresponding Lagrangian is given by

$$L(x, \lambda) = f(x) - \lambda x. \tag{3.134}$$

The point $x^0 = -1$ is clearly a solution of this program. The saddlepoint problem is that of finding a $\lambda^0 \geq 0$ such that

$$L(x^0, \lambda) \geq L(x^0, \lambda^0) \geq L(x, \lambda^0). \tag{3.135}$$

In our case we have

$$-1 + \lambda \geq -1 + \lambda^0 \geq f(x) - \lambda^0 x. \tag{3.136}$$

In order to satisfy the left-hand inequality we need $\lambda^0 = 0$, but then the right-hand inequality does not hold, since we can take $f(x) = 0$. It seems that the property of $x^0 = -1$ being a local but nonglobal maximum causes the lack of equivalence between the solutions of the mathematical program and its corresponding saddlepoint problem.

Let us turn now to the differentiable case. There are several extensions of the necessary optimality conditions of Theorem 2.29, in which the concave, convex, and affine functions are replaced by more general concave ones. Some of these extensions are given below. Consider the mathematical program

P
$$\max f(x) \tag{3.137}$$

$$g_i(x) \leq 0, \qquad i = 1, \ldots, m, \tag{3.138}$$

$$h_j(x) = 0, \qquad j = 1, \ldots, p, \tag{3.139}$$

where f, g_i, h_j are assumed to be differentiable functions defined on an open convex subset C of R^n. To establish necessary optimality conditions for P some type of constraint qualifications are needed, similar to the strong consistency condition for concave programs.

Definition 3.17 (Mangasarian, 1969). Program P is said to satisfy the generalized strong consistency condition if the g_i are pseudoconvex, the h_j are pseudomonotonic and there exists a point x^0 such that $g_i(x^0) < 0$, $i = 1, \ldots, m$ and $h_j(x^0) = 0$, $j = 1, \ldots, p$.

We then have the following theorem.

Theorem 3.46. Suppose that P satisfies the generalized strong consistency condition and x^* is an optimal solution of P. Then there exist vectors λ^* and μ^* such that

$$\nabla_x L(x^*, \lambda^*, \mu^*) \equiv \nabla f(x^*) - \sum_{i=1}^{m} \lambda_i^* \nabla g_i(x^*) - \sum_{j=1}^{p} \mu_j^* \nabla h_j(x^*) = 0, \tag{3.140}$$

$$\lambda_i^* g_i(x^*) = 0, \qquad i = 1, \ldots, m, \tag{3.141}$$

$$\lambda^* \geq 0. \tag{3.142}$$

For a proof of this theorem the reader is referred to Martos (1975).

It is interesting to note that no concavity of any type is assumed for the objective function f in the above theorem. Also, the pseudoconvexity condition for all the g_i can be weakened to only those for which $g_i(x^*) = 0$.

Arrow and Enthoven (1961) consider necessary optimality conditions in a program P without equality constraints and in which the inequality constraints are defined by quasiconvex functions:

Theorem 3.47. Suppose that the g_i, $i = 1, \ldots, m$ are quasiconvex functions and that P has no equality constraints. Also suppose that there exists a point $x^0 \in R^n$ such that

$$g_i(x^0) < 0, \qquad i = 1, \ldots, m \tag{3.143}$$

and for each i let either g_i be convex, or for each feasible x let $\nabla g_i(x) \neq 0$. Let x^* be an optimal solution of P. Then there exist vectors λ^* and μ^* satisfying (3.140)-(3.142).

Turning to sufficient optimality conditions we now show that Theorem 2.30 can be quite simply extended to generalized concave functions. The next theorem is due to Mangasarian (1965).

Theorem 3.48. Suppose that f is pseudoconcave, the g_i are quasiconvex, and the h_j are quasimonotonic. If there exists an x^* feasible for P and λ^*, μ^* satisfying (3.140)-(3.142) then x^* is an optimal solution of P.

Proof. Let x be any feasible point and let $I = \{i\colon g_i(x^*) = 0, i = 1, \ldots, m\}$. Then $g_i(x) \leq g_i(x^*)$ for $i \in I$ and it follows from the quasiconvexity of the g_i that

$$(x - x^*)^T \nabla g_i(x^*) \leq 0. \tag{3.144}$$

Since $\lambda_i^* \geq 0$ and $\lambda_i^* = 0$ for $i \notin I$ we have

$$(x - x^*)^T \sum_{i=1}^{m} \lambda_i^* \nabla g_i(x^*) \leq 0. \tag{3.145}$$

Similarly,

$$(x - x^*)^T \sum_{j=1}^{p} \mu_j^* \nabla h_j(x^*) = 0. \tag{3.146}$$

It follows that

$$(x - x^*)^T \left\{ \sum_{i=1}^{m} \lambda_i^* \nabla g_i(x^*) + \sum_{j=1}^{p} \mu_j^* \nabla h_j(x^*) \right\} \leq 0 \tag{3.147}$$

and by (3.138),

$$(x - x^*)^T \nabla f(x^*) \leq 0. \tag{3.148}$$

Since f is pseudoconcave, we have $f(x^*) \geq f(x)$. □

For an extension of Theorem 3.48 to the nondifferentiable case, see Diewert (1981).

Another extension of Theorem 2.30 for a quasiconcave objective function was obtained by Ferland (1972):

Theorem 3.49. Suppose that f is quasiconcave on a convex set S having a nonempty interior and such that $C \subset S$, the g_i, $i = 1, \ldots, m$ are quasiconvex on C and P has no equality constraints. Let x^* be feasible, and let x^*, λ^* satisfy (3.140)–(3.142). If (i) there is an $x \in S$ such that $(x - x^*)^T \nabla f(x^*) < 0$ or (ii) $\nabla f(x^*) \neq 0$ and f is twice differentiable on S, then x^* is a solution of P.

In view of the existence of various types of generalized concave functions as defined in this chapter, it should be clear to the reader that the theory of optimality conditions for mathematical programs involving such functions is far from complete. A thorough investigation of the properties of such programs is yet to be accomplished. Results obtained so far apply to very special types of programs. Concave duality results presented in the preceding chapter have been extended to more general types of functions by Crouzeix (1977, 1981) and Greenberg and Pierskalla (1973). Duality results for fractional programs, a special case of generalized concave programs, will be presented in Chapter 7.

References

ARROW, K. J., and ENTHOVEN, A. C. (1961), Quasi-concave programming, *Econometrica* **29**, 779-800.

AVRIEL, M. (1972), *r*-Convex functions, *Math. Programming* **2**, 309-323.

AVRIEL, M. (1976), *Nonlinear Programming: Analysis and Methods*, Prentice Hall, Englewood Cliffs, New Jersey.

AVRIEL, M., and SCHAIBLE, S. (1978), Second order characterizations of pseudoconvex functions, *Math. Programming* **14**, 170-185.

AVRIEL, M., DIEWERT, W. E., SCHAIBLE, S., and ZIEMBA, W. T. (1981), Introduction to concave and generalized concave functions, in *Generalized Concavity in Optimization and Economics*, Edited by S. Schaible and W. T. Ziemba, Academic Press, New York.

BARTLE, R. G. (1976), *The Elements of Real Analysis*, 2nd ed., Wiley, New York.

BERGE, C. (1963), *Topological Spaces*, Oliver and Boyd, Edinburgh.

CROUZEIX, J. P. (1976), Conjugacy in quasiconvex analysis, in *Convex Analysis and Its Applications*, Edited by A. Auslender, Springer-Verlag, Berlin.

CROUZEIX, J. P. (1977), Contributions à l'étude des fonctions quasiconvexes, These de Doctorat, Universite de Clermont-Ferrand II, France.

CROUZEIX, J. P. (1980), On second order conditions for quasiconvexity, *Math. Programming* **18**, 349-352.

CROUZEIX, J. P. (1981), A duality framework in quasiconvex programming, in *Generalized Concavity in Optimization and Economics*, Edited by S. Schaible and W. T. Ziemba, Academic Press, New York.

CROUZEIX, J. P., and FERLAND, J. A. (1981), Criteria for quasiconvexity and pseudo-convexity and their relationships, in *Generalized Concavity in Optimization and Economics*, Edited by S. Schaible and W. T. Ziemba, Academic Press, New York.

DIEWERT, W. E. (1981), Alternative characterizations of six kinds of quasiconcavity in the nondifferentiable case with applications to nonsmooth programming, in *Generalized Concavity in Optimization and Economics*, Edited by S. Schaible and W. T. Ziemba, Academic Press, New York.

DIEWERT, W. E., AVRIEL, M., and ZANG, I. (1977), Nine kinds of quasiconcavity and concavity, Discussion Paper 77-31, Department of Economics, University of British Columbia, Vancouver, N.C., Canada.

DIEWERT, W. E., AVRIEL, M., and ZANG, I. (1981), Nine kinds of quasiconcavity and concavity, *J. Econ. Theory* **25**, 397-420.

ELKIN, R. M. (1968), Convergence theorems for Gauss-Seidel and other minimization algorithms, Ph.D. dissertation, University of Maryland, College Park.

FENCHEL, W. (1951), Convex cones, sets and functions, Mimeographed lecture notes, Princeton University, Princeton, New Jersey.

FENCHEL, W. (1956), Über konvexe Funktionen mit vorgeschriebenen Niveaumanning-faltigkeiten, *Math. Z.* **63**, 496-506.

FERLAND, J. A. (1971), Quasi-convex and pseudo-convex functions on solid convex sets, Ph.D. dissertation, Stanford University, Stanford, California.

FERLAND, J. A. (1972), Mathematical programming problems with quasi-convex objective functions, *Math. Programming* **3**, 296–301.
DE FINETTI, B. (1949), Sulle stratificazioni convesse, *Ann. Math. Pura Appl.* **30**, 173–183.
GERENCSÉR, L. (1973), On a close relation between quasi-convex and convex functions and related investigations, *Math. Operationsforsh. Statist.* **4**, 201–211.
GINSBERG, W. (1973), Concavity and quasiconcavity in economics, *J. Econ. Theory* **6**, 596–605.
GREENBERG, H. J., and PIERSKALLA, W. P. (1971), A review of quasi-convex functions, *Oper. Res.* **19**, 1553–1570.
GREENBERG, H. J., and PIERSKALLA, W. P. (1973), Quasi-conjugate functions and surrogate duality, *Cah. Cent. Etud. Rech. Oper.* **15**, 437–448.
HANSON, M. A. (1964), Bounds for functionally convex optimal control problems, *J. Math. Anal. Appl.* **8**, 84–89.
KANNAI, Y. (1977), Concavifiability and constructions of concave utility functions, *J. Math. Econ.* **4**, 1–56.
KARAMARDIAN, S. (1967), Duality in mathematical programming, *J. Math. Anal. Appl.* **20**, 344–358.
KATZNER, D. W. (1970), *Static Demand Theory*, Macmillan, New York.
MANGASARIAN, O. L. (1965), Pseudo-convex functions, *J. SIAM Control Ser. A* **3**, 281–290.
MANGASARIAN, O. L. (1969), *Nonlinear Programming*, McGraw-Hill, New York.
MARTOS, B. (1965), The direct power of adjacent vertex programming methods, *Manage. Sci.* **12**, 241–252.
MARTOS, B. (1969), Subdefinite matrices and quadratic forms, *Siam J. Appl. Math.* **17**, 1215–1223.
MARTOS, B. (1975), *Nonlinear Programming Theory and Methods*, North-Holland, Amsterdam.
MCFADDEN, D. (1978), Convex analysis, in *Production Economics: A Dual Approach to Theory and Applications*, Vol. 1, Edited by M. Fuss and D. McFadden, North-Holland, Amsterdam.
NEWMAN, D. J. (1965), Location of the maximum on unimodal surfaces, *J. Assoc. Comp. Mach.* **12**, 395–398.
ORTEGA, J. M., and RHEINBOLDT, W. C. (1970), *Iterative Solution of Nonlinear Equations in Several Variables*, Academic Press, New York.
OTANI, K. (1983), A characterization of quasiconcave functions, *J. Econ. Theory* **31**, 194–196.
PONSTEIN, J. (1967), Seven kinds of convexity, *SIAM Rev.* **9**, 115–119.
SCHAIBLE, S. (1971), Beiträge zur quasikonvexen Programming, Doctoral Dissertation, Köln, Germany.
STOER, J., and WITZGALL, C. (1970), *Convexity and Optimization in Finite Dimensions* I, Springer-Verlag, Berlin.
THOMPSON, W. A., and PARKE, D. W. (1973), Some propèrties of generalized concave functions, *Oper. Res.* **21**, 305–313.
TUY, H. (1964), Sur les inégalités linéaires, *Colloq. Math.* **13**, 107–123.
WILDE, D. J. (1964), *Optimum Seeking Methods*, Prentice-Hall, Englewood Cliffs, New Jersey.
ZANG, I. (1981), Concavifiability of C^2 functions: A unified exposition, in *Generalized Concavity in Optimization and Economics*, Edited by S. Schaible and W. T. Ziemba, Academic Press, New York.

4

Application of Generalized Concavity to Economics

In this chapter, we indicate where the main types of generalized concave functions are used in economics. At the same time, we shall study in some detail three models of constrained choice that are of great interest in the economics literature. The three models together cover the fundamentals of modern (competitive) microeconomic theory.

The first constrained choice model that we shall study in Sections 4.1–4.3 is the model where a producer (or consumer) minimizes the cost of producing or achieving a given output (or utility) level. The second and third constrained maximization problems to be studied are the consumer's utility maximization problem (Sections 4.4–4.8) and the producer's profit maximization problem (Section 4.9).

Section 4.10 shows how the theory of concave programming, studied above in Chapter 2, may be combined with economic duality theory to yield some interesting theorems in the pure theory of trade or in the theory of economic planning.

Theorems 4.2, 4.3, and 4.4 in Section 4.1 below deal with rather technical continuity issues. These theorems can be omitted by the applications-oriented reader.

Some of the material in this chapter appeared in Diewert (1981) in condensed form.

4.1. The Cost Function

One of the fundamental paradigms in economics is the one that has a producer competitively minimizing costs subject to his technological constraints. "Competitive" means that the producer takes input prices as fixed and unchanging during the given period of time irrespective of the producer's demand for those inputs.

We assume that only one output is produced using n inputs and that the producer's technology can be summarized by a production function F: $y = F(x)$, where $y \geq 0$ is the maximal amount of output that can be produced during a period given the nonnegative vector of inputs $x = (x_1, x_2, \ldots, x_n) \geq 0$. We further assume that the cost of purchasing one unit of input i is $p_i > 0$, and that the positive vector of input prices that the producer faces is $p = (p_1, p_2, \ldots, p_n) > 0$. For $y \geq 0$, $p > 0$, the *producer's cost function* C is defined as the solution to the following constrained minimization problem:

$$C(y, p) = \min_x \{p^T x: F(x) \geq y, x \geq 0\}. \tag{4.1}$$

Of course, the minimum in (4.1) may not exist. However, if we impose the following very weak regularity condition on the production function F, it can be shown that C will be well defined as a minimum (at least for $p > 0$):

Assumption 1. F is a real-valued nonnegative function defined for all nonnegative input vectors $x \geq 0$ and F is *upper semicontinuous*—that is, $U(F, y) = \{x \in R^n: F(x) \geq y, x \geq 0\}$ is closed for every y in the range of F (see Definition 3.4). In the economics literature this property is also called *continuity from above.*

Assumption 1 is very weak from an empirical point of view, since it cannot be contradicted by a finite set of data on the inputs and outputs of a producer.

Let Y be the set of all feasible output levels—that is, $Y = \{y \in R: y \geq 0, U(F, y) \neq \varnothing\}$. We can now state the following theorem.

Theorem 4.1. If F satisfies Assumption 1 above, then C is well defined by (4.1) for $y \in Y$ and $p > 0$. Moreover, C has the following eight properties:

(a) If $y \in Y$, $p > 0$, then $C(y, p) > 0$ (nonnegativity).
(b) If $y \in Y$, $p > 0$, $\lambda > 0$, then $C(y, \lambda p) = \lambda C(y, p)$ (linear homogeneity in prices for fixed output).
(c) If $y \in Y$, $0 < p^1 \leq p^2$, $p^1 \neq p^2$, then $C(y, p^1) \leq C(y, p^2)$ (nondecreasing in prices for fixed output).
(d) If $y \in Y$, $p^1 > 0$, $p^2 > 0$, $0 \leq \lambda \leq 1$, then $C(y, \lambda p^1 + (1 - \lambda)p^2) \geq \lambda C(y, p^1) + (1 - \lambda)C(y, p^2)$ (concavity in prices for fixed output).

(e) If $y \in Y$, then $C(y, p)$ is continuous in p for $p > 0$ (continuity in prices for fixed output).

(f) If $p > 0$, $y^0 \in Y$, $y^1 \in Y$, $y^0 < y^1$, then $C(y^0, p) \leq C(y^1, p)$ (nondecreasing in output for fixed prices).

(g) If $p > 0$, $\alpha \in R$, then $\{y: C(y, p) \leq \alpha\}$ is a closed set (lower semicontinuity in output for fixed prices).

(h) If $\sup_y \{y: y \in Y\} = +\infty$, then $C(y, p) \to +\infty$ as $Y \to +\infty$ for every $p > 0$; and $C(0, p) = 0$ for every $p > 0$.

Proof. We first show that the minimum in (4.1) exists, i.e., that the cost function C is well defined as a minimum. Let $y \in Y$. Then there exists $\bar{x}$ such that $F(\bar{x}) \geq y$. Since $p > 0$, the set $\{x: p^T x \leq p^T \bar{x}; x \geq 0; F(x) \geq y\}$ is bounded and closed (using Assumption 1). Since $\bar{x}$ belongs to this set, it is also nonempty. Thus $\min_x \{p^T x: p^T x \leq p^T \bar{x}; x \geq 0; F(x) \geq y\}$ exists and is equal to $C(y, p)$.

(a) The nonnegativity of $C(y, p)$ follows from the nonnegativity of x and the positivity of p, so that $p^T x \geq 0$ if $p > 0$ and $x \geq 0$.

(b) Let $y \in y$, $p > 0$ and $\lambda > 0$. Then,

$$C(y, \lambda p) = \min_x \{\lambda p^T: F(x) \geq y, x \geq 0\} \tag{4.2}$$

$$= \lambda \min_x \{p^T x: F(x) \geq y, x \geq 0\} \qquad \text{(since } \lambda > 0) \tag{4.3}$$

$$= \lambda C(y, p). \tag{4.4}$$

(c) Let $y \in Y$, $0 < p^1 \leq p^2$, $p^1 \neq p^2$. Then,

$$C(y, p^2) = \min_x \{(p^2)^T x: F(x) \geq y, x \geq 0\} \tag{4.5}$$

$$= (p^2)^T x^2 \qquad \text{(say where } x^2 \geq 0) \tag{4.6}$$

$$\geq (p^1)^T x^2 \qquad \text{(since } p^2 \geq p^1 \text{ and } x^2 \geq 0) \tag{4.7}$$

$$\geq \min_x \{(p^1)^T x: F(x) \geq y, x \geq 0\} \qquad \text{(since } x^2 \text{ is feasible)} \tag{4.8}$$

$$= C(y, p^1). \tag{4.9}$$

(d) Let $y \in Y$, $p^1 > 0$, $p^2 > 0$, and $0 \le \lambda \le 1$. Then,

$$C(y, \lambda p^1 + (1-\lambda)p^2) = \min_x \{(\lambda p^1 + (1-\lambda)p^2)^T x\colon F(x) \ge y, x \ge 0\} \tag{4.10}$$

$$= (\lambda p^1 + (1-\lambda)p^2)^T x^0$$

$$\text{for some } x^0 \text{ such that } F(x^0) \ge y \tag{4.11}$$

$$= \lambda((p^1)^T x^0) + (1-\lambda)((p^2)^T x^0) \tag{4.12}$$

$$\ge \lambda C(y, p^1) + (1-\lambda) C(y, p^2) \tag{4.13}$$

since x^0 is feasible for the cost minimization problems with p^1 and p^2, but it is not necessarily optimal; i.e., $(p^i)^T x^0 \ge C(y, p^i)$ for $i = 1, 2$.

(e) By (a) and (d) above, $C(y, p)$ is a nonnegative concave function in p for each $y \in y$ over the positive orthant $\{p\colon p > 0\}$, an open set. Hence, by Proposition 2.9 $C(y, p)$ is continuous in p for each $y \in Y$.

(f) Let $p > 0$, $y^0 \in Y$, $y^1 \in Y$, $y^0 < y^1$. Then,

$$C(y^1, p) = \min_x \{p^T x\colon F(x) \ge y^1 > y^0, x \ge 0\} \tag{4.14}$$

$$\ge \min_x \{p^T x\colon F(x) \ge y^0, x \ge 0\} \tag{4.15}$$

since the feasible set is now larger. Hence,

$$C(y^1, p) \ge C(y^0, p). \tag{4.16}$$

(g) Let $y^k \in Y$ for $k = 1, 2, \dots$, $\lim_{k\to\infty} y^k = y^0$, $p > 0$ and $C(y^k, p) \le \alpha$ for each k. We wish to show that $C(y^0, p) \le \alpha$. Let $x^k > 0$ solve the cost minimization problem with y^k; i.e., $F(x^k) \ge y^k$, $p^T x^k = C(y^k, p) \le \alpha$ for $k = 1, 2, \dots$. Since the x^k belong to the compact set $\{x\colon p^T x \le \alpha, x \ge 0\}$, there is a convergent subsequence of the x^k that converges to $x^0 \ge 0$. Note that

$$p^T x^0 \le \alpha. \tag{4.17}$$

Since F is upper semicontinuous, its hypograph, $H(F) = \{(x, y)\colon x \in R^n, y \in R, x \ge 0, y \le F(x)\}$ is a closed set (for a proof of this equivalence, see, for example, Rockafellar, 1970). Since $F(x^k) \ge y^k$, we have $(x^k, y^k) \in H(F)$ for each k. Since $H(F)$ is closed, also $(x^0, y^0) \in H(F)$, i.e., $F(x^0) \ge y^0$. Thus,

$$C(y^0, p) = \min_x \{p^T x\colon F(x) \ge y^0\} \le p^T x^0 \tag{4.18}$$

since x^0 is feasible for the problem. Hence,

$$C(y^0, p) \le \alpha. \tag{4.19}$$

(h) If $\bar{y} = \sup_x \{F(x): x \ge 0\} = +\infty$, then since $F(x)$ is real valued for finite x, it is evident that at least one component of x must tend to infinity if we are to produce an infinite output level. Since $p > 0$, $C(y, p) \to \infty$ as $y \to \infty$. Also,

$$C(0, p) = \min_x \{p^T x: F(x) \ge 0\} < p^T 0 = 0 \tag{4.20}$$

since 0 is feasible for the minimization problem and hence using property (a), $C(0, p) = 0$. □

From the perspective of concavity and its generalizations, the interesting thing to note is the concavity in prices, property (d). This is our first example showing how (generalized) concavity properties arise naturally in economics.

Our proof of Theorem 4.1 combines various results that have appeared in the economics literature. The proof of property (a) is taken from McFadden (1966) and the proof of property (d) is taken from McKenzie (1956–1957). Under somewhat stronger assumptions on the production function F, Shephard (1953) obtained properties (a)–(d). Uzawa (1964) deduced properties (e) and (f), and Shephard (1970) obtained (g). Property (h) is a technical property of little intrinsic interest.

To an economist, properties (a), (b), (c) and (f) will be intuitively obvious. To a mathematician the concavity property (d) will be obvious, since $C(y, p)$ can be recognized as a *support function* for the upper-level set $U(F, y) = \{x \in R^n: F(x) \ge y, x \ge 0\}$, the set of input vectors that can produce at least the output level y.

However, the continuity properties (e) and (g) are not so intuitively obvious, so we will make some additional observations about them. The reader who is not too concerned about these rather delicate continuity problems is advised to omit this material and turn to the paragraph following (4.53).

The careful reader will have noticed that we have defined the cost function $C(y, p)$ by (4.1) for $p > 0$. If we attempt to define $C(y, p)$ by (4.1) for $p \ge 0$, we find that it may be necessary to change "minimum" in (4.1) to "infimum."

Example 4.1. Let $n = 2$, $x_1 \ge 0$, $x_2 \ge 0$, and $F(x_1, x_2) = 2(x_1)^{1/2}(x_2)^{1/2}$. First notice that $F(x_1, 0) = F(0, x_2) = 0$, hence $C(0, p) = 0$. For $y > 0$ all feasible x must be positive. The reader can verify that for $x > 0$ we have

$\nabla F(x) \neq 0$ and the function F satisfies (3.83). It follows then from Corollary 3.20 that F is quasiconcave for $x > 0$ and that (4.1) satisfies the conditions of Theorem 3.48—that is, the Karush-Kuhn-Tucker necessary optimality conditions are sufficient for problem (4.1). Moreover, since F is increasing and $p > 0$, it follows that $F(x) = y$ must hold at an optimal solution. Hence the Karush-Kuhn-Tucker conditions take the form

$$p_1 = \lambda (x_1)^{-1/2}(x_2)^{1/2} \tag{4.21}$$

$$p_2 = \lambda (x_1)^{1/2}(x_2)^{-1/2} \tag{4.22}$$

$$y = 2(x_1)^{1/2}(x_2)^{1/2} \quad \text{and} \quad \lambda \geq 0. \tag{4.23}$$

Solving (4.21)-(4.23) for x_1, x_2, and λ as functions of y, p_1, and p_2, where $y \geq 0$, $p_1 > 0$, and $p_2 > 0$, the following solution functions are obtained:

$$x_1(y, p_1, p_2) = \tfrac{1}{2}(p_2/p_1)^{1/2}y, \tag{4.24}$$

$$x_2(y, p_1, p_2) = \tfrac{1}{2}(p_1/p_2)^{1/2}y \tag{4.25}$$

and

$$\lambda(y, p_1, p_2) = (p_1 p_2)^{1/2}. \tag{4.26}$$

The cost function may now be calculated using the solution functions (input demand functions) (4.24) and (4.25) as follows:

$$C(y, p) = p_1 x_1(y, p_1, p_2) + p_2 x_2(y, p_1, p_2) = (p_1)^{1/2}(p_2)^{1/2}y. \tag{4.27}$$

Let $p_2 = 1$ and let p_1 approach 0. From (4.24) and (4.25), the demand for input 1 approaches ∞, while the demand for input 2 approaches 0. However, $C(y, 0, 1)$ is not well defined as a minimum in this example, although it is well defined as an infimum, i.e., $\inf_x \{0x_1 + 1x_2 : 2(x_1)^{1/2}(x_2)^{1/2} \geq y\} = 0$ for all $y > 0$.

Rather than changing the word "minimum" in (4.1) to "infimum," we prefer to define $C(y, p)$ only for p strictly positive as in (4.1), and then use the Fenchel (1953) closure operation to extend the domain of definition of $C(y, p)$ from $p > 0$ to $p \geq 0$. The Fenchel closure operation works as follows: fix $y \in Y$ and define $g(p) = C(y, p)$ for $p > 0$. Form the hypograph of g, $H(g) = \{(c, p): c \leq g(p), p > 0\}$, take its closure in R^{n+1}, $\bar{H}(g)$ say, and now define the Fenchel closure of g, $\bar{g}$ say, over $p \geq 0$ by $\bar{g}(p) = \max_c \{c: (c, p) \in \bar{H}(g)\}$. It turns out that $\bar{g}$ coincides with g for $p > 0$ and

$\bar{g}$ will always be upper semicontinuous (unfortunately, the extension of g may not be finite valued; i.e., we may have $\bar{g}(p) = -\infty$ if p is a boundary point, as the example $g(p) = -1/p$, $p > 0$ shows). Although $\bar{g}$ constructed by the Fenchel closure operation will always be upper semicontinuous, it need not be lower semicontinuous, and hence continuous in general. However, in our particular situation, it turns out that $g(p) = C(y, p)$ does have a continuous extension to the nonnegative orthant, as the following theorem shows.

Theorem 4.2 (Gale, Klee, and Rockafellar, 1968; Rockafellar, 1970). Let g be a concave function of n variables defined over the positive orthant $P = \{p \in R^n: p > 0\}$. Suppose g is bounded from below over every bounded subset of P. Then the Fenchel extension of g to the nonnegative orthant $\bar{P} = \{p: p \geq 0\}$ is continuous over $\bar{P}$.

Proof. Let $\bar{g}$ be the Fenchel extension of g to $\bar{P}$. By construction, $\bar{g}$ is upper semicontinuous over $\bar{P}$, and the boundedness property on g implies that $\bar{g}$ is a finite-valued function and hence bounded on every bounded subset of $\bar{P}$. Proposition 2.9 shows that $\bar{g}$ is continuous over P, so we need only show that $\bar{g}$ is lower semicontinuous at boundary points of P.

Since the closure of a convex set is convex, using Proposition 2.5, it can be seen that $\bar{g}$ is a concave function over $\bar{P}$.

Let p^0 be a boundary point of P. Then any point in the square of diameter $2\delta > 0$ centered at p^0 and intersected with the nonnegative orthant may be written as

$$p = t_0 p^0 + \sum_{i=1}^{m} t_i(p^0 + q^i) \tag{4.28}$$

$$= \left(1 - \sum_{i=1}^{m} t_i\right)p^0 + \sum_{i=1}^{m} t_i(p^0 + q^i), \qquad 0 \leq t_i \leq \delta \leq 1/m \tag{4.29}$$

where the points q^i are positive or negative unit vectors. For example, if p^0 were 0, then $m = n$ and $q^i \equiv e^i$, a unit vector with a one in component i and zeros elsewhere. Let $1 \geq \varepsilon > 0$ be given, and define

$$\alpha = \max_i \{|\bar{g}(p^0 + q^i) - \bar{g}(p^0)|: i = 1, 2, \ldots, m\}, \tag{4.30}$$

$$\alpha^* = \max\{\alpha, 1\}, \tag{4.31}$$

$$\delta = \varepsilon/\alpha^* m. \tag{4.32}$$

Using the concavity of $\bar{g}$ on a p of the form defined by (4.29) yields the following inequality:

$$\bar{g}(p) \geq \left(1 - \sum_{i=1}^{m} t_i\right)\bar{g}(p^0) + \sum_{i=1}^{m} t_i\bar{g}(p^0 + q^i) \tag{4.33}$$

$$= \bar{g}(p^0) + \sum_{i=1}^{m} t_i[\bar{g}(p^0 + q^i) - \bar{g}(p^0)]. \tag{4.34}$$

By the definition of α^*

$$\bar{g}(p) \geq \bar{g}(p^0) - \sum_{i=1}^{m} t_i\alpha^*. \tag{4.35}$$

Since $0 \leq t_i \leq \delta$ for each i

$$\bar{g}(p) \geq \bar{g}(p^0) - m\delta\alpha^* \tag{4.36}$$

and by using (4.32)

$$\bar{g}(p) \geq \bar{g}(p^0) - \varepsilon. \tag{4.37}$$

The above inequality establishes that $\bar{g}$ is lower semicontinuous at the boundary point p^0. □

The following example shows that Theorem 4.2 is not necessarily true if the domain of definition set P is not polyhedral.

Example 4.2 (Fenchel, 1953; Rockafellar, 1970). Let

$$P = \{(p_1, p_2): p_1 \in R, p_2 \in R, (p_2)^2 < p_1\} \tag{4.38}$$

and

$$g(p_1, p_2) = -(p_2)^2/p_1. \tag{4.39}$$

Then,

$$\bar{P} = \{(p_1, p_2): p_1 \in R, p_2 \in R, (p_2)^2 \leq p_1\} \tag{4.40}$$

$$\bar{g}(p_1, p_2) = \begin{cases} -(p_2)^2/p_1 & \text{if } p_2 \neq 0, (p_2)^2 \leq p_1, \\ 0 & \text{if } p_2 = 0, p_1 = 0. \end{cases} \quad (4.41)$$

By calculating the matrix of second-order derivatives of g, it can be verified that $\bar{g}$ is indeed a concave function over the interior of $\bar{P}$. Hence it follows (see the proof of Theorem 4.2) that $\bar{g}$ is concave over $\bar{P}$. However, $\bar{g}$ is discontinuous at $p^0 = (0, 0)$. Let (p_1, p_2) approach $(0, 0)$ along the line $p_2 = 0$, $p_1 > 0$, which belongs to $\bar{P}$. Then

$$\lim_{p \to 0} \bar{g}(p, 0) = \lim_{p_1 \to 0} -(0)^2/p_1 = 0. \quad (4.42)$$

Now let (p_1, p_2) approach $(0, 0)$ along the boundary of $\bar{P}$; i.e., let $p_1 = (p_2)^2$, where $p \neq 0$. Then

$$\lim_{p_2 \to 0} \bar{g}((p_2)^2, p_2) = \lim_{p_2 \to 0} -(p_2)^2/(p_2)^2 = -1. \quad (4.43)$$

Equations (4.41) and (4.42) show that $\bar{g}$ is discontinuous at $(0, 0)$.

Actually, Gale, Klee, and Rockafellar (1968) show that Theorem 4.2 is valid if the domain of definition set $\bar{P}$ is *polyhedral*, i.e., $\bar{P} = \bigcap_{i=1}^{I} H_i$, where each H_i is a *halfspace* (a set of the form $\{p: p^T a^i \leq b_i\}$, where a^i is a vector and b_i is a scalar). The reader should be able to modify our proof of Theorem 4.2 to deal with this more general situation.

We are not quite through with our discussion of continuity problems. The reader will recall that we proved the lower semicontinuity property of the cost function, property (g), by a direct argument. It is also possible to prove this property using a very general theorem due to Berge (1963). Since this general theorem will be useful later, we find it convenient to pause momentarily and state this theorem and a closely related one due to Debreu. We do not prove these theorems, since the proofs do not involve generalized concavity.

In the following definitions, let S denote a subset of R^m, T a subset of R^n, $\{s^k\}$ a sequence of points belonging to S, and $\{t^k\}$ a sequence of points belonging to T.

Definition 4.1. ϕ is called a correspondence from S into T if for every $s \in S$ there exists a nonempty image set $\phi(s)$ that is a subset of T.

Definition 4.2. A correspondence ϕ is called upper hemicontinuous at the point $s^0 \in S$ if

$$\lim_{k \to \infty} \{t^k\} = t^0, \qquad s^k \in \phi(t^k), \qquad \lim_{k \to \infty} \{s^k\} = s^0 \Rightarrow s^0 \in \phi(t^0). \quad (4.44)$$

A correspondence ϕ is called lower hemicontinuous at the point $s^0 \in S$ if $\lim_{k\to\infty}\{s^k\} = s^0$, $t^0 \in \phi(s^0)$ imply that there exists a sequence $\{t^k\}$ such that $t^k \in \phi(s^k)$ and $\lim_{k\to\infty}\{t^k\} = t^0$. A correspondence is called continuous at $s^0 \in S$ if it is both upper and lower hemicontinuous at s^0.

Berge (1963) shows that ϕ is an upper hemicontinuous correspondence over all of S if $G(\phi) = \{(s, t)\colon s \in S, t \in \phi(s)\}$ is a closed set in $S \times T$.

Theorem 5.3 (Upper Hemicontinuous Maximum Theorem, Berge, 1963). Let $f(s, t)$ be an upper semicontinuous function defined over $S \times T$, where T is a compact subset of R^n. Suppose that ϕ is an upper hemicontinuous correspondence from S into T. Then the function g given for $s \in S$ by

$$g(s) = \max_x \{f(s, t)\colon t \in T, t \in \phi(s)\} \tag{4.45}$$

is well defined and is upper semicontinuous over S.

Theorem 4.4 (Maximum Theorem, Berge, 1963; Debreu, 1952, 1959). Let f be a continuous real-valued function defined over $S \times T$, where T is a compact subset of R^n. Let ϕ be a correspondence from S into T and let ϕ be continuous over S. Define the function g on S by (4.45) and the (set of maximizers) correspondence ψ from S into subsets of T by $\psi(s) = \{t\colon t \in \phi(s)$ and $f(s, t) = g(s)\}$. Then the function g is continuous over S and the correspondence ψ is upper hemicontinuous over S.

Now let us see how Theorem 4.3 may be used in order to establish property (g) for the cost function. Let $y^1 < y^2$ with $y^1 \in Y$, $y^2 \in Y$, $p > 0$ and suppose x^2 solves the minimization problem $\min_x \{p^T x\colon F(x) \geq y^2\} = C(y^2, p)$. For y, $y^1 \leq y \leq y^2$, define the following correspondences:

$$A(y) = \{x\colon F(x) \geq y, x \geq 0\}, \tag{4.46}$$

$$B(y) = \{x\colon x \geq 0, \text{and } p^T x \leq p^T x^2\} \tag{4.47}$$

and

$$\phi(y) = A(y) \cap B(y). \tag{4.48}$$

Since F is upper semicontinuous, the correspondence $A(y)$ is upper hemicontinuous. The correspondence $B(y)$ is a constant compact set (using the positivity of the price vector (p)) and hence is also upper hemicontinuous. It is easy to show that the intersection of two upper hemicontinuous correspondences is also upper hemicontinuous. Thus $\Phi(y)$ is an upper hemicontinuous correspondence which is contained in a compact set. For $y^1 \le y \le y^2$,

$$C(y, p) = -\max_x \{-p^T x \colon F(x) \ge y, x \ge 0\} \tag{4.49}$$

$$= -\max_x \{-p^T x \colon x \in A(y)\} \tag{4.50}$$

$$= -\max_x \{-p^T x \colon x \in A(y), x \ge 0, p^T x \le p^T x^2\} \tag{4.51}$$

(since x^2 is feasible for the maximization problem)

$$= -\max_x \{-p^T x \colon x \in A(y) \text{ and } x \in B(y)\} \tag{4.52}$$

$$= -\max_x \{-p^T x \colon x \in \Phi(y)\}. \tag{4.53}$$

By Theorem 4.3, $\max_x \{-p^T x \colon x \in \Phi(y)\}$ is upper semicontinuous for $y^1 \le y \le y^2$, and hence since the negative of an upper semicontinuous function is lower semicontinuous, $C(y, p)$ is lower semicontinuous in y on every compact subset of Y, hence it is lower semicontinuous on Y.

So much for the continuity properties of the cost function. What is surprising is that the cost function C has so many properties in spite of the fact that we have assumed next to nothing about the underlying production function F. These properties of the cost function have some important empirical consequences. For example, economists often have data on cost, output, and input prices for a firm during period t, C^t, y^t, and p^t, respectively. A functional form for the firm's cost function C^*, say, is assumed and the unknown parameters that characterize C^* are estimated by minimizing the sum of squared errors $(e^t)^2$, where the error in period t is defined by $e^t = C^t - C^*(y^t, p^t)$, $t = 1, 2, \ldots, T$.

Frequently, the functional form for C^* is assumed to be linear:

$$C^*(y, p) = a_0 + \sum_{i=1}^{n} a_i p_i + a_{n+1} y \tag{4.54}$$

where the a_i are the unknown parameters to be estimated. However, if the firm's production function satisfies the very weak regularity condition in

Assumption 1 above and if the firm is competitively minimizing costs during the T periods, then by Theorem 4.1, property (a) implies that the firm's true cost function must be linearly homogeneous in prices p. Thus if C^* is the firm's true cost function, then we must have $a_0 = 0 = a_{n+1}$. But then C^* does not depend on the output level y, which is completely unrealistic from an economic point of view. If we assume that $C^*(y, p)$ is a quadratic function in (y, p), we encounter similar difficulties. For examples of functional forms for cost functions that are consistent with Theorem 4.1 but at the same time can approximate an arbitrary (differentiable) cost function to the second order, see Diewert (1974, 1982) and Lau (1974, 1978).

It is obvious that the production function F determines the cost function C. In the following section we ask if the process can be reversed—that is, given a cost function satisfying properties (a)-(h), can we determine the underlying production function?

4.2. Duality between Cost and Production Functions

Suppose we have a production function F satisfying Assumption 1. We remind the reader that the upper-level sets of F for $y \in Y$ are given by $U(F, y) = \{x: F(x) \geq y, x \geq 0\}$. It is easy to see that a family of upper level sets (or production possibilities sets) $U(F, y)$ completely determines a production function F. Also, the producer's cost function C can be defined by (4.1) or equivalently by $C(y, p) = \min_x \{p^T x: x \in U(F, y)\}$ for $y \in Y$, $p > 0$. So far, we addressed the problem of determining $C(y, p)$ given $F(x)$ or $U(F, y)$. Now we address the opposite problem: given a cost function $C(y, p)$, does there exist a compatible family of production possibilities sets $U(F, y)$, and hence via the above construction a production function F, for which the corresponding cost function obtained via definition (4.1) is the original cost function that we started with?

Suppose that the set Y and the cost function $C(y, p)$ are given. For $y \in Y$, $p > 0$, define the isocost plane for output level y and input price vector p as $\{x: p^T x = C(y, p)\}$. From the definitions of $C(y, p)$ and $U(F, y)$, it is obvious that the set $U(F, y)$ must lie "above" this isocost plane and be tangent to it; i.e., $U(F, y) \subset \{x: p^T x \geq C(y, p)\}$, where $\{x: p^T x \geq C(y, p)\}$ is a supporting half-space for the true production possibilities set $U(F, y)$. Thus we may use the cost function in order to form an *outer approximation* $M(y)$ to the true set $U(F, y)$:

$$M(y) = \bigcap_{p>0} \{x: p^T x \geq C(y, p)\}. \tag{4.55}$$

Since $U(F, y) \subset \{x: p^T x \geq C(y, p)\}$ for every $p > 0$, $U(F, y) \subset M(y)$, where by (4.55), $M(y)$ is the intersection of the supporting halfspaces to the set $U(F, y)$. The set $M(y)$ is often called (e.g., McFadden, 1966) the *free disposal convex hull* of $U(F, y)$.

Since $M(y)$ is the intersection of a family of convex sets, $M(y)$ is also a convex set. $M(y)$ also has the following (free disposal) property: $x^0 \in M(y)$, $x^0 \leq x^1$ implies that $x^1 \in M(y)$. Thus if we want $U(F, y)$ to coincide with $M(y)$ for each $y \in Y$, then $U(F, y)$ must be a convex set with the free disposal property for every $y \in Y$. However, if the family of sets $U(F, y)$ is convex for every $y \in Y$, then it must be the case that the production function F is quasiconcave (recall Definition 3.1). Similarly, if $U(F, y)$ satisfies the free disposal property for every $y \in Y$, then F must be a nondecreasing function. Thus if we wish to use the cost function in order to determine precisely the underlying production function (rather than an outer approximation to it), we must assume that the production function satisfies the following two additional properties:

Assumption 2. F is a quasiconcave function.

Assumption 3. F is nondecreasing—that is, if $x^0 \leq x^1$, $x^0 \neq x^1$, then $F(x^0) \leq F(x^1)$.

We now ask whether, given a cost function C satisfying properties (a)-(h), there exists a corresponding production function F satisfying Assumptions 1-3 that is consistent with this given cost function. The next theorem answers this question affirmatively. First, however, we need some introductory material on supergradients of concave functions. Let g be a concave function defined over a convex set $S \subset R^n$.

Definition 4.3. The vector z^0 is called a supergradient to the concave function g at the point $p^0 \in S$ if

$$g(p) \leq g(p^0) + z^{0T}(p - p^0) \qquad \text{for every } p \in S. \tag{4.56}$$

Rockafellar (1970) shows that if p^0 belongs to the interior of S, then g has at least one supergradient at p^0. Moreover, if g has only one supergradient vector z^0 at p^0, then g is differentiable at p^0 and $\nabla g(p^0) = z^0$. Essentially, the set of supergradients, denoted by $\partial g(p^0)$, is the set of all normals of the nonvertical supporting hyperplanes to the boundary point $(p^0, g(p^0))$ of the hypograph of g (see Figure 2.5).

If g is positively linearly homogeneous, and p^0 is an interior point of S, then we may let p in (4.56) equal λp^0 for λ close to unity. Using $g(\lambda p^0) = \lambda g(p^0)$, (4.56) becomes

$$(\lambda - 1)g(p^0) \leq (\lambda - 1)z^{0T}p^0. \tag{4.57}$$

Hence if $\lambda > 1$, $g(p^0) \leq z^{0T}p^0$ while if $\lambda < 1$, (4.57) yields the inequality $g(p^0) \geq z^{0T}p^0$. Hence $g(p^0) = z^{0T}p^0$. Thus when g is positively linearly homogeneous in addition to being concave, (4.56) becomes

$$g(p) \leq z^{0T}p \qquad \text{for every } p \in S \text{ and } g(p^0) = z^{0T}p^0. \tag{4.58}$$

In other words, z^0 is a supergradient to a linearly homogeneous concave function g at the interior point p^0 of S if z^0 satisfies (4.58).

Theorem 4.5. Suppose that a set of outputs is Y defined by $Y = \{y \in R: 0 \leq y \leq \bar{y}\}$ or $Y = \{y \in R: 0 \leq y < \bar{y}\}$ and given a cost function $C(y, p)$ satisfying properties (a)–(h) on $Yx\{p \in R^n: p > 0\}$. Then the production function F, defined in terms of the cost function for $x \geq 0$ by

$$F(x) = \max_y \{y \in Y: C(y, p)^T \leq p^T x \text{ for every } p > 0\}, \tag{4.59}$$

is well defined and it satisfies Assumptions 1–3. Moreover, if we define the cost function that corresponds to the F defined by (4.59) as C^0, that is, for $y \in Y$ and $p > 0$,

$$C^0(y, p) = \min_x \{p^T x: F(x) \geq y, x \geq 0\}, \tag{4.60}$$

then C^0 equals the originally given cost function C.

Proof. For each $p > 0$, define the following correspondence:

$$\psi_p(x) = \{y: \in Y, C(y, p) \leq p^T x\}. \tag{4.61}$$

For $x \geq 0$, property (g) implies that $\psi_p(x)$ is an upper hemicontinuous correspondence in x for each $p > 0$. $\psi_p(x)$ is nonempty, since by property (h), $0 \in \psi_p(x)$. Property (h) also implies that $\psi_p(x)$ is bounded from above and hence using (g), $\psi_p(x)$ is a compact set. For $x \geq 0$, define the intersection correspondence Φ by

$$\Phi(x) = \bigcap_{p>0} \psi_p(x) = \{y: y \in y, C(y, p) \leq p^T x \text{ for every } p > 0\}. \tag{4.62}$$

Note that $\Phi(x)$ is nonempty, since $0 \in \psi_p(x)$ for every p. From (4.62) and the properties of the $\psi_p(x)$, it is evident that Φ is an upper hemicontinuous correspondence for $x \geq 0$, and for each $x \geq 0$ $\Phi(x)$ is a compact set. It follows that the maximum in (4.59) exists. Thus $F(x)$ is well defined. Moreover, by Theorem 4.3, F is upper semicontinuous for $x \geq 0$.

To show that F defined by (4.59) is quasiconcave, let $x^1 \geq 0$, $x^2 \geq 0$, $0 < \lambda < 1$, $F(x^1) \geq y$ and $F(x^2) \geq y$. Then by (4.59) and property (f), we have

$$C(y, p) \leq p^T x^1 \qquad \text{for every } p > 0 \tag{4.63}$$

and

$$C(y, p) \leq p^T x^2 \qquad \text{for every } p > 0. \tag{4.64}$$

Hence $C(y, p) = \lambda C(y, p) + (1 - \lambda)C(y, p) \leq p^T(\lambda x^1 + (1 - \lambda)x^2)$ for every $p > 0$, and hence using (4.59), $F(\lambda x^1 + (1 - \lambda)x^2) \geq y$ since y is feasible for the maximization problem.

To show that F is nondecreasing, let $x^1 \geq x^0 \geq 0$, $x^1 \neq x^0$. Then, using (4.61), $p > 0$, and property (f), we have $\psi_p(x^0) \subset \psi_p(x^1)$ for every $p > 0$, and hence

$$\Phi(x^0) \subset \Phi(x^1). \tag{4.65}$$

Thus

$$F(x^1) = \max_y \{y: y \in \Phi(x^1)\} \geq \max_y \{y: y \in \Phi(x^0)\} = F(x^0). \tag{4.66}$$

Finally, we need to show that C^0 defined by (4.60) coincides with C. Let $y^0 \in Y$, $p^0 > 0$. We wish to show that $C^0(y^0, p^0) = C(y^0, p^0)$. By properties (b) and (d), $C(y^0, p)$ is a linearly homogeneous concave function of p for $p > 0$. Hence, there exists at least one supergradient vector x^0 to this function at the point $p^0 > 0$. Using (4.57), x^0 satisfies

$$C(y^0, p) \leq p^T x^0 \qquad \text{for every } p > 0 \quad \text{and} \quad C(y^0, p^0) = p^{0T} x^0. \tag{4.67}$$

Using property (a) we see that $x^0 \geq 0$. Thus, by (4.59) and (4.67) we obtain

$$F(x^0) = \max_y \{y \in y: C(y, p) \leq p^T x^0 \text{ for every } p > 0\} \geq y^0. \tag{4.68}$$

Thus,

$$C^0(y^0, p^0) = \min \{(p^0)^T x: F(x) \geq y^0, x \geq 0\} \tag{4.69}$$

$$\leq (p^0)^T x \qquad \text{[because } x^0 \text{ is feasible by (4.68)]} \tag{4.70}$$

$$= C(y^0, p^0) \qquad \text{[using (4.67)]}. \tag{4.71}$$

Using (4.59) and (4.60), we also have

$$C^0(y^0, p^0) = \min_x \{(p^0)^T x: \max_y \{y \in Y: C(y, p) \leq p^T x \text{ for every } p > 0\} \geq y^0\} \tag{4.72}$$

$$= \min_x \{(p^0)^T x: x \geq 0, C(y^0, p) \leq p^T x \text{ for every } p > 0\}$$

$$\text{(using property 6)} \tag{4.73}$$

$$\geq \min_x \{(p^0)^T x: x \geq 0, C(y^0, p^0) \leq (p^0)^T x\} \tag{4.74}$$

$$= C(y^0, p^0). \tag{4.75}$$

From (4.71) and (4.75), it follows that $C^0(y^0, p^0) = C(y^0, p^0)$. □

Corollary 4.6 (Gorman, 1968). Let C satisfy properties (a)–(h) and define the corresponding production function F by (4.59). Let $p^0 > 0$ and $y^0 \in Y$. Then the solution set to the cost minimization problem $\min_x \{(p^0)^T x: F(x) \geq y^0, x \geq 0\}$ is $\partial_p C(y^0, p^0)$, the set of supergradients to the concave function $g(p) = C(y^0, p^0)$ at the point p^0.

Proof. In the proof of Theorem 4.5, we have shown that if $x^0 \in \partial_p C(y^0, p^0)$, then x^0 solves the cost minimization problem (4.60). If $x \geq 0$ but $x \notin \partial_p C(y^0, p^0)$, then two cases are possible: (i) We have $C(y^0, \bar{p}) > \bar{p}^T x$ for at least one $\bar{p} > 0$ and hence $F(x) = \max_y \{y \in Y: C(y, p) \leqq p^T x$ for every $p > 0\} < y^0$, so in this case x is not even feasible for the cost minimization problem (4.60) and hence cannot solve it. (ii) $C(y^0, p) \leq p^T x$ for every $p > 0$ and $C(y^0, p^0) < (p^0)^T x$. But in this case, x cannot be optimal in (4.60). □

Corollary 4.7 (Shephard, 1953). Let C satisfy properties (a)–(h), $p^0 > 0$, $y^0 \in Y$ and suppose in addition that $C(y^0, p)$ is differentiable with respect to input prices at $p = p^0$; i.e., the gradient vector $\nabla_p C(y^0, p^0)$ exists. Then the solution x^0 to the cost minimization problem $\min_x \{(p^0)^T x: F(x) \geq y^0\}$, where F is defined by (4.59), is unique and is equal to the vector of first-order partial derivatives of C with respect to the components of the input price vector, i.e.,

$$x^0 = \nabla_p C(y^0, p^0). \tag{4.76}$$

Proof. If a concave function is differentiable at p^0, say, then the set of supergradients of the function at this point reduces to the gradient vector (see Figure 2.4). Hence the present corollary is implied by the previous corollary. □

In the following section, we shall indicate some applications of Corollary 4.7, or *Shephard's lemma* as it is known in the economics literature.

The careful reader may have noticed that our proof of Theorem 4.5 did not use properties (c) (nondecreasing in prices) and (e) (continuous in prices) of the cost function C. This is due to the fact that property (d) implies (e), and (c) is implied by properties (a), (b), and (d). Thus if $y \in Y$, $p^1 > 0$, $p^2 > 0$, then

$$C(y, p^1 + p^2) = 2C(y, (1/2)p^1 + (1/2)p^2) \quad \text{[by (4.3)]} \tag{4.77}$$

$$\geq 2\{(1/2)C(y, p^1) + (1/2)C(y, p^2)\} \quad \text{[by (4.5)]} \tag{4.78}$$

$$= C(y, p^1) + C(y, p^2) \tag{4.79}$$

$$\geq C(y, p^1) \quad \text{[by (4.2)].} \tag{4.80}$$

The general case where $p^2 \geq 0$ follows by a continuity argument.

Theorem 4.5 is an example of an *economics duality theorem*. In this case, the duality is between cost and production functions satisfying certain regularity conditions: given one of these functions, the other is completely determined and vice versa. The first theorems of this type were established by Shephard (1953, 1970). The regularity conditions that we have used in Theorem 4.5 are somewhat weaker than those appearing in the existing literature (and we do not think that they can be weakened much further). From a mathematical point of view, the above duality theorem rests on Minkowski's (1911) theorem: every closed convex set can be characterized as the intersection of its supporting half-spaces. Typically, the proofs of duality theorems in economics are straightforward with the exception of the proofs of continuity properties. The reader will recall that we required Berge's Maximum Theorem 4.3 in order to prove Theorem 4.5.

Similar duality theorems between cost and production functions have been established in the literature under a variety of regularity conditions. In the economics literature, see Diewert (1971, 1974, 1982), McFadden (1966, 1978), Samuelson (1953-1954), Shephard (1953, 1970), and Uzawa (1964), and for local duality theorems, see Diewert (1982), Diewert and Blackorby (1979), and Epstein (1981). In the mathematics literature, see Crouzeix (1977) and Fenchel (1953).

4.3. Generalized Concavity and Producer Theory

We have already seen why concavity plays a central role in economic theory. From Theorem 4.1, the cost function $C(y, p)$ is concave in p, no matter what the functional form for the producer's production function F is (provided that F is upper semicontinuous, an empirically harmless assumption). It is now possible to see why quasiconcavity also plays a central role in economic theory: from an empirical point of view, it is harmless to assume that the producer's production function is quasiconcave (and nondecreasing), provided that the producer is competitively minimizing costs. Why is this? Assume that the true production function F satisfies only Assumption 1 and construct F's cost function C using (4.1). Then use C to construct the function F^*, say, by (4.59). It follows from Theorem 4.5 that the cost function that corresponds to F^* is also C. Thus the original production function F and the derived (from the cost function) production function F^* both generate the same cost function, and so we might as well assume that F^* is the true production function. However, F^* is quasiconcave (and nondecreasing) even though the original F need not be.

One common application of Theorem 4.5 (see Diewert, 1971, 1974, or Lau, 1974, 1978) is an easy method of generating a theoretically valid system of cost minimizing input demand functions $x(y, p)$: simply postulate a functional form for the cost function C that is consistent with properties (a)–(h), and in addition is differentiable with respect to input prices, and then estimate the unknown parameters occurring in the cost function by minimizing some function of the errors e^t in the equation system

$$x^t = \nabla_p C(y^t, p^t) + e^t, \qquad t = 1, 2, \ldots, T, \tag{4.81}$$

where x^t, y^t, and p^t are observed data on inputs, output, and input prices, respectively, during period t. The alternative to the above method is to assume a functional form for the production function F and then solve the cost minimization problem (4.1) directly for the system of cost minimizing demand function's $x(y, p)$, assuming that the solution set is unique, and then estimate the unknown parameters. The problem with this more direct method is that the unknown parameters that characterize the production function F usually occur in the demand system $x(y, p)$ in a highly nonlinear manner. On the other hand, if we use Theorem 4.5 in order to generate the system of demand functions $x(y, p)$, we can choose our functional form for C in such a way that the system of equations (4.81) is linear in the unknown parameters, and thus linear regression techniques can be applied to the estimation problem.

A second equally important application of Theorem 4.5 can be given. Suppose that C satisfies properties (a)-(h) and, in addition, is twice continuously differentiable with respect to input prices at the point $y^0 \in Y$, $p^0 > 0$. Then by Theorem 4.5, the input demand functions are given by $x(y^0, p) = \nabla_p C(y^0, p)$ for p close to p^0. Now differentiate each demand function $x_i(y^0, p^0) = \partial C(y^0, p^0)/\partial p_i$ with respect to input prices and denote the resulting matrix of partial derivatives as $\nabla_p x(y^0, p^0)$. Then,

$$\nabla_p x(y^0, p^0) = \nabla^2_{pp} C(y^0, p^0) \tag{4.82}$$

where $\nabla^2_{pp} C(y^0, p^0) = [\partial^2 C(y^0, p^0)/\partial p_i \partial p_j]$ denotes the matrix of second partial derivatives of C with respect to the components of p. By property (d), $C(y^0, p)$ is concave in p. By Theorem 2.14, $\nabla^2_{pp} C(y^0, p^0)$ is a negative semidefinite matrix, so that in particular, we must have

$$\partial x_i(y^0, p^0)/\partial p_i = \partial^2 C(y^0, p^0)/\partial p_i^2 \le 0, \qquad i = 1, 2, \ldots, n, \tag{4.83}$$

and at least one component must be negative. The inequalities in (4.83) have a very simple economic interpretation: If the price of input i increases, then the cost minimizing demand for input i needed to produce a fixed output level y^0 will not increase; i.e., demand for input functions do not slope upwards with respect to that input's price.

If $n = 2$, so that there are only two inputs, then we may deduce that

$$\frac{\partial x_1(y^0, p^0)}{\partial p_2} = \frac{\partial^2 C(y^0, p^0)}{\partial p_1 \, \partial p_2} \qquad \text{[by differentiating (4.76)]} \tag{4.84}$$

$$= \frac{\partial^2 C(y^0, p^0)}{\partial p_2 \, \partial p_1} \qquad \text{(by interchanging the order of differentiation)} \tag{4.85}$$

$$= \frac{\partial x_2(y^0, p^0)}{\partial p_1} \qquad \text{[by using (4.76) again].} \tag{4.86}$$

Using

$$p_1 \frac{\partial^2 C}{\partial p_2 \, \partial p_1} + p_2 \frac{\partial^2 C}{\partial p_2 \, \partial p_2} = 0, \tag{4.87}$$

which is implied by property (b) and Euler's theorem on homogeneous functions, we obtain

$$\frac{\partial x_1(y^0, p^0)}{\partial p_2} = \frac{\partial x_2(y^0, p^0)}{\partial p_1} = \frac{-p_2}{p_1} \frac{\partial^2 C(y^0, p^0)}{\partial p_2 \, \partial p_2} \tag{4.88}$$

$$\ge 0 \qquad \text{[by using (4.83)].} \tag{4.89}$$

The equality (4.86) is a Samuelson (1947), Hicks (1946) *symmetry restriction.*

All of the above results were obtained (under somewhat stronger hypotheses) by Hicks (1946), and Samuelson (1947). The present derivation is taken from Diewert (1982) and McFadden (1966, 1978). The above results illustrate a second major application of the duality between cost and production functions: duality theory usually enables us to derive theoretical theorems about the solutions to various economic optimization problems in a comparatively effortless manner.

One question that we have not yet resolved is: under what conditions on the production function F will the cost function C be differentiable with respect to input prices? The following theorem provides a sufficient condition.

Theorem 4.8. Let F be a nonnegative, continuous, nondecreasing, and strictly quasiconcave function defined over the nonnegative orthant in R^n. Define the cost function C that corresponds to F by (4.1). Then for every $y^0 \in Y$ satisfying $y^0 > F(0)$ and $p^0 > 0$, $\nabla_p C(y^0, p^0)$ exists; i.e., C is differentiable with respect to input prices.

Proof. Let $y^0 > F(0)$, $p^0 > 0$ and let $x^1 \geq 0$ solve

$$\min_x \{(p^0)^T x \colon F(x) \geq y^0\} = C(y^0, p^0). \tag{4.90}$$

From Corollary 4.7 it can be seen that $x^1 = \nabla_p C(y^0, p^0)$ if x^1 is the unique solution to (4.90). Hence suppose that x^1 is not the unique solution to (4.90), i.e., there exists $x^2 \neq x^1$ such that

$$F(x^2) \geq y^0, \qquad (p^0)^T x^2 = C(y^0, p^0). \tag{4.91}$$

Since F is continuous, we must have $F(x^1) = y^0$ and $F(x^2) = y^0$ for if $F(x^1) > y^0$ or $F(x^2) > y^0$, then x^1 and x^2 could not solve the cost minimization problem (4.90); i.e., there would exist $0 \leq \lambda < 1$ such that $F(\lambda x^1) = y^0$ and $(p^0)^T \lambda x^1 < C(y^0, p^0)$, etc. Since F is strictly quasiconcave, using Definition 3.8, we have

$$F(\tfrac{1}{2}x^1 + \tfrac{1}{2}x^2) > \min\{F(x^1), F(x^2)\} = y^0. \tag{4.92}$$

However, (4.92) and the continuity of F imply that there exists a λ^0 such that

$$F[\lambda^0(x^1 + x^2)/2] = y^0 \quad \text{and} \quad 0 \leq \lambda^0 < 1. \tag{4.93}$$

Since

$$(p^0)^T\lambda^0(x^1 + x^2)/2 = \lambda^0\{(p^0)^Tx^1 + (p^0)^Tx^2\}/2 \tag{4.94}$$

$$= \lambda^0 C(y^0, p^0) \tag{4.95}$$

$$< C(y^0, p^0) \qquad [\text{since } C(y^0, p^0) > 0]. \tag{4.96}$$

Hence, (4.93) and (4.96) show that x^1 and x^2 cannot be the solutions to the cost minimization problem (4.90). This contradiction rules out our supposition, and hence the solution to (4.90) must be unique. □

Thus strict quasiconcavity (and continuity) of the production function F provides a sufficient condition for the differentiability of the cost function C with respect to input prices. However, the following example shows that strict quasiconcavity of F is not necessary.

Example 4.3. Let $a_1 > 0$, $a_2 > 0$, $n = 2$ and define $F(x_1, x_2) = \min\{x_1/a_1, x_2/a_2\}$. This production function is known in the economics literature as a Leontief or fixed coefficient production function. Note that it is not strictly quasiconcave and it is not differentiable since whenever $x_1/a_1 = x_2/a_2$, the two-sided directional derivatives of F, $D_vF(x) = \lim_{t\to 0}[F(x + tv) - F(x)]/t$, do not exist for all directions v. However, the cost function that corresponds to F is $C(y, p_1, p_2) = (a_1p_1 + a_2p_2)y$, which has partial derivatives of all orders.

Note that a simple, necessary and sufficient condition for the differentiability of $C(y^0, p^0)$ with respect to p at some point $y^0 \in Y$ and $p^0 > 0$ is that the solution to $\min_x\{(p^0)^Tx = F(x) \geq y^0\}$ be unique (where F satisfies Assumption 1).

It should also be noted that the hypothesis of continuity is required in Theorem 4.6 since Example 4.4 below exhibits a (discontinuous) strictly quasiconcave function that does not have a differentiable cost function.

Example 4.4. Let $n = 2$ and define F as follows (assume $x_1 > 0$ and $x_2 > 0$):

$$F(x_1, x_2) = \begin{cases} x_1x_2 & \text{for } x_1x_2 < 1 \text{ and for } \frac{1}{2} < x_1 < 2 \text{ and } x_2 = 1/x_1, \\ 1 + x_1x_2 & \text{for } x_1x_2 \geq 1 \text{ and } x_1 + x_2 \geq \frac{5}{2}, \\ \dfrac{x_2 - x_1 - 2(x_1)^{-1} + (5/2)}{(5/2) - x_1 - (x_1)^{-1}} & \text{for } x_1x_2 > 1 \text{ and } x_1 + x_2 < \frac{5}{2} \end{cases} \tag{4.97}$$

Level curves of the function are shown in Figure 4.1. It can be verified that $U(F, 2) = \{x\colon F(x) \geq 2, x \geq 0\}$ is

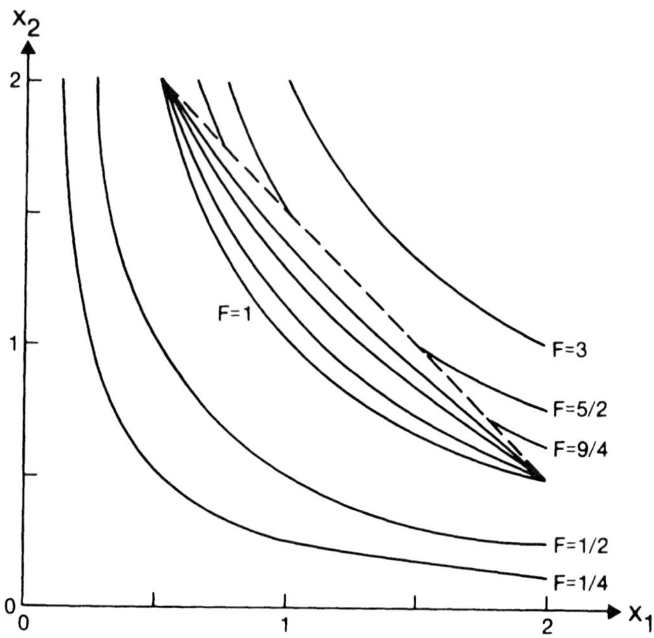

Figure 4.1. The function (4.97).

$$U(F, 2) = \{(x_1, x_2): x_1 > 0, x_2 > 0, x_1x_2 \geq 1 \text{ and } x_1 + x_2 \geq \tfrac{5}{2}\}. \tag{4.98}$$

Thus if $p_1 = p_2 = 1$, $C(2, (p_1, p_2)) = \frac{5}{2}$, and the set of solutions to the cost minimization problem is $\{(x_1, x_2): \frac{1}{2} \leq x_1 \leq 2 \text{ and } x_2 = \frac{5}{2} - x_1\}$. Hence C is not differentiable with respect to p_1 and p_2 at $y = 2, p_1 = 1, p_2 = 1$. Moreover, it can be verified that F is strictly quasiconcave and upper semicontinuous over the positive orthant.

Let us consider another interesting question: under what conditions on F will C be twice differentiable with respect to prices with a negative definite Hessian matrix $\nabla^2_{pp}C(y^0, p^0)$ [which would imply that the inequalities in (4.83) are strict]. Unfortunately, $\nabla^2_{pp}C(y^0, p^0)$ cannot be negative definite for any F, since property (b) on C (linear homogeneity in prices) implies that

$$\nabla^2_{pp}C(y^0, p^0)p^0 = 0 \tag{4.99}$$

so that p^0 is an eigenvector of $\nabla^2_{pp}C(y^0, p^0)$ that corresponds to a zero eigenvalue. However, we can ask under what conditions will $\nabla^2_{pp}C(y^0, p^0)$ be negative definite in the subspace orthogonal to $p^0 > 0$, and we can also ask what does the last condition on C imply about F?

Suppose C satisfies properties (a)–(h), y^0 is in the interior of Y, $p^0 > 0$, C is twice continuously differentiable in a neighborhood around (y^0, p^0) with $x^0 = \nabla_p C(y^0, p^0) > 0$ and $\nabla^2_{pp} C(y^0, p^0)$ has the following property:

$$z^T p^0 = 0, \qquad z \neq 0 \;\Rightarrow\; z^T \nabla^2_{pp} C(y^0, p^0) z < 0. \tag{4.100}$$

First we show that (4.100) implies that the inequalities in (4.83) hold strictly, provided that $n \geq 2$. By property (d), $\nabla^2_{pp} C(y^0, p^0)$ is negative semidefinite, while property (b) implies (4.99), i.e., that p^0 is an eigenvector of $\nabla_{pp} C(y^0, p^0)$ that corresponds to a zero eigenvalue. Since every vector $z \in R^n$ satisfying $z \neq kp^0$ for some $k \in R$ must be of the form $z = w + \alpha p^0$, where $w \neq 0$ is in the subspace orthogonal to p^0, it follows from (4.99) and (4.100) that $\nabla^2_{pp} C(y^0, p^0)$ satisfies the following property:

$$k \in R, \qquad z \neq kp^0 \;\Rightarrow\; z^T \nabla^2_{pp} C(y^0, p^0) z < 0. \tag{4.101}$$

From (4.83), $\partial^2 C(y^0, p^0)/\partial p_i^2 \leq 0$. Suppose that $\partial^2 C(y^0, p^0)/\partial p_i^2 = 0$. Then $e_i^T \nabla^2_{pp} C(y^0, p^0) e_i = \partial^2 C(y^0, p^0)/\partial p_i^2 = 0$, where e_i is the ith unit vector. Since $p^0 > 0$, $e_i \neq kp^0$ for any scalar k, we have contradicted (4.101). Thus our supposition is false and

$$\partial x_i(y^0, p^0)/\partial p_i = \partial^2 C(y^0, p^0)/\partial p_i^2 < 0 \qquad \text{for } i = 1, 2, \ldots, n. \tag{4.102}$$

Thus assumption (4.100) on C is sufficient to imply that the ith input demand function $x_i(y, p)$ has a negative slope with respect to the ith price p_i for all (y, p) in a neighborhood of (y^0, p^0).

The meaning of (4.100) will become clearer if we note that (4.101) is equivalent to the following expression [assuming the other conditions in (4.100) hold]:

$$z^T \nabla_p C(y^0, p^0) = 0, \qquad z \neq 0 \;\Rightarrow\; z^T \nabla^2_{pp} C(y^0, p^0) z < 0. \tag{4.103}$$

Suppose now that (4.101) holds. Let $z^T \nabla_p C(y^0, p^0) = 0$, $z \neq 0$, where $\nabla_p C(y^0, p^0) = x^0 > 0$ by an assumption made above (4.100). If $z = kp^0$ for some $k \neq 0$, then $z^T \nabla_p C(y^0, p^0) < 0$ if $k < 0$ or $z^T \nabla_p C(y^0, p^0) > 0$ if $k > 0$ since $p^0 > 0$, which contradicts $z^T \nabla_p C(y^0, p^0) = 0$. Thus $z \neq kp^0$ for any scalar k and (4.101) implies $z^T \nabla_{pp} C(y^0, p^0) z < 0$. Thus (4.103) holds. The proof that (4.103) implies (4.101) is similar.

From Proposition 3.45, it can be seen that (4.103) implies that $C(y, p)$ is strongly pseudoconcave with respect to p for (y, p) close to (y^0, p^0). We now ask: what does local strong pseudoconcavity in prices of the cost function C imply about the corresponding dual production function F

defined by (4.59)? Using the material in Blackorby and Diewert (1979), it can be shown that if C satisfies (4.100), then F satisfies Assumptions 1-3 and the following additional properties [where $x^0 = \nabla_p C(y^0, p^0) > 0$]:

(i) F is twice continuously differentiable in a neighborhood around x^0 with $\nabla F(x^0) > 0$; (4.104)

(ii) $z^T \nabla F(x^0) = 0, \quad z \Rightarrow 0$ implies $z^T \nabla^2 F(x^0) z < 0.$ (4.105)

Using Proposition 3.45, it can be seen that the differentiability properties of F and (4.105) imply that F is strongly pseudoconcave in a neighborhood of x^0.

It can also be shown that if a production function F satisfies Assumptions 1-3, and properties (i) and (ii), then the corresponding cost function C defined by (4.1) satisfies properties (a)-(h) and (4.103). Thus local strong pseudoconcavity of the production function is equivalent to local strong pseudoconcavity of the cost function in the twice continuously differentiable case. This result may be found in Blackorby and Diewert (1979). We have shown how concavity, quasiconcavity, strict quasiconcavity, and strong pseudoconcavity arise in the context of the producer's cost minimization problem. We conclude this section by indicating a few additional economic applications of cost functions.

First, we note that the producer's cost minimization problem can be given an interpretation in the context of consumer theory: Interpret F as the consumer's *utility function* [if $F(x^1) > F(x^2)$, then the consumer prefers the commodity vector x^1 to x^2], interpret $p > 0$ as a vector of commodity prices, and interpret $y \in Y$ as a utility level rather than as an output level. Then the cost minimization problem (4.1), $\min_x \{p^T x: F(x) \geqq y\}$ can be interpreted as the problem of minimizing the cost or expenditure needed to achieve a certain utility level $y \in Y$ given that the consumer faces the vector of commodity prices $p > 0$. In the economics literature, the resulting cost function $C(y, p)$ is often called an *expenditure function*; see, e.g., Blackorby and Diewert (1979).

Two applications of expenditure functions can readily be given. The expenditure function plays a central role in the economic theory of the cost of living index. Let the consumer face the commodity price vectors $p^1 > 0$ and $p^2 > 0$ during periods 1 and 2, respectively, and let F be the consumer's utility function and C be the corresponding expenditure function defined by (4.1). Then the Konyus (1939) *cost of living index* P corresponding to $x > 0$ (the reference vector of quantities), is defined as

$$P(p^1, p^2, x) = C[F(x), p^2]/C[F(x), p^1]. \qquad (4.106)$$

P is interpreted as follows: pick a reference indifference surface indexed by the quantity vector $x > 0$. Then $P(p^1, p^2, x)$ is the minimum cost of achieving the standard of living indexed by x when the consumer faces commodity prices p^2 relative to the minimum cost of achieving the same standard of living when the consumer faces period 1 prices p^1. Thus P can be interpreted as a level of prices in period 2 relative to a level of prices in period 1. Note that the mathematical properties of P are completely determined by the mathematical properties of F and C.

The expenditure function also plays a central role in *cost-benefit analysis*, which makes use of the *consumer surplus* concept. Let F, C, p^1, and p^2 be defined as in the previous paragraph and let x^1 and x^2 be the consumer's observed consumption vectors in periods 1 and 2, respectively. Hicks (1941–1942) has used the expenditure function in order to define two closely related measures of consumer surplus or welfare change: Hicks' *compensating variation in income* is defined as $C[F(x^2), p^2] - C[F(x^1), p^2]$ while Hicks' *equivalent variation in income* is defined as $C[F(x^2), p^1] - C[F(x^1), p^1]$. If the sign of either of these consumer surplus measures is positive (negative), then the consumer's utility or welfare has increased (decreased) going from period 1 to period 2. Note that the mathematical properties of these consumer surplus measures are again completely determined by the mathematical properties of F and C.

4.4. The Consumer's Utility Maximization Problem

Another fundamental paradigm in economics is the one where a consumer's preferences over alternative (nonnegative) consumption vectors x are represented by a continuous *utility function* F and the consumer chooses his optimal consumption vector by maximizing utility $F(x)$ with respect to the consumption vector x, subject to a budget constraint of the form $\bar{p}^T x \le y$, where $\bar{p} > 0$ is a vector of positive commodity prices that he faces and $y > 0$ is the amount of income that he can spend on the commodities during the period under consideration. Formally, the consumer's utility maximization problem is defined for $p > 0$ as

$$G(p) = \max_x \{F(x): p^T x \le 1, x \ge 0\}, \tag{4.107}$$

where we have replaced the consumer's original budget constraint, $\bar{p}^T x \le y$, say, by the equivalent budget constraint $p^T x \le 1$, where $p = (p_1, p_2, \ldots, p_n) = (\bar{p}_1/y, \bar{p}_2/y, \ldots, \bar{p}_n/y)$ denotes a vector of normalized prices (the original commodity price vector divided by the consumer's positive income). In what follows, the price vector p is always interpreted as a vector of normalized commodity prices.

Note that we have not only defined the consumer's constrained utility maximization problem in the right-hand side of (4.107), but we have also defined the function G as the maximized utility level as a function of the normalized prices that the consumer faces. The function G is known in the economics literature as the *indirect utility function.*

It turns out to be technically more convenient to assume that the utility function F is continuous rather than just being upper semicontinuous as in Section 4.1. Formally, we now make the following assumption:

Assumption 4. F is a real-valued continuous function defined for all nonnegative consumption vectors $x \geq 0$.

The properties of G are given in the following theorem.

Theorem 4.9. Let F satisfy Assumptions 2, 3, and 4. Then G is well defined by (4.107) for $p > 0$ and satisfies

$$G \text{ is a continuous function for all } p > 0, \tag{4.108}$$

G is nonincreasing

$$[\text{if } 0 < p^0 \leq p^1, p^0 \neq p^1 \text{ then } G(p^0) \geq G(p^1)], \tag{4.109}$$

$$G \text{ is a quasiconvex function for all } p > 0 \tag{4.110}$$

and if for $x > 0$ we define

$$F^0(x) = \min_p \{\bar{G}(p)\colon p^T x \leq 1, p \leq 0\}, \tag{4.111}$$

where $\bar{G}$ is the extension of G, using the Fenchel closure operation, to the nonnegative orthant, then F^0 is continuous for all $x > 0$ and has a continuous extension to the nonnegative orthant $x \geq 0$. Moreover, the function F^0 defined by (4.110) coincides with F.

Proof. If $p > 0$, the constraint set in (4.107), given by $\{x\colon x \geq 0, p^T x \leq 1\}$, is compact and since F is continuous, the maximum in (4.107) exists.

For $\varepsilon > 0$ denote

$$P(\varepsilon) = \{p\colon p_i \geq \varepsilon, i = 1, \ldots, n\}. \tag{4.112}$$

We now show that for every $\varepsilon > 0$ the function $G(p)$ is continuous on $P(\varepsilon)$, from which the continuity for all $p > 0$ follows. First note that for a given $\varepsilon > 0$, a feasible x in (4.107) must satisfy

$$\varepsilon \sum_{i=1}^{n} x_i \leq p^T x \leq 1. \tag{4.113}$$

Hence $x \geq 0$ and $\sum_{i=1}^{n} x_i \leq 1/\varepsilon$ hold. Regardless of p and for a given ε all feasible x are in a compact set. Moreover, $\phi(p) = \{x: p^T x \leq 1\}$ is a continuous correspondence on $P(\varepsilon)$. Hence the continuity of G follows using Theorem 4.4.

To show that G is nonincreasing let $0 < p^0 \leq p^1$, $p^0 \neq p^1$. Then

$$G(p^1) = \max_x \{F(x): (p^1)^T x \leq 1, x \geq 0\} \tag{4.114}$$

$$\leqq \max_x \{F(x): (p^0)^T x \leq 1, x \geqq 0\} \tag{4.115}$$

since $0 < p^0 \leq p^1$ and so the set of feasible x's has increased. It follows that

$$G(p^1) \leq G(p^0). \tag{4.116}$$

To prove (4.110), let $0 < \lambda < 1$, $p^1 > 0$, $p^2 > 0$, and $G(p^i) \leqq u$ for $i = 1, 2$. Define the sets

$$H^i = \{x: (p^i)^T x \leq 1, x \geq 0\}, \qquad i = 1, 2 \tag{4.117}$$

and

$$H^\lambda = \{x: [\lambda p^1 + (1-\lambda)p^2]^T x \leq 1, x \geq 0\}. \tag{4.118}$$

Then,

$$G(\lambda p^1 + (1-\lambda)p^2) = \max_x \{F(x): x \in H^\lambda\} \tag{4.119}$$

$$\leq \max_x \{F(x): x \in H^1 \cup H^2\} \tag{4.120}$$

since $H^\lambda \subset H^1 \cup H^2$. Hence

$$G(\lambda p^1 + (1-\lambda)p^2) \leq u \tag{4.121}$$

since $\max_x \{F(x): x \in H^i\} = G(p^i) \leq u$ for $i = 1, 2$.

Since G is a continuous quasiconvex function for all $p > 0$, we may use the Fenchel closure operation to extend G to the nonnegative orthant; i.e, define the epigraph of G as Epi $G = \{(u, p): u \geq G(p), p > 0\}$, take the closure of this set, $\overline{\text{Epi } G}$ say, and define $\bar{G}(p) = \min_u \{u: (x, p) \in \overline{\text{Epi } G}\}$ for $p \geq 0$. The resulting $\bar{G}$ will be lower semicontinuous over the nonnegative orthant and so by a result due to Berge (1963, p. 76), for $x > 0$, the minimum in (4.122) below exists:

$$F^0(x) = \min_p \{\bar{G}(p): p^T x \leq 1, p \geq 0\}. \tag{4.122}$$

A minor complication that causes no difficulties should be noted: the extended function $\bar{G}$ defined above can take on the value $+\infty$ if p is on the boundary of the nonnegative orthant.

Property (4.111) will follow if we can show that the F^0 defined by (4.122) equals the original F over the positive orthant. Let $x^0 > 0$ and define $u^0 = F(x^0)$. There are two cases to consider:

Case (i). x^0 is on the boundary of the set $L(u^0) = \{x: F(x) \geq u^0\}$. By Assumption 2 on F, $L(u^0)$ is a convex set and since x^0 is on the boundary of this closed, convex set, at least one supporting hyperplane $p^0 \neq 0$ exists; i.e., $L(u^0)$ is contained in the half-space $\{x: (p^0)^T x \geq (p^0)^T x^0\}$. Using Assumption 3, it can be seen that we must have $p^0 \geq 0$. Let us normalize p^0 so that it satisfies

$$p^0 \geq 0, \qquad (p^0)^T x^0 = 1. \tag{4.123}$$

Since $\{x: F(x) \geq u^0\} \subset \{x: (p^0)^T x \geq (p^0)^T x^0 = 1\}$ and $F(x^0) = u^0$,

$$G(p^0) = \max_x \{F(x): (p^0)^T x \leq 1, x \geq 0\} = F(x^0) = u^0. \tag{4.124}$$

Case (ii). x^0 is in the interior of $L(u^0)$ (this can happen if F is constant over an open set). In this case, we can still find a p^0 that satisfies (4.123) and (4.124). If $\max_x \{F(x): x \geq 0\} = F(x^0)$, then any p^0 satisfying (4.123) will do. If there exists an $x \geq 0$, $x \neq 0$, such that $F(x) > F(x^0)$, then define the convex set $M(u^0) = \{x: F(x) > F(x^0) = u^0, x \geq 0\}$ and take its closure, $\bar{M}(u^0)$. Let $\lambda > 1$; solve $\min_\lambda \{\lambda: \lambda x^0 \in \bar{M}(u^0)\}$. Thus $\lambda^0 x^0$ is on the boundary of the closed convex set $\bar{M}(u^0)$ and so a supporting hyperplane $\tilde{p}^0 > 0$ must exist. Divide this $\tilde{p}^0$ by a scalar so that (4.123) is satisfied for the scaled p^0. Again, we find that (4.124) is also satisfied for this p^0. Now pick any p that satisfies

$$p \geq 0, \qquad p^T x^0 \leq 1. \tag{4.125}$$

Then,

$$\bar{G}(p) = \sup_x \{F(x): p^T x \leq 1, x \geq 0\} \tag{4.126}$$

$$\geq F(x^0) \tag{4.127}$$

since x^0 is feasible for the maximization problem by (4.125). Thus,

$$F^0(x^0) = \min_p \{\bar{G}(p): p^T x^0 \leq 1, p \geq 0\} \tag{4.128}$$

$$= F(x^0) \tag{4.129}$$

using (4.123), (4.124), and (4.127). □

A few comments on Theorem 4.9 are in order. Note that the indirect utility function G as defined by (4.107) will satisfy (4.108), (4.109), and (4.110) provided that the direct utility function F satisfies Assumption 4; i.e., Assumptions 2 and 3 were not required in order to derive (4.108)-(4.110). However, if we wish to use the indirect function G in order to construct a corresponding direct function F^0 via (4.122) then F^0 will not coincide with the original F unless F is also quasiconcave and nondecreasing. However, as in the cost function case, it is completely harmless to assume that F satisfies Assumptions 2 and 3 (unless the consumer can actually affect the prices he pays for his purchases).

Theorem 4.8 is another example of an *economics duality theorem*; i.e., direct functions F satisfying Assumptions 2, 3, and 4 are completely characterized by indirect functions G satisfying (4.108)-(4.111). Our proofs of properties (4.108)-(4.111) are taken from Diewert (1974). For references to other duality theorems in the economics literature, see Blackorby and Diewert (1979) or Diewert (1974). For similar duality theorems in the mathematical programming literature, see Crouzeix (1977, 1981b, 1982).

From the perspective of the role of generalized concavity in economics, the interesting point to notice about the above theorem is that the indirect utility function G is *quasiconvex* (i.e., $-G$ is *quasiconcave*) irrespective of the properties of the direct utility function F (provided only that F is continuous).

In the following sections, we indicate how other types of generalized concave functions arise naturally in the context of consumer theory.

4.5. Semistrict Quasiconcavity and Consumer Theory

Suppose the utility function F satisfies Assumption 1 and we define the consumer's utility maximization problem by (4.107) for every (normalized) commodity price vector $p > 0$. By a theorem due to Berge (1963) stating that the maximum of an upper semicontinuous function over a compact set exists, the indirect utility function $G(p)$ is well defined as a maximum in (4.107). Suppose that $G(p) < \sup_x \{F(x): x \geq 0\}$ and denote the solution set of optimal consumption vectors for (4.107) as $\beta(p)$. The correspondence $\beta(p)$ is the *demand correspondence.* If it is single valued, then it becomes the *demand function.* If F is constant over a neighborhood of a solution point (in economics terminology, some indifference surfaces are thick), then $\beta(p)$ can contain points x such that $p^T x < 1$; i.e., the consumer's entire budget is not spent. From the viewpoint of economic applications, it is useful to be able to rule out this kind of behavior. The following theorem gives sufficient conditions for this.

Theorem 4.10. Suppose that F satisfies Assumption 1 and is semistrictly quasiconcave (recall Definition 3.11). Then F satisfies the following property:

$$p > 0, \qquad G(p) < \sup_x \{F(x): x \geq 0\}, \qquad x \in \beta(p) \Rightarrow p^T x = 1. \qquad (4.130)$$

Proof. Suppose there exists p^0 such that

$$p^0 > 0, \qquad G(p^0) < \sup_x \{F(x): x \geq 0\}, x^0 \in \beta(p^0) \quad \text{and} \quad (p^0)^T x^0 < 1. \qquad (4.131)$$

Then there exists $\bar{x} \geq 0$ such that $G(p^0) = F(x^0) < F(\bar{x})$ with $(p^0)^T \bar{x} > 1$. By the semistrict quasiconcavity of F,

$$F(x^0) < F(\lambda x^0 + (1 - \lambda)\bar{x}) \qquad \text{for } 0 \leq \lambda < 1. \qquad (4.132)$$

Since $(p^0)^T x < 1$ and $(p^0)^T \bar{x} > 1$, there exists $0 < \lambda^0 < 1$ such that $(p^0)^T[\lambda^0 x^0 + (1 - \lambda^0)x] = 1$. Thus $\lambda^0 x^0 + (1 - \lambda^0)\bar{x}$ is feasible for the $G(p^0)$ maximization problem, $\max_x \{F(x): (p^0)^T x < 1, x > 0\}$. But (4.132) shows that this feasible solution gives a higher level of utility than $F(x^0) = G(p^0)$, a contradiction to the definition of $G(p^0)$. Thus our supposition (4.131) is false and (4.130) follows. □

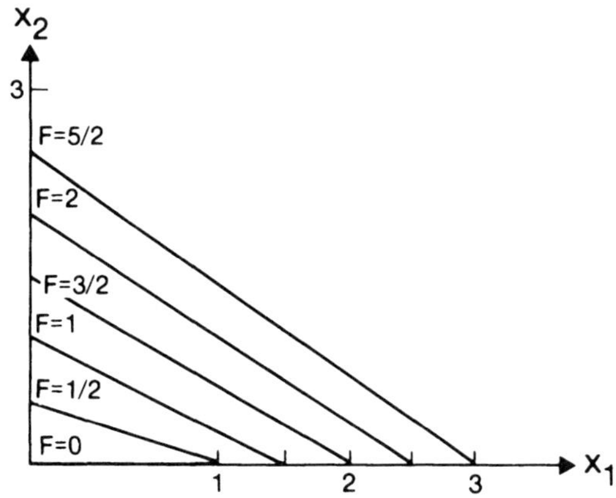

Figure 4.2. The function (4.133).

We note that semistrict quasiconcavity of F is not a necessary condition for (4.130); the following function of two variables satisfies Assumptions 1-4 as well as (4.130), but it is not semistrictly quasiconcave (consider the behavior of the function along the x_1 axis):

Example 4.5.

$$F(x_1, x_2) = \begin{bmatrix} 0 & \text{if } 0 \le x_1 \le 1, \quad x_2 = 0 \\ t & \text{if } x_1 \ge 0, \quad x_2 \ge 0, \quad t > 0, \quad tx_1 + (1+t)x_2 = t(1+t). \end{bmatrix} \tag{4.133}$$

The level surfaces (or indifference curves) of the above function are a family of nonparallel straight lines (see Figure 4.2). This is also an example of a quasiconcave function that cannot be transformed into a concave function (see Chapter 8 for a detailed study of this problem).

4.6. Strict Quasiconcavity and Consumer Theory

In most applications of consumer theory in economics, we require not only that the property (4.130) hold, but also that the set of optimal consumption vectors $\beta(p)$ reduce to a singleton. Thus we are interested in finding conditions on the direct utility function F or the indirect utility function G that will ensure that the consumer's demand correspondence, the set of maximizers $\beta(p)$ to (4.107), is an ordinary function, at least for a region of prices.

Theorem 4.11. Suppose the direct utility function F satisfies Assumptions 2 and 4 and for $p^0 > 0$, $G(p^0) < \sup_x \{F(x): x \geq 0\}$, $x^0 \in \beta(p^0)$ and F is strictly quasiconcave over $N_\delta(x^0) = \{x: (x - x^0)^T(x - x^0) < (\delta)^2, x \geq 0\}$, the open ball of radius $\delta > 0$ around x^0. Then the following condition holds: $\beta(p)$ is a continuous function on a neighborhood $N(p^0)$ of normalized prices around p^0, with $p^T\beta(p) = 1$ for $p \in N(p^0)$.

Proof. Let p^0 and x^0 be defined as above and let $\bar{x}$ be such that $G(p^0) = F(x^0) < F(\bar{x})$. Suppose that $(p^0)^Tx^0 < 1$. Then, since F is quasiconcave, for $0 \leq \lambda \leq 1$, $F(\lambda x^0 + (1 - \lambda)\bar{x}) \geq F(x^0)$. By the local strict quasiconcavity of F, there exists a $0 < \lambda^0 < 1$ such that

$$F(\lambda x^0 + (1 - \lambda)\bar{x}) > F(x^0) \qquad \text{for } \lambda^0 \leq \lambda < 1. \tag{4.134}$$

For λ^1 sufficiently close to 1, $(p^0)^T(\lambda^1 x^0 + (1 - \lambda^1)\bar{x}) \leq 1$, which implies along with (4.134) that $G(p^0) \geq F(\lambda^1 x^0 + (1 - \lambda^1)\bar{x}) > F(x^0)$, a contradiction. Thus our supposition is false, and thus $(p^0)^Tx^0 = 1$.

Now suppose there exists $\tilde{x} \neq x^0$ such that $\tilde{x} \in \beta(p^0)$. Then the quasiconcavity and local strict quasiconcavity of F again imply a contradiction. Thus $\beta(p^0)$ is the singleton x^0 with $(p^0)^Tx^0 = 1$. By Theorem 4.4, $\beta(p)$ is an upper hemicontinuous correspondence for $p > 0$. Hence for p sufficiently close to p^0, $\beta(p) \cap N_\delta(x^0)$ is not empty. Strict quasiconcavity on F in $N_\delta(x^0)$ will imply as before that $\beta(p)$ is single valued and $p^T\beta(p) = 1$ for p close to p^0. When β is single valued, upper hemicontinuity reduces to continuity. □

Versions of the above theorem are common in the economics literature (e.g., Debreu, 1959) except that strict quasiconcavity is generally assumed to hold globally rather than locally.

Another set of conditions that are sufficient to ensure that the consumer's demand correspondence $\beta(p)$ is single valued and continuous around $p^0 > 0$ can be phrased in terms of the indirect utility function G. Sufficient conditions are that G satisfies (4.108)-(4.111) and the following condition: G is once continuously differentiable on a neighborhood around p^0 with $\nabla G(p^0) \neq 0$. If the above conditions on G are satisfied, then it can be shown (see Diewert, 1974, or Roy, 1947) that $\beta(p) = [p^T\nabla G(p)]^{-1}\nabla G(p)$ for p close to p^0 (see also Theorem 4.11 below).

Neither set of sufficient conditions for the single valuedness and continuity of $\beta(p)$ implies the other. It appears to be difficult to find necessary and sufficient conditions for this problem, just as it was difficult to find necessary and sufficient conditions for the differentiability of the cost function with respect to input prices.

4.7. Strong Pseudoconcavity and Consumer Theory

Suppose now that we not only want the consumer's demand correspondence $\beta(p)$ to be single valued and continuous for a neighborhood of prices p, but also we want the demand functions to be once continuously differentiable. As usual, it is difficult to find necessary and sufficient conditions on F for the above property to hold. However, the following theorem gives sufficient conditions.

Theorem 4.12 (Blackorby and Diewert, 1979). Suppose that the direct utility function F satisfies Assumptions 2, 3, and 4. Furthermore, let $x^0 \in R^n$, $x^0 > 0$ be such that F is twice continuously differentiable in a neighborhood of x^0 with $\nabla F(x^0) > 0$, and

$$z^T z = 1, \qquad z^T \nabla F(x^0) = 0 \Rightarrow z^T \nabla^2 F(x^0) z < 0. \tag{4.135}$$

Then the indirect utility function G defined by (4.107) satisfies (4.108)-(4.110). Moreover, if we define $p^0 = [x^{0T} \nabla F(x^0)]^{-1} \nabla F(x^0)$, then G is twice continuously differentiable in a neighborhood of p^0 with $\nabla G(p^0) < 0$ and

$$z^T z = 1, \qquad z^T \nabla G(p^0) = 0 \quad \Rightarrow \quad z^T \nabla^2 G(p^0) z > 0. \tag{4.136}$$

Moreover, the solution to the utility maximization problem (4.107) reduces to $\beta(p) = [p^T \nabla G(p)]^{-1} \nabla G(p)$ for p close to p^0 and is once continuously differentiable there.

Proof. Since F satisfies Assumptions 2, 3, and 4, by Theorem 4.8, G satisfies (4.108)-(4.111). Let $x^0 > 0$ satisfy the hypotheses above and define $p^0 > 0$ by

$$p^0 = \nabla F(x^0)/(x^0)^T \nabla F(x^0). \tag{4.137}$$

For this p^0 consider the following utility maximization problem:

$$G(p^0) = \max_x \{F(x) : (p^0)^T x \le 1, x \ge 0\}. \tag{4.138}$$

The Karush-Kuhn-Tucker (Karush, 1939; Kuhn and Tucker, 1951), first-order necessary conditions given by Theorem 3.47 for x^* to solve (4.138) are

$$\frac{\partial F(x^*)}{\partial x_i} - \lambda^* p^0 = \begin{cases} 0 & \text{if } x^* > 0 \\ \le 0 & \text{otherwise,} \end{cases} \tag{4.139}$$

$$\lambda^*(1 - (p^0)^T x^*) = 0, \tag{4.140}$$

$$\lambda^* \geq 0. \tag{4.141}$$

Define $\lambda^*(x^0)^T \nabla F(x^0) > 0$, and using (4.137), note that $x^* = x^0$ and λ^* satisfy (4.139)-(4.141). If we make use of the equality $\nabla F(x^0) = \lambda^* p^0$, it can be verified that assumption (4.135) implies that the second-order sufficient conditions for x^* to be a locally unique solution to the constrained maximization problem (4.138) are satisfied. (See Fiacco, 1976.) Note that replacing the inequality constraint $(p^0)^T x \leq 1$ in (4.138) by an equality constraint $(p^0)^T x = 1$, then (4.135) implies that the second-order sufficient conditions for an equality constrained maximum are satisfied by x^*; see Hancock (1917) or Afriat (1971) for these conditions.

We now show that x^0 is the (globally) unique solution to the constrained maximization problem (4.138). Suppose there exists $\bar{x} \geq 0$ such that $\bar{x} \neq x^0$, $F(\bar{x}) \geq F(x^0)$, and $(p^0)^T \bar{x} \leq 1$. By the quasiconcavity of F, for $0 < \lambda < 1$, $F(\lambda x + (1 - \lambda) x^0) \geq F(x^0)$. Note also that $\lambda \bar{x} + (1 - \lambda) x^0 \geq 0$ and $(p^0)^T [\lambda \bar{x} + (1 - \lambda) x^0] \leq 1$. Hence $\lambda \bar{x} + (1 - \lambda) x^0$ is also a solution to (4.138) for $0 < \lambda < 1$. But this contradicts the strict local maximizing property of x^0 for λ sufficiently close to 0. Hence our supposition is false and x^0 is the unique solution to (4.138).

It was shown in Fiacco (1976) that for p in a neighborhood of p^0, there exist once differentiable functions $x(p)$ and $\lambda(p)$ such that $x(p^0) = x^*$ and $\lambda(p^0) = \lambda^*$ and $x(p)$ locally solves (4.138) when p^0 is replaced by p. Repeating our argument in the previous paragraph, we may show that $x(p)$ is the globally unique solution to (4.138) when p replaces p^0. These functions $x(p)$ and $\lambda(p)$ are obtained by using the Implicit Function theorem on the system (4.139)-(4.141) treated as a system of equations, since $x^* > 0$ and $\lambda^* > 0$.

Differentiating (4.139)-(4.141) with respect to the components of p, we obtain the following system of equations:

$$\begin{bmatrix} \nabla^2 F(x^*) & -p^0 \\ -(p^0)^T & 0 \end{bmatrix} \begin{bmatrix} \nabla_p x(p^0) \\ \nabla_p \lambda(p^0)^T \end{bmatrix} = \begin{bmatrix} \lambda^* I \\ (x^*)^T \end{bmatrix} \tag{4.142}$$

where $\nabla_p x(p^0)$ is the $n \times n$ partial derivative matrix which has i, j element $\partial x_i(p^0)/\partial p_j$ and where $\nabla_p \lambda(p^0)$ is the n-vector of partial derivatives of λ with respect to p. Using (4.135) and $\nabla F(x^*) = \lambda^* p^0$, it can be seen that $\nabla^2 F(x^*)$ is negative definite in the subspace orthogonal to p^0; i.e., if $z^T p^0 = 0$, $z \neq 0$, then $z^T \nabla^2 F(x^*) z < 0$. Hence by Finsler's (1937) theorem (see also Proposition 8.24), there exists $k^* > 0$ such that

$$A = \nabla^2 F(x^*) - k^* p^0 (p^0)^T \tag{4.143}$$

is a negative definite symmetric matrix. Using elementary row operations, it can be verified that we may use A defined by (4.143) in order to construct the inverse of the matrix that appears in the left-hand side of (4.142):

$$\begin{bmatrix} \nabla^2 F(x^*) & -p^0 \\ -(p^0)^T & 0 \end{bmatrix}^{-1}$$
$$= \begin{bmatrix} A^{-1} - A^{-1}p^0[(p^0)^T A^{-1} p^0]^{-1}(p^0)^T A^{-1} & -A^{-1}p^0[(p^0)^T A^{-1} p^0]^{-1} \\ -[(p^0)^T A^{-1} p^0]^{-1}(p^0)^T A^{-1} & -k^* - [(p^0)^T A^{-1} p^0]^{-1} \end{bmatrix} \tag{4.144}$$

Define the $n \times n$ symmetric matrix D by

$$D = A^{-1} - A^{-1}p^0[(p^0)^T A^{-1} p^0]^{-1}(p^0)^T A^{-1}. \tag{4.145}$$

We wish to prove that D is negative definite in the subspace orthogonal to p^0 (obviously $Dp^0 = 0$) or more precisely, we wish to prove the following inequality:

$$z \neq \lambda p^0 \quad \text{for any } \lambda \in R \quad \text{implies} \quad z^T D z < 0. \tag{4.146}$$

The inequality in (4.146) is equivalent to

$$z^T A^{-1} z - (z^T A^{-1} p^0)^2 [(p^0)^T A^{-1} p^0]^{-1} < 0 \tag{4.147}$$

or

$$(z^T A^{-1} z)[(p^0)^T A^{-1} p^0] < (z^T A^{-1} p^0)^2 \tag{4.148}$$

since $(p^0)^T A^{-1} p^0 < 0$, or

$$z^T(-A^{-1})(p^0)^T(-A^{-1})p^0 > [z^T(-A^{-1})p^0]^2 \tag{4.149}$$

and the last inequality is true for $z \neq \lambda p^0$ by the Cauchy-Schwarz inequality (using the fact that $-A^{-1}$ is a positive definite symmetric matrix). A more detailed alternative derivation of this material starting at (4.142) may be found in Diewert and Woodland (1977). Equation (4.144) may be substituted into (4.142) in order to obtain explicit formulas for the matrix of derivatives $\nabla_p x(p^0)$ and the vector of derivatives $\nabla_p \lambda(p^0)$.

For p close to p^0, let us use Fiacco's solution functions $x(p)$ and $\lambda(p)$ in order to evaluate the optimized Lagrangian:

$$L(x(p), \lambda(p)) = F(x(p)) + \lambda(p)[1 - p^T x(p)] \tag{4.150}$$

$$= F(x(p)) \qquad [\text{since } p^T x(p) = 1] \tag{4.151}$$

$$= G(p). \tag{4.152}$$

The functions $x(p)$ and $\lambda(p)$ will satisfy the first-order conditions (4.139)-(4.141) for p close to p^0; i.e., we have

$$\nabla_x F(x(p)) - \lambda(p)p = 0 \tag{4.153}$$

$$1 - p^T x(p) = 0. \tag{4.154}$$

It follows from (4.150)-(4.152) that for p close to p^0 we have

$$\nabla_p G(p) = \nabla_p x(p)^T \nabla_x F(x(p)) + [1 - p^T x(p)]\nabla_p \lambda(p) - \lambda(p)x(p) - \lambda(p)\nabla_p x(p)^T p \tag{4.155}$$

$$= -\lambda(p)x(p) \qquad [\text{using } (4.153)\text{-}(4.154)] \tag{4.156}$$

$$< 0 \tag{4.157}$$

since $x(p) > 0$ and $\lambda(p) > 0$ for p close to p^0. Thus $\nabla_p G(p)$ exists for p close to p^0 and is negative, as asserted. Equation (4.156) may be rewritten as $x(p) = -\nabla_p G(p)/\lambda(p)$. Since $p^T x(p) = 1$, $-p^T \nabla_p G(p)/\lambda(p) = 1$ or

$$\lambda(p) = -p^T \nabla_p G(p). \tag{4.158}$$

If we substitute (4.158) into (4.156) we obtain the unique solution for p close to p^0 to the utility maximization problem in terms of the derivatives of the indirect utility function [Roy's (1947) identity]:

$$x(p) = \beta(p) = \nabla G(p)/p^T \nabla G(p). \tag{4.159}$$

Since the functions $\lambda(p)$ and $x(p)$ are once continuously differentiable for p close to p^0, the matrix of second-order partial derivatives of G may be calculated by differentiating (4.156):

$$\nabla^2_{pp} G(p) = -x(p)\nabla_p \lambda(p)^T - \lambda(p)\nabla_p x(p). \tag{4.160}$$

Finally, we have to establish (4.136). Let $z^T z = 1$ and $z^T \nabla G(p^0) = 0$. Using (4.156) we have $-z^T \lambda^* x^* = 0$ or $z^T x^* = 0$. Hence $z \neq \lambda p^0$ for any λ since $z \neq 0$ and $(p^0)^T x = 1$. Using (4.160) when $p = p^0$, and solving (4.142) for $\nabla_p x(p^0)$ employing (4.144), we find that using $z^T x^* = 0$:

$$z^T \nabla^2_{pp} G(p^0) z = -(\lambda^*)^2 z^T \{A^{-1} - A^{-1} p^0 [(p^0)^T A^{-1} p^0]^{-1} (p^0)^T A^{-1}\} z \qquad (\text{since } z^T x^* = 0) \tag{4.161}$$

$$= -(\lambda^*)^2 z^T D z \qquad \text{[by Definition (4.145)]} \tag{4.162}$$

$$> 0 \qquad \text{[using (4.146) and } \lambda^* > 0]. \tag{4.163}$$

□

Corollary 4.13. Suppose the indirect utility function $\bar{G}$ satisfies (4.108)-(4.111). Then the direct utility function F defined by $F(x) = \min_p \{\bar{G}(p): p^T x \leq 1, p \geq 0\}$ satisfies Assumptions 2, 3, and 4. Moreover, if $\bar{G}$ is twice continuously differentiable in a neighborhood of p^0 with $\nabla G(p^0) < 0$, and (4.136) holds for $\bar{G}$, and we define $x^0 = \nabla \bar{G}(p^0)/(p^0)^T \nabla \bar{G}(p^0)$, then the direct utility function F is twice continuously differentiable in a neighborhood of x^0 with $\nabla F(x^0) > 0$ and (4.135) holds.

Proof. (Blackorby and Diewert, 1979). The proof is similar to the proof of the preceding theorem, except a minimization problem has replaced a maximization problem, and various inequalities are reversed. □

Note that our differentiability assumptions and (4.135) imply that F is *strongly pseudoconcave* in a neighborhood around x^0, while (4.136) implies that G is *strongly pseudoconvex* around p^0; see Proposition 3.46.

Thus sufficient conditions for the local continuous differentiability of a consumer's system of demand functions can be obtained by assuming that the direct utility function F is strongly pseudoconcave locally, or equivalently, by assuming that the indirect function G is strongly pseudoconvex locally. For additional material on the differentiability properties of quasiconcave functions and the relationship between primal and dual differentiability, see Crouzeix (1981a, 1982).

4.8. Pseudoconcavity and Consumer Theory

In order to obtain a system of demand functions consistent with utility maximizing behavior, economists frequently postulate a functional form for the direct utility function F and then they attempt to solve algebraically the utility maximization problem (4.107) for an explicit solution. As part of this exercise, economists often assume that necessary conditions for x^0 to solve (4.107) are also sufficient.

In the case where F is once continuously differentiable, if x^0 is a solution to (4.107) when normalized prices $p > 0$ prevail, then it can be verified that the following necessary conditions must be satisfied:

$$v^T \nabla F(x^0) \leq 0 \qquad \text{for every direction } v \in S(x^0, p) \tag{4.164}$$

where the set of feasible directions is defined as $S(x^0, p) = \{v: v^T v = 1$ and there exists a $t > 0$ such that $x^0 + tv \geq 0$ and $p^T(x^0 + tv) \leq 1\}$. The meaning of (4.164) is that for all feasible directions, F cannot increase locally.

It can be shown that (4.164) is equivalent to the following more familiar Karush-Kuhn-Tucker conditions; see Theorem 3.47: If $x^0 \geq 0$ solves (4.107), then there exists a multiplier λ^0 such that

$$\nabla F(x^0) - \lambda^0 p \leq 0, \qquad \lambda^0 \geq 0, \qquad x^{0T}[\nabla F(x^0) - \lambda^0 p] = 0, \tag{4.165}$$

$$p^T x^0 - 1 \leq 0, \qquad \lambda^0[p^T x^0 - 1] = 0. \tag{4.166}$$

It follows from Theorem 3.48 that the necessary conditions (4.165)-(4.166) are also sufficient if F is pseudoconcave (see Definition 3.13); i.e., if F is pseudoconcave and x^0 satisfies (4.165) and (4.166), then x^0 is a solution to the consumer's utility maximization problem. Similarly, if F is strictly pseudoconcave (see Definition 3.13) and x^0 satisfies (4.165) and (4.166), then, since F is also strictly quasiconcave and by Proposition 3.29, we have that x^0 is the unique solution to (4.107).

There are several other similar applications of the concept of strict pseudoconcavity to economics; e.g., see Donaldson and Eaton (1981).

The final type of generalized concave function that occurs in consumer theory that we would like to mention is the class of quasiconcave functions that may be transformed into concave functions by a monotonically increasing function of one variable. This class of functions is important in welfare economics and game theory where we often wish to maximize a convex combination of individual utility functions. The resulting sum function is not even quasiconcave in general unless the individual utility functions are concave. This concavifiability question will be studied in detail in Chapter 8.

A problem that is related to the issue of concavifiability is the following one: given that the direct utility function is concave, what does this imply about the corresponding cost and indirect utility functions C and G? On the other hand, if G is convex, what does this imply about the corresponding F and C? Answers to these questions can be found in Crouzeix (1977), Diewert (1978), and Fenchel (1953).

Another interesting problem is the following: given that the n components of the vector $x \in R^n$ can be partitioned into m partitions, $m \leq n$ and $F(x) = F(x_1, \ldots, x_m) = \sum_{i=1}^m f^i(x_i)$, where $x_i \in S^i$, an open convex set such that S, the domain of F, is the Cartesian product of the S^i, and F is quasiconcave, what does this imply about the functions f^i? For answers to this question and some economic applications (particularly to consumer choice under uncertainty), see Blackorby, Davidson, and Donaldson (1977), Debreu and Koopmans (1982), and Yaari (1977).

4.9. Profit Maximization and Comparative Statics Analysis

Consider the model of producer behavior presented in Section 4.1 above, where the producer minimized costs subject to an output constraint. In this section, we make a stronger assumption about producer behavior, namely, that he competitively maximizes profits. Denote the production function by F, an input vector by $x > 0$, an input price vector by $\bar{p} > 0$, and a scalar price of output by $\bar{p}_0 > 0$. A competitive producer takes all prices as fixed and chooses x in order to maximize profits, $\bar{p}_0 F(x) - \bar{p}^T x$. Thus we define the producer's competitive profit maximization problem as

$$\pi(p) = \sup_x \{F(x) - p^T x: x \geq 0\}, \tag{4.167}$$

where we have defined the vector of normalized input prices as $p = (\bar{p}_1/\bar{p}_0, \bar{p}_2/\bar{p}_0, \ldots, \bar{p}_n/\bar{p}_0)$. It is obvious that maximizing profits is equivalent to solving (4.166) since the objective function in (4.167) is simply profits divided by the positive price of output $\bar{p}_0$. The maximized value in (4.167) is defined to be the producer's (normalized) *profit function* π (see Jorgenson and Lau, 1974a, 1974b or Lau, 1978 for additional details on this model).

Mathematicians will recognize π as the (negative of the) *conjugate function* to F (see Fenchel, 1953 or Rockafellar, 1970).

Recall how we used the cost function C to define outer approximations to the true level sets of the production function F. Now we show how the profit function π can be used in order to provide another outer approximation to the true production function.

For $p > 0$, define the following half-space in R^{n+1}: $H(p) = \{(y, x) \in R \times R^n: y - p^T x \leq \pi(p)\}$. Then an outer approximation to the true production possibilities set $S = \{(y, x): y \leq F(x), x \geq 0\}$ can be defined as $\hat{S} = \bigcap_{p>0} H(p)$. Assuming that $\pi(p) < +\infty$ for at least one $p > 0$, and since $\hat{S}$ is closed, then for a given $x \geq 0$ the set $\{y \in R: (y, x) \in \hat{S}\}$ is bounded from above. It follows that an outer approximation to F can be defined for $x \geq 0$ as

$$\hat{F}(x) = \max_y \{y: (y, x) \in S\}. \tag{4.168}$$

Since $\hat{S}$ is a convex set, it can be seen that the approximating production function $\hat{F}$ is a *concave function.* Moreover, for any input price vector $p^0 > 0$ such that there exists an $x \geq 0$ that attains the sup in (4.167), we have $\max_x \{\hat{F}(x) - (p^0)^T x: x \geq 0\} = \max_x \{F(x) - (p^0)^T x: x \geq 0\} = F(x^0) - (p^0)^T x = \pi(p^0)$. Hence from an empirical point of view, it is harmless to assume that the producer's production function is concave, provided that the producer is competitively maximizing profits. The present situation is analogous to the result we obtained in Section 4.1, where we indicated that the production function could be assumed to be quasiconcave provided that the producer was competitively minimizing costs; we now obtain the stronger concavity result because we assume the stronger hypothesis of competitive profit-maximizing behavior.

For references to duality theorems between production functions, production possibilities sets, and profit functions, see Diewert (1974), Gorman (1968), Lau (1978), and McFadden (1966, 1978).

However, our main concern in this section is not to show how concavity arises naturally in the content of production theory, but to show how *strong concavity* arises naturally in the content of comparative statics theory.

A typical application of the Hicks (1946) and Samuelson (1947) comparative statics analysis can be illustrated by using the profit maximization model (4.167). The analysis given below follows Lau (1978). Given a positive vector of (normalized) input prices $p^0 > 0$, it is assumed that $x^* > 0$ is the unique solution to the profit maximization problem $\max_x \{F(x) - (p^0)^T x: x \geq 0\} = \pi(p^0)$. [A sufficient condition for this is that F be globally concave and locally strictly concave around a point $x^* > 0$ such that $\nabla F(x^*) = p^0$.] It is also assumed that F is twice continuously differentiable in a neighborhood of x^*. Since $x^* > 0$ solves the profit maximization problem, the nonnegativity constraints $x \geq 0$ are not binding and hence the following first-order necessary conditions must be satisfied:

$$\nabla F(x^*) - p^0 = 0. \tag{4.169}$$

Since F is assumed to be twice continuously differentiable, the following second-order necessary conditions must also be satisfied (e.g., see Avriel, 1976):

$$z \neq 0 \quad \Rightarrow \quad z^T \nabla^2 F(x^*) z \leq 0; \tag{4.170}$$

i.e., the matrix of second-order partial derivatives of F evaluated at x^* must be negative semidefinite. Comparative statics analysis proceeds by assuming that F satisfies the following stronger conditions at x^* (second-order sufficient conditions); see for example, Samuelson (1947):

$$x \neq 0 \quad \Rightarrow \quad z^T \nabla^2 F(x^*) z < 0; \tag{4.171}$$

i.e., $\nabla^2 F(x^*)$ is a symmetric negative definite matrix, and hence $[\nabla^2 F(x^*)]^{-1}$ exists. Note that (4.171) implies that F is *strongly concave* in a neighborhood of x^*; see Proposition 2.19.

The reason for assuming (4.171) is that we may now apply the Implicit Function theorem to (4.171) and deduce the existence of a unique once continuously differentiable solution to the profit maximization problem (4.167); $x(p)$, say, for p close to p^0, and the $n \times n$ matrix of partial derivatives of the functions $[x_1(p), \ldots, x_n(p)]^T = x(p)$ evaluated at p^0 is

$$\nabla x(p^0) = [\partial x_i(p^0)/\partial p_j] = [\nabla^2 F(x^*)]^{-1}. \tag{4.172}$$

Since $\nabla^2 F(x^*)$ is a negative definite matrix, so is its inverse $[\nabla^2 F(x^*)]^{-1}$. Hence the input demand functions satisfy $\partial x_i(p^0)/\partial p_i < 0$, $i = 1, 2, \ldots, n$ (the diagonal elements of a negative definite matrix are negative). Thus if the price of the ith input increases, the demand for the ith input decreases. Hence the assumption of local strong concavity leads to economically meaningful theorems.

In the mathematical programming literature, the terms "stability theory" or "perturbation theory" are used in place of the economist's term "comparative statics analysis."

We have considered three models of economic behavior in this chapter: the producer's cost minimization problem, the consumer's utility maximization problem, and the producer's profit maximization problem; and we have shown how nine different kinds of generalized concavity and convexity arise in the context of these models.

4.10. Concave Programs and Economics

In this section, we shall indicate how Theorem 2.28, the Karlin–Uzawa Saddle Point theorem, can be used in conjunction with the economics duality theorems explained earlier in this chapter in order to establish some interesting results in economics.

The economic model we wish to consider is that of a small open economy which faces positive fixed world prices $w = (w_1^0, w_2^0, \ldots, w_m^0)^T > 0$ for m internationally traded goods. The production sector of this economy also has available to it positive amounts $v = (v_1^0, v_2^0, \ldots, v_n^0)^T > 0$ of n domestic inputs. An interesting *planning problem* for this country is the maximization of the net value of internationally traded goods subject to the n domestic resource constraints. Below, we shall examine this problem in some detail and we shall seek to determine the dependence of this planning problem on the vector of world prices w^0 and the vector of aggregate domestic resource availability v^0.

We assume that the entire production sector in the economy consists of $K \leq n$ constant returns to scale sectors. The technology of the kth sector is represented by a set S^k of feasible net outputs (inputs and imports are indexed negatively). Thus if $(x, y) \in S^k$, then the nonpositive vector $x = (x_1, x_2, \ldots, x_n)^T$ of domestic inputs and the vector $y = (y_1, y_2, \ldots, y_m)^T$ of net exports are producible by sector k. Technically, we assume that each set S^k has the following representation:

$$S^k = \{(x, y): x = a^k z_k, y = b^k z_k, (a^k, b^k) \in C^k, z_k \geq 0\}, \qquad k = 1, \ldots, K \tag{4.173}$$

where z_k is a nonnegative sector k scalar scale variable and C^k is a nonempty, closed convex subset of R^{n+m} that represents the sector k unit scale production possibilities. Our representation of the economy's production possibilities is broad enough to encompass both the usual neoclassical smooth production functions and the activity analysis framework. In the latter case, the sets C^k are the convex hulls of a finite number of points or activities.

Define the sector k *unit profit function* π^k, which is dual to the unit scale production possibilities set C^k by

$$\pi^k(p, w) = \max_{a,b} \{p^T a + w^T b: (a, b) \in C^k\}, \qquad k = 1, \ldots, K, \tag{4.174}$$

where $p > 0$ is a positive vector of domestic prices and $w > 0$ is a positive vector of world prices. Note that C^k determines π^k. Using the outer approximation idea defined by (4.168) we can show that π^k determines C^k. This is why we say that π^k is dual to C^k: each can represent the technology of sector k. The unit profit functions π^k have mathematical properties that are similar to those of the normalized profit function defined by (4.167) or to the price properties of the negative of the cost function defined by (4.1). In particular, it can be shown that the π^k are (positively) linearly homogeneous and convex functions of their price arguments [modify the proof of properties (b) and (d), respectively; changing a min to a max reverses the sense of inequalities in property (d)]. Moreover, if $\pi^k(p, w)$ is differentiable with respect to its components for some k, then by a result analogous to Corollary 4.7, the solution to (4.174) is unique for this k and is given by $a^k = \nabla_p \pi^k(p, w)$, $b^k = \nabla_w \pi^k(p, w)$. For additional material on unit profit functions, see Diewert and Woodland (1977).

Given the world price vector w^0 and the positive vector of domestic resource availabilities v^0, the economy's production planning problem is

$$G(w^0, v^0) = \max_{y, y^1, \ldots, y^K, x^1, \ldots, x^K}$$

$$\times \left\{ (w^0)^T y : y = \sum_{k=1}^{K} y^k; -\sum_{k=1}^{K} x^k \le v^0; (y^k, x^k) \in S^k, k = 1, \ldots, K \right\}, \quad (4.175)$$

where we have defined the economy's *national product function* G as the optimized value of the objective function in the programming problem defined in (4.175). The first set of constraints in (4.175) serves to define y, the economy's net output vector of internationally traded goods; the second set of constraints requires that (nonnegative) aggregate input demand, $-\sum_{k=1}^{K} x^k$, be equal to or less than the vector of aggregate supply availabilities, v^0; the last set of constraints merely requires that the net supply vector y^k and the (nonpositive) domestic input demand vector x^k for sector k belong to the sector k production possibilities set S^k. We may eliminate the first set of constraints by substitution and rewrite (4.175) as

$$G(w^0, v^0) = \max_{y^1, \ldots, y^K, x^1, \ldots, x^K}$$

$$\times \left\{ \sum_{k=1}^{K} (w^0)^T y^k : \sum_{k=1}^{K} x^k + v^0 \ge 0; (y^k, x^k) \in S^k, k = 1, \ldots, K \right\}. \quad (4.176)$$

It is evident that the above primal programming problem and the equivalent problem (4.175) are concave programming problems of the type studied at the end of Chapter 2. We assume that a finite optimal solution, y^*, $y^{1*}, \ldots, y^{K*}$, $x^{1*}, \ldots, x^{K*}$ exists for (4.175) and that a feasible solution for (4.175) exists that satisfies the resource inequality constraints strictly. Under these conditions, we may rewrite (4.175) using Theorem 2.28 as follows:

$$G(w^0, v^0) = \max_{y^1, \ldots, y^K, x^1, \ldots, x^K} \min_{p} \left\{ \sum_{k=1}^{K} (w^0)^T y^k + p^T \left(\sum_{k=1}^{K} x^k + v^0 \right) : \right.$$

$$\left. (y^k, x^k) \in S^k, k = 1, \ldots, K; p \ge 0 \right\} \quad (4.177)$$

$$= \max_{a^1, \ldots, a^K, b^1, \ldots, b^K, z} \min_{p} \left\{ \sum_{k=1}^{K} (w^0)^T b^k z_k + p^T \left(\sum_{k=1}^{K} a^k z_k + v^0 \right) : \right.$$

$$(a^k, b^k) \in C^k, k = 1, \ldots, K; z \in R^k, z \geq 0, p \geq 0\Big\}$$

(using the constant returns to scale representation of the S^k)

$$(4.178)$$

$$= \max_{z\geq 0} \min_{p\geq 0} \left\{ \sum_{k=1}^{K} \pi^k(p, w^0) z_k + p^T v^0 \right\} \tag{4.179}$$

where the last line follows upon performing the maximization with respect to the (a^k, b^k) and using the definition of the unit profit functions π^k, (4.174).

Note the simplicity of the unconstrained max-min problem (4.179) compared to the primal constrained maximization problem (4.175): Problem (4.179) involves only $K + n$ variables, whereas (4.175) involves $(K + 1)m + Kn$ variables, where K is the number of sectors, m is the number of internationally traded goods, and n is the number of domestic inputs.

There is yet an additional way of rewriting the primal planning problem that gives some additional insight into the structure of the problem. Assume that a nonnegative domestic price vector $\bar{p} \geq 0$, $\bar{p} \neq 0$ exists such that $\pi^k(\bar{p}, w^0) < 0$ for $k = 1, \ldots, K$: i.e., so that each sector makes negative unit profits at the prices $\bar{p}$, w^0. Then we may apply Theorem 2.28 in a different way to the max-min problem (4.179). Rewrite (4.179) as

$$G(w^0, v^0) = -\min_{z\geq 0} \max_{p\geq 0} \left\{ -\sum_{k=1}^{K} \pi^k(p, w^0) z_k - p^T v^0 \right\} \tag{4.180}$$

$$= -\max_{p\geq 0} \{-p^T v^0 : -\pi^k(p, w^0) \geq 0, k = 1, \ldots, K\} \tag{4.181}$$

(applying Theorem 2.28)

$$= \min_{p\geq 0} \{p^T v^0 : \pi^k(p, w^0) \leq 0, k = 1, \ldots, K\}. \tag{4.182}$$

Thus if z^*, p^* solves the max-min problem (4.179) then p^* solves the convex programming problem (4.182) and z^* is a vector of optimal Karush-Kuhn-Tucker multipliers for the constrained minimization problem (4.182). Moreover, the equality of the primal representation for the national production function (4.175) to the dual representation (4.182) yields

$$G(w^0, v^0) = (w^0)^T y^* = (p^*)^T v^0. \tag{4.183}$$

Thus $(w^0)^T y^*$, the maximal value of the country's net output valued at international prices is equal to the value of the country's endowment of inputs $(p^*)^T v$ valued at the optimal dual prices p^*. Note that the dual problem (4.182) asks us to find optimal shadow or accounting prices p^* that will minimize the value of the country's resource endowments subject to each industry making nonpositive profits. Note also that the primal programming problem (4.179) reduces to a linear program if the sectoral unit scale production possibilities sets C^k are single points and (4.179) reduces to the linear programming dual to the primal linear program. Thus the equality between (4.175) and (4.179) generalizes the usual duality theory for linear programs, where an optimal pricing problem is equivalent to a direct resource allocation problem.

We shall now explore the sensitivity properties of the national product function $G(w, v)$ with respect to the vector of international prices w and the endowment vector v.

Define the set

$$S(v) = \left\{ y: y = \sum_{k=1}^{K} y^k;\ \sum_{k=1}^{K} x^k + v \geq 0;\ (y^k, z^k) \in S^k, k = 1, \ldots, K \right\}.$$

The convexity of the sets S^k implies the convexity of the set $S(v^0)$. By the definition of $G(w, v)$ given by (4.175) and the definition of the set $S(v)$, it can be seen that

$$G(w, v^0) = \max_y \{w^T y: y \in S(v^0)\}. \tag{4.184}$$

Thus $G(w, v^0)$ regarded as a function of the world price vector w is obviously the (upper) support function for the convex set $S(v^0)$ and hence $G(w, v^0)$ is a *convex function* of w. Alternatively, this convexity property may be established by the method of proof of property (d) in Theorem 4.1 as follows. Let $w^1 > 0$, $w^2 > 0$ be m-dimensional vectors, let $0 < \lambda < 1$, and let y^λ solve (4.184) when world prices $\lambda w^1 + (1 - \lambda)w^2$ prevail. Then

$$G(\lambda w^1 + (1 - \lambda)w^2, v^0) = \max_y \{[\lambda w^1 + (1 - \lambda)w^2]^T y: y \in S(v^0)\} \tag{4.185}$$

$$= [\lambda w^1 + (1 - \lambda)w^2]^T y^\lambda \tag{4.186}$$

$$= \lambda (w^1)^T y^\lambda + (1 - \lambda)(w^2)^T y^\lambda \tag{4.187}$$

$$\leq \lambda \max_y \{(w^1)^T y: y \in S(v^0)\}$$

$$+ (1 - \lambda) \max_y \{(w^2) y: y \in S(v^0)\} \tag{4.188}$$

$$= \lambda G(w^1, v^0) + (1 - \lambda) G(w^2, v^0) \tag{4.189}$$

since y^λ is feasible (but not necessarily optimal) for the two maximization problems.

If $G(w, v^0)$ is differentiable with respect to the components of the world price vector w at $w = w^0$, then the solution y^* to (4.184) is unique and is equal to the gradient vector of $G(w^0, v^0)$ with respect to the components of w; i.e., we have

$$\nabla_w G(w^0, v^0) = y^*. \tag{4.190}$$

To prove the above result, define for $w > 0$,

$$f(w) = G(w, v^0) - w^T y^* \geq 0, \tag{4.191}$$

where y^* is any solution to (4.184) when $w = w^0$. The inequality in (4.191) follows from the definition (4.184) of $G(w, v^0)$ as a constrained maximization problem and the feasibility of y^* for the problem. When $w = w^0$ we have

$$f(w^0) = G(x^0, v^0) - (w^0)^T y^* = 0 \tag{4.192}$$

since by assumption, y^* is a solution to $\max\{(w^0)^T y: y \in S(v^0)\} = G(w^0, v^0)$. It follows from (4.191) and (4.192) that the function f is globally minimized over nonnegative w vectors at $w = w^0 > 0$. Since $G(w^0, v^0)$ is differentiable with respect to w by assumption, the first-order necessary conditions for a local minimum yield (4.190). If we take any other solution, $\tilde{y}^*$, say, to (4.175), we may repeat the above argument and find that $\tilde{y}^* = \nabla_w G(w^0, v^0) = y^*$. Hence the solution to (4.175) is unique if $G(w^0, v^0)$ is differentiable with respect to w. Thus if G is differentiable with respect to w, the economy's system of net supply functions for internationally traded goods, $y(w^0, v^0)$, may be obtained by differentiating the national product function with respect to the vector of international prices, i.e., $y(w^0, v^0) = \nabla_w G(w^0, v^0)$. The result is an economy-wide counterpart to Shephard's lemma (Corollary 4.7).

There is also a counterpart to Corollary 4.6: The solution set y to the primal maximization problem (4.175) is the set of subgradients to the convex function $g(w) = G(w, v^0)$ at the point w^0. Hence if the solution set Y is a singleton consisting of y^* only, then $G(w^0, v^0)$ is differentiable with respect to w and $\nabla_w G(w^0, v^0) = y^*$.

Finally, there is a counterpart to the results on the cost function discussed in Section 4.3. Suppose $G(w^0, v^0)$ is twice continuously differentiable with respect to the components of w. Then differentiating (4.190) with respect to w tells us that the $m \times m$ matrix of derivatives of the net supply functions for internationally traded goods with respect to the components of the international price vector is equal to the Hessian matrix of the national product function with respect to the components of the international price vector, i.e.,

$$\nabla_w y(w^0, v^0) = \nabla^2_{ww} G(w^0, v^0). \tag{4.193}$$

Since G is convex and linearly homogeneous in w, the above matrices satisfy the following restrictions:

$$\nabla^2_{ww} G(w^0, v^0) \text{ is a symmetric positive semidefinite matrix,} \tag{4.194}$$

and

$$[\nabla^2_{ww} G(w^0, v^0)]w^0 = 0. \tag{4.195}$$

Specifically, (4.194) implies that $\partial y_i(w^0, v^0)/\partial w_i \geq 0$ for $i = 1, \ldots, m$ (the supply curve for internationally traded good i as a function of the price w_i does not slope downwards), and also the *Samuelson symmetry conditions*:

$$\partial y_i(w^0, v^0)/\partial w_j = \partial y_j(w^0, v^0)/\partial w_i, \qquad 1 \leq i \leq j \leq m, \tag{4.196}$$

see Samuelson (1953–1954).

We now turn to the properties of $G(w, v)$ with respect to the n components of the resource vector v. Consider the convex programming problem (4.182) and note that the set of feasible domestic accounting or shadow prices is $P(w^0) = \{p: \pi^k(p, w^0) \leq 0, k = 1, \ldots, K; p \geq 0\}$. Thus we may rewrite (4.182) for a general $v > 0$ as

$$G(w^0, v) = \min_p \{p^T v: p \in P(w^0)\}. \tag{4.197}$$

The convexity of the functions $\pi^k(p, w^0)$ in p implies that the set of feasible prices $P(w^0)$ in (4.197) is a convex set. Thus $G(w^0, v)$ regarded as a function of the economy's endowment vector of domestic inputs v is obviously the (lower) support function for the convex set $P(w^0)$, and hence $G(w^0, v)$ is a concave function of v. Alternatively, this concavity property may be established by modifying the proof of (4.189).

If $G(w^0, v)$ is differentiable with respect to the components of the endowment vector v at v^0, then we can show that the solution p^* to (4.182) is unique and is equal to the gradient vector of $G(w^0, v^0)$ with respect to the components of v; i.e., we have

$$\nabla_v G(w^0, v^0) = p^*. \tag{4.198}$$

The proof of (4.198) is analogous to the proof of (4.190).

Result (4.198) enables us to provide an economic interpretation for the vector of accounting prices $p^* = (p_1^*, \ldots, p_n^*)^T$: $p_j^* = \partial G(w^0, v^0)/\partial v_j$ is the marginal increase in national product due to a marginal change in the economy's endowment of resource j. Put another way, if G is differentiable with respect to v, the economy's system of inverse demand functions for domestic goods, $p(w^0, v^0)$, may be obtained by differentiating the national product function with respect to the endowment vector, i.e., $p(w^0, v^0) = \nabla_v G(w^0, v^0)$.

There is also another counterpart to Corollary 4.6: The solution set P to the dual minimization problem (4.182) is the set of supergradients to the concave (in v) function $G(w^0, v)$ at the point v^0. Hence if the solution set to (4.182) is a singleton consisting of p^* only, then $G(w^0, v^0)$ is differentiable with respect to v and $\nabla_v G(w^0, v^0) = p^*$.

Suppose now that $G(w, v)$ is twice continuously differentiable with respect to the components of v. Then differentiating (4.198) with respect to v tells us that the $n \times n$ matrix of derivatives of the inverse demand functions for domestic inputs is equal to the Hessian matrix of the national product function with respect to the components of the domestic endowment vector; i.e.,

$$\nabla_v p(w^0, v^0) = \nabla^2_{vv} G(w^0, v^0). \tag{4.199}$$

Since G is concave in v, the above matrices satisfy the following condition:

$$\nabla^2_{vv} G(w^0, v^0) \text{ is a negative semidefinite symmetric matrix.} \tag{4.200}$$

Since G is linearly homogeneous in v, we have the additional restrictions implied by Euler's theorem:

$$[\nabla_v p(w^0, v^0)]v^0 = [\nabla^2_{vv} G(w^0, v^0)]v^0 = 0. \tag{4.201}$$

It follows from (4.200) that $\partial p_j(w^0, v^0)/\partial v_j \leq 0$ for $j = 1, \ldots, n$ (as the endowment of the jth domestic input increases, the corresponding jth input price cannot increase; i.e., the inverse demand curve for the jth domestic good does not slope upwards). The negative semidefiniteness of $\nabla^2_{vv} G(w^0 v^0)$ also implies the following Samuelson symmetry condition:

$$\partial p_i(w^0, v^0)/\partial v_j = \partial p_j(w^0, v^0)/\partial v_i, \qquad i \leq i < j \leq n, \tag{4.202}$$

see Samuelson (1953–1954).

As in the earlier sections of this chapter, again the reader can see how functions with convexity and concavity properties arise naturally in economics and how these curvature properties may be translated into results that have economic content. One additional result can be obtained if we assume that $G(w^0, v^0)$ is jointly twice continuously differentiable with respect to all of its arguments. Upon differentiating (4.198) with respect to w and differentiating (4.190) with respect to v, we obtain the following *Samuelson reciprocity relations*:

$$\nabla_w p(w^0, v^0) = [\nabla_v y(w^0, v^0)]^T = \nabla^2_{vw} G(w^0, v^0) \tag{4.203}$$

or

$$\partial p_j(w^0, v^0)/\partial w_i = \partial y_i(w^0, v^0)/\partial v_j, \qquad j = 1, \ldots, n; \qquad i = 1, \ldots, m; \tag{4.204}$$

see Samuelson (1953–1954). Thus the rate of change of the jth input price with respect to a change in the ith internationally traded good is equal to the rate of increase in the net supply of the ith internationally traded good to a change in the endowment of domestic good j. This is not an intuitively obvious result.

We conclude this chapter with an application of Theorem 2.30. Assume that the unit profit functions $\pi^k(p^*, w^0)$, $k = 1, \ldots, K$, are differentiable with respect to p and that a strictly positive solution $p^* > 0$ and $z^* > 0$ exists for the max–min problem (4.179). Then the following Karush–Kuhn–Tucker conditions for (4.179) or (4.182) will be satisfied:

$$-\sum_{k=1}^{K} \nabla_p \pi^k(p^*, w^0) z_k = v \tag{4.205}$$

$$\pi^k(p^*, w^0) = 0, \qquad k = 1, \ldots, K. \tag{4.206}$$

Define a certain endowment set v^0 that corresponds to this initial equilibrium as follows:

$$V^0 = \left\{ v\colon v = -\sum_{k=1}^{K} \nabla_p \pi^k(p^*, w^0) z_k\,;\ z_k \geq 0,\ k = 1, \ldots, K \right\}. \tag{4.207}$$

Then we have the following theorem.

Theorem 4.14 (Diewert and Woodland, 1977; McKenzie, 1955; Samuelson, 1953–1954). The initial domestic price solution p^* to the dual minimization problem (4.182) will continue to solve (4.197) [which is (4.182) except that v replaces v^0] for all endowment vectors $v \in V^0$.

Proof. Let $v \in V^0$. Then there exists $z = (z_1, \ldots, z_K)^T \geq 0$ such that

$$-\sum_{k=1}^{K} \nabla_p \pi^k(p^*, w^0) z_k = v. \tag{4.208}$$

Equations (4.206) and (4.207) and the nonnegativity of z show that the Karush–Kuhn–Tucker conditions for (4.197) are satisfied. Hence by Theorem 2.30, p^* solves (4.197). □

Theorem 4.14 is known in the economics literature as a *factor price equalization theorem*; all endowment vectors in the cone V^0 are consistent with a competitive equilibrium that has the same vector of domestic factor prices p^*.

In summary, this chapter considered four models of economic behavior: A producer's cost minimization problem, a consumer's utility maximization problem, a producer's profit maximization problem, and an open economy's national product maximization problem. We have shown how the various kinds of generalized concavity and convexity arise in the context of these models, and we have developed various sensitivity theorems associated with these models.

References

AFRIAT, S. N. (1971), Theory of maxima and the method of Lagrange, *SIAM J. Appl. Math.* **29**, 343-357.

BERGE, C. (1963), *Topological Spaces*, Macmillan, New York.

BLACKORBY, C., DAVIDSON, R., and DONALDSON, D. (1977), A homiletic exposition of the expected utility hypothesis, *Economica* **44**, 351-458.

BLACKORBY, C., and DIEWERT, E. W. (1979), Expenditure functions, local duality and second order approximations, *Econometrica* **47**, 579-601.

CROUZEIX, J.-P. (1977), Contributions à l'étude des fonctions quasiconvexes, Thèse presentee a l'université de Clermont II, France.

CROUZEIX, J.-P. (1981a), Continuity and differentiability properties of quasiconvex functions on R^n, in *Generalized Concavity in Optimization and Economics*, Edited by S. Schaible and W. T. Ziemba, Academic Press, New York.

CROUZEIX, J.-P. (1981b), A duality framework in quasiconvex programming, in *Generalized Concavity in Optimization and Economics*, Edited by S. Schaible and W. T. Ziemba, Academic Press, New York.

CROUZEIX, J.-P. (1982), Duality and differentiability in economics, mimeo, Department de Mathématiques Appliquées, Université de Clermont II, B.P. 45, 63170 Aubiere, France.

DEBREU, G. (1952), A social equilibrium existence theorem, *Proc. Nat. Acad. Sci.* **38**, 886-893.

DEBREU, G. (1959), *Theory of Value*, Wiley, New York.

DEBREU, G., and KOOPMANS, T. C. (1982), Additively decomposed functions, *Math. Programming* **24**, 1-38.

DIEWERT, W. E. (1971), An application of the Shephard duality theorem: A generalized Leontief production function, *J. Pol. Econ.* **79**, 481-507.

DIEWERT, W. E. (1974), Applications of duality theory, in *Frontiers of Quantitative Economics*, Vol. II, Edited by M. D. Intriligator and D. A. Kendrick, North-Holland, Amsterdam.

DIEWERT, W. E. (1978), Hicks' aggregation theorem and the existence of a real value added function, in *Production Economics: A Dual Approach to Theory and Applications*, Vol. 2, Edited by M. Fuss and D. McFadden, North-Holland, Amsterdam.

DIEWERT, W. E. (1981), Generalized concavity and economics, in *Generalized Concavity in Optimization and Economics*, Edited by S. Schaible and W. T. Ziemba, Academic Press, New York.

DIEWERT, W. E. (1982), Duality approaches to microeconomic theory, in *Handbook of Mathematical Economics*, Vol. II, Edited by K. J. Arrow and M. D. Intriligator, North-Holland, Amsterdam.

DIEWERT, W. E., and WOODLAND, A. D. (1977), Frank Knight's theorem in linear programming revisited, *Econometrica* **45**, 375-398.

DONALDSON, D., and EATON, B. C. (1981), Patience, more than its own reward: A note on price discrimination, *Can. J. Econ.* **14**, 93-105.

EPSTEIN, L. (1981), Generalized duality and integrability, *Econometrica* **49**, 655-678.

FENCHEL, W. (1953), Convex cones, sets and functions, lecture notes, Department of Mathematics, Princeton University, Princeton, New Jersey.

FIACCO, A. V. (1976), Sensitivity analysis for nonlinear programming using penalty methods, *Math. Programming* **10**, 287-311.

FINSLER, P. (1937), Über das Vorkommen definiter und semidefiniter Formen in scharen quadratischen Formen, *Comm. Math. Helv.* **9**, 188–192.

GALE, D., KLEE, V. L., and ROCKAFELLAR, R. T. (1968), Convex functions on convex polytopes, *Proc. Am. Math. Soc.* **19**, 867–873.

GORMAN, W. M. (1968), Measuring the quantities of fixed factors, in *Value, Capital and Growth: Papers in Honour of Sir John Hicks*, Edited by J. N. Wolfe, Aldine, Chicago, Illinois.

HANCOCK, H. (1917), *Theory of Maxima and Minima*, Ginn, Boston.

HICKS, J. R. (1941–1942), Consumers' surplus and index-numbers, *Rev. Econ. Stud.* **9**, 126–137.

HICKS, J. R. (1946), *Value and Capital*, Clarendon Press, Oxford.

JORGENSON, D. W., and LAU, L. J. (1974a), The duality of technology and economic behavior, *Rev. Econ. Stud.* **41**, 181–200.

JORGENSON, D. W., and LAU, L. J. (1974b), Duality and differentiability in production, *J. Econ. Theory* **9**, 23–42.

KARLIN, S. (1959), *Mathematical Methods and Theory in Games, Programming and Economics*, Vol. 1, Addison-Wesley, Reading, Massachusetts.

KARUSH, W. (1939), Minima of functions of several variables with inequalities as side conditions, M.S. dissertation, Department of Mathematics, University of Chicago.

KONYUS, A. A. (1939), The problem of the true index of the cost of living, *Econometrica* **7**, 10–29.

KUHN, H. W., and TUCKER, A. W. (1951), Nonlinear programming, in *Proceedings of the Second Berkeley Symposium on Mathematical Statistics and Probability*, Edited by J. Neyman, University of California Press, Berkeley, California.

LAU, L. J. (1974), Applications of duality theory: Comments, in *Frontiers of Quantitative Economics*, Vol. II, Edited by M. D. Intriligator and D. A. Kendrick, North-Holland, Amsterdam.

LAU, L. J. (1978), Applications of profit functions, in *Production Economics: A Dual Approach to Theory and Applications*, Vol. 1, Edited by M. Fuss and D. McFadden, North-Holland, Amsterdam.

MCFADDEN, D. (1966), Cost, revenue and profit functions: A cursory review, Working Paper No. 86, IBER, University of California, Berkeley, California.

MCFADDEN, D. (1978), Convex analysis, in *Production Economics: A Dual Approach to Theory and Applications*, Vol. 1, Edited by M. Fuss and D. McFadden, North-Holland, Amsterdam.

MCKENZIE, L. W. (1955), Equality of factor prices in world trade, *Econometrica* **23**, 239–257.

MCKENZIE, L. W. (1956–1957), Demand theory without a utility index, *Rev. Econ. Stud.* **24**, 185–189.

MINKOWSKI, H. (1911), *Theorie der Konvexen Körper in besondere Begründing ihres Oberflächenbegriffs*, Gesammelte Abhandlungen II, B. G. Teubner, Leipzig, Germany.

ROCKAFELLAR, R. T. (1970), *Convex Analysis*, Princeton University Press, Princeton, New Jersey.

ROY, R. (1947), La distribution du revenu entre les divers biens, *Econometrica* **15**, 205–225.

SAMUELSON, P. A. (1947), *Foundations of Economic Analysis*, Harvard University Press, Cambridge, Massachusetts.

SAMUELSON, P. A. (1953–1954), Prices of factors and goods in general equilibrium, *Rev. Econ. Stud.* **21**, 1–20.

SHEPHARD, R. W. (1953), *Cost and Production Functions*, Princeton University Press, Princeton, New Jersey.

SHEPHARD, R. W. (1970), *Theory of Cost and Production Functions*, Princeton University Press, Princeton, New Jersey.

UZAWA, H. (1964), Duality principles in the theory of cost and production, *Int. Econ. Rev.* **5**, 216–220.

YAARI, M. E. (1977), A note on separability and quasiconcavity, *Econometrica* **45**, 1183–1186.

5

Special Functional Forms I: Composite Functions, Products, and Ratios

In this chapter and in the next one we focus our attention on the generalized concavity properties of special families of functions. The study of composite functions, products, and ratios in this chapter is followed by a study of quadratic functions in Chapter 6. In both chapters we want to derive conditions under which functions belonging to the above families are generalized concave.

The need for readily verifiable criteria is quite obvious. We saw in Chapter 3 that generalized concave programs have many properties in common with concave programs. They often can be solved by methods of concave programming or extensions of such methods. In order to identify a given optimization problem as a generalized concave program, one has to have suitable criteria to verify that the objective function and constraint functions are generalized concave. Furthermore, we saw in Chapter 4 that generalized concave functions are used in economic theory; hence, again, we need practical criteria to establish the generalized concavity of the functions involved.

Unfortunately, the standard definitions in Chapter 3 are often not easily applicable to decide whether a given function is generalized concave or not. Even for functions in two variables it may be quite difficult to verify the defining inequalities of quasiconcavity or pseudoconcavity, for example. By restricting ourselves to specific classes of functions, having a certain algebraic structure, we can hope to find more practical characterizations. And this indeed is so.

Early works in this area had concentrated on finding verifiable criteria for rather special types of functions, for example, special ratios or products. Seldom were these criteria related to each other. Also, the criteria were established by different methods. More recent works tried to use a unified approach by which all known, and many new, results can be derived (Mangasarian, 1970; Schaible, 1971, 1972, 1981a, 1981b). We shall follow this approach here. The main idea is to look at functions, such as products, ratios, or quadratic functions, as composite functions.

In Chapter 3 we already saw that under certain conditions composite functions inherit quasiconcavity and pseudoconcavity from their "parent" functions. In the following we derive several other results. These criteria will then be used in the latter part of the chapter to obtain criteria for the generalized concavity of products and ratios of convex and/or concave functions. This analysis will also provide the basis for establishing criteria for quadratic functions in the next chapter.

5.1. Composite Functions

In this section we derive sufficient conditions for the generalized concavity of composite functions. Although they will be applied later to obtain characterizations for other classes of functions, they are of independent interest in the applications of generalized concavity.

In Propositions 5.1–5.5 quasiconcavity and semistrict quasiconcavity of these functions is analyzed. Following these results, the pseudoconcave and strictly pseudoconcave counterpart for the same functions are treated in Propositions 5.6–5.10. Most of the criteria of this section were shown independently by Mangasarian (1970) and Schaible (1971).

We start with a criterion for the quasiconcavity of the composition of two functions by restating Proposition 3.2 here, and by complementing it by a corresponding result for semistrict quasiconcavity. The proof of the latter case is similar to the proof of the former result, and is omitted here.

Throughout this chapter C will denote a convex set in R^n.

Proposition 5.1. Let ϕ be a (semistrictly) quasiconcave function defined on $C \subset R^n$ and let f be a (increasing) nondecreasing function on $D \subset R$, containing the range of ϕ. Then the composite function $f\phi(x)$ is also (semistrictly) quasiconcave on C.

From this result we can derive the following Corollary.

Corollary 5.2. If ϕ is either a positive or negative (semistrictly) quasiconcave function on $C \subset R^n$, then $1/\phi$ is (semistrictly) quasiconvex on C.

Proof. We can write $1/\phi(x) = -f\phi(x)$ with $f(y) = -1/y$. Since $f'(y) = 1/(y)^2 > 0$ for $y \neq 0$, f is increasing over the range of ϕ, and, applying Proposition 5.1, $f\phi(x)$ is (semistrictly) quasiconcave on C if $\phi(x)$ is so. Hence, $1/\phi(x) = -f\phi(x)$ is (semistrictly) quasiconvex on C. □

Hence, the reciprocal of an either positive or negative (semistrictly) quasiconcave function is (semistrictly) quasiconvex. Moreover, the reciprocal of an either positive or negative quasiconvex function is quasiconcave as we can also easily see from Corollary 5.2. Note that there is no such similar symmetric relationship between concave and convex functions. Although it is true that the reciprocal of a positive concave function is convex, the reverse does not hold. The reader can verify this by considering, for example, the convex function $\phi(x) = \exp(x)$.

In Proposition 5.1 we considered the composition of only two functions. We now turn to more general composite functions. In Chapter 2 we already saw that under certain concavity assumptions the composition of m functions is again concave (see Proposition 2.16). We will now extend this result to the quasiconcave case:

Proposition 5.3. Let $\phi_1, \ldots, \phi_m$ be concave functions on $C \subset R^n$ and let f be a nondecreasing (semistrictly) quasiconcave function on $D \subset R^m$. Suppose that D contains the range of $\phi = (\phi_1, \ldots, \phi_m)$. Then the composite function $f(\phi_1(x), \ldots, \phi_m(x))$ is (semistrictly) quasiconcave on C.

Proof. We prove here the semistrictly quasiconcave case only. The proof for the quasiconcave case is similar. Let $x^1 \in C$, $x^2 \in C$ such that $f\phi(x^1) \neq f\phi(x^2)$. Since $\phi_1, \ldots, \phi_m$ are concave and f is nondecreasing, we have for every λ, $0 < \lambda < 1$

$$f(\phi(\lambda x^1 + (1-\lambda)x^2)) \geq f(\lambda\phi(x^1) + (1-\lambda)\phi(x^2)). \tag{5.1}$$

By the semistrict quasiconcavity of f it follows that

$$f(\lambda\phi(x^1) + (1-\lambda)\phi(x^2)) > \min\{f\phi(x^1), f\phi(x^2)\}. \tag{5.2}$$

Combining (5.1) and (5.2) we obtain the result. □

Note that for $m = 1$ $f\phi(x)$ is quasiconcave if $\phi(x)$ is only quasiconcave rather than concave. This follows from Proposition 5.1. However, for $m > 1$ the concavity of ϕ_i cannot be replaced by quasiconcavity. Otherwise, the sum of quasiconcave functions ϕ_i would always be quasiconcave, since $\sum_{i=1}^m \phi_i(x) = f(\phi_1(x), \ldots, \phi_m(x))$, where $f(y_1, \ldots, y_m) = \sum_{i=1}^m y_i$ is nondecreasing and (quasi)concave. However, the sum of quasiconcave functions is generally *not* quasiconcave. For example, $f(x_1, x_2) = (x_1)^3 + (x_2)^2$ is not quasiconcave on the positive orthant of R^2, since the upper-level sets are not convex there.

In applications, some of the functions ϕ_i may be convex rather than concave. In this case a result similar to Proposition 5.3 holds provided $f(y_1, \ldots, y_m)$ is nonincreasing rather than nondecreasing in the relevant y_i. With these modifications the proof of the following proposition is similar to the proof of Proposition 5.3 and will be omitted here.

Proposition 5.4. Let ϕ_i be concave functions on $C \subset R^n$ for $i \in I_1$, $I_1 \subset \{1, \ldots, m\}$ and let ϕ_i be convex functions on C for $i \in I_2$, $I_2 = \{1, \ldots, m\} \backslash I_1$. Let f be (semistrictly) quasiconcave on $D \subset R^m$, where D contains the range of $\phi = (\phi_1, \ldots, \phi_m)$. Furthermore, assume that $f(y_1, \ldots, y_m)$ is nondecreasing in y_i for $i \in I_1$ and nonincreasing in y_i for $i \in I_2$. Then $f\phi(x)$ is (semistrictly) quasiconcave on C.

In Proposition 2.17 it was shown that concavity is invariant under linear transformations of the domain. The next proposition extends this result to (semistrictly) quasiconcave functions.

Proposition 5.5. Let f be a (semistrictly) quasiconcave function on a convex set $D \subset R^m$, A a real $m \times n$ matrix, and $b \in R^m$. Let C be a convex set in R^n such that $Ax + b \in D$ for all $x \in C$. Then $f(Ax + b)$ is (semistrictly) quasiconcave on C.

Proof. Let $\phi(x) = Ax + b$ in Proposition 5.3. Then the proof of Proposition 5.3 is valid without assuming monotonicity of f. This is so since (5.1) holds as an equality because of the linearity of $\phi(x)$. □

Let us now turn to differentiable functions. Essentially we shall see that semistrict quasiconcavity properties established in Propositions 5.1–5.5 can be replaced by pseudoconcavity. In addition, we shall also derive criteria for strictly pseudoconcave functions. The pseudoconcave counterpart of Proposition 5.1 is the following proposition.

Proposition 5.6. Let ϕ be a (strictly) pseudoconcave function on $C \subset R^n$ and let f be differentiable on $D \subset R$, containing the range of ϕ. Assume that $f'(y) > 0$ on D. Then $F(x) = f\phi(x)$ is (strictly) pseudoconcave on C.

Proof. We only prove the strictly pseudoconcave case. Let $x^1 \in C$, $x^2 \in C$, such that $x^1 \neq x^2$ and $(x^2 - x^1)^T \nabla F(x^1) \leq 0$. Since $\nabla F(x^1) = f'(\phi(x^1)) \nabla \phi(x^1)$ and $f'(y) > 0$, we have $(x^2 - x^1)^T \nabla \phi(x^1) \leq 0$. This yields $\phi(x^2) < \phi(x^1)$ by the strict pseudoconcavity of ϕ (see Definition 3.13). Hence $f\phi(x^2) < f\phi(x^1)$. □

From Proposition 5.6 we can derive the next result following the idea of the proof of Corollary 5.2:

Corollary 5.7. If ϕ is either a positive or negative (strictly) pseudoconcave function on $C \subset R^n$, then $1/\phi$ is (strictly) pseudoconvex on C.

The next result is the pseudoconcave counterpart of Proposition 5.3.

Proposition 5.8. Let $\phi_1, \ldots, \phi_m$ be differentiable concave functions on $C \subset R^n$ and let f be pseudoconcave on $D \subset R^m$, where D contains the range of $\phi = (\phi_1, \ldots, \phi_m)$. Assume that $\nabla f(y) \geq 0$ on D. Then the composite function $F(x) = f\phi(x)$ is pseudoconcave on C. Moreover, if f is strictly pseudoconcave, at least one function ϕ_i is strictly concave and $\partial f/\partial y_i > 0$, then $F(x) = f\phi(x)$ is strictly pseudoconcave on C.

Proof. Let $x^1 \in C$, $x^2 \in C$ such that $(x^2 - x^1)^T \nabla F(x^1) \leq 0$. Since

$$\frac{\partial F}{\partial x_j} = \sum_{i=1}^{m} \frac{\partial f}{\partial y_i} \frac{\partial \phi_i}{\partial x_j} \tag{5.3}$$

we have

$$0 \geq (x^2 - x^1)^T \nabla F(x^1) = \sum_{j=1}^{n} \sum_{i=1}^{m} \frac{\partial f}{\partial y_i} \frac{\partial \Phi_i(x^1)}{\partial x_j} (x_j^2 - x_j^1) \tag{5.4}$$

$$= \sum_{i=1}^{m} \frac{\partial f}{\partial y_i} \sum_{j=1}^{n} (x_j^2 - x_j^1) \frac{\partial \phi_i(x^1)}{\partial x_j} \tag{5.5}$$

$$\geq \sum_{i=1}^{m} \frac{\partial f}{\partial y_i} [\phi_i(x^2) - \phi_i(x^1)]$$

$$= [\phi(x^2) - \phi(x^1)]^T \nabla_y f(\phi(x^1)) \tag{5.6}$$

because of the concavity of ϕ_i and since $\partial f/\partial y_i \geq 0$. Hence

$$[\phi(x^2) - \phi(x^1)]^T \nabla_y f(\phi(x^1)) \leq 0. \tag{5.7}$$

Since f is pseudoconcave, (5.7) implies $f(\phi(x^2)) \leq f(\phi(x^1))$, i.e., $F(x^2) \leq F(x^1)$.

In the strictly pseudoconcave case we assume $x^1 \neq x^2$. We see that (5.7) becomes a strict inequality implying $\phi(x^1) \neq \phi(x^2)$. Then, the strict pseudoconcavity of f yields $f(\phi(x^2)) < f(\phi(x^1))$. □

Note that strict pseudoconcavity of f alone is not sufficient to ensure strict pseudoconcavity of F. By appropriate modifications in the proof of Proposition 5.8 we obtain the following proposition.

Proposition 5.9. Let ϕ_i be differentiable concave functions on $C \subset R^n$ for $i \in I_1$, $I_1 \subset \{1, \ldots, m\}$ and let ϕ_i be differentiable convex functions on C for $i \in I_2$, $I_2 = \{1, \ldots, m\} \backslash I_1$. Let f be a pseudoconcave function on $D \subset R^m$, containing the range of $\phi = (\phi_1, \ldots, \phi_m)$. Assume that $\partial f/\partial y_i \geq 0$ for $i \in I_1$ and $\partial f/\partial y_i \leq 0$ for $i \in I_2$. Then $F(x) = f\phi(x)$ is pseudoconcave on C. If at least one ϕ_i is strictly concave (strictly convex) for $i \in I_1 (i \in I_2)$, $\partial f/\partial y_i > 0\ (<0)$ and f is strictly pseudoconcave, then $F(x) = f\phi(x)$ is strictly pseudoconcave on C.

Finally, an analogous result to Proposition 5.5 is the following proposition.

Proposition 5.10. Let f be a pseudoconcave function on a convex set $D \subset R^m$, A a real $m \times n$ matrix, and $b \in R^m$. Let C be a convex set in R^n such that $Ax + b \in D$ for all $x \in C$. Then $f(Ax + b)$ is pseudoconcave on C. If f is strictly pseudoconcave and A is a nonsingular $n \times n$ matrix then $f(Ax + b)$ is strictly pseudoconcave.

In this section we presented sufficient conditions for the generalized concavity of certain composite functions. To some extent these conditions are also necessary. The interested reader is referred to Bereanu (1972).

In the remaining part of this chapter and also in the next chapter, we shall make use of the above results.

5.2. Products and Ratios

We now examine products and ratios of concave and/or convex functions and will see that they are often generalized concave in some sense. Following the approach in Schaible (1971, 1972) we consider these functions as composite functions and apply the criteria derived in the preceding section.

First we introduce the function

$$f(y, z) = \prod_{i=1}^{k} (y_i)^{a_i} \Big/ \prod_{j=1}^{l} (z_j)^{b_j} \tag{5.8}$$

where $a_i > 0$, $b_j > 0$ are given numbers. This function is considered on the domain

$$D = \{(y, z) \in R^k \times R^l : y \geq 0, z > 0\}. \tag{5.9}$$

Let D^I denote the interior of D, i.e.,

$$D^I = \{(y, z) \in R^k \times R^l : y > 0, z > 0\}. \tag{5.10}$$

In addition, let G be an arbitrary open convex set in D. We then have the following Lemma.

Lemma 5.11. (a) Suppose that $l = 1$. If $b_1 > \sum_{i=1}^k a_i$, then $f(y, z)$ is strictly pseudoconcave on D^I, and if $b_1 = \sum_{i=1}^k a_i$, then $f(y, z)$ is pseudoconcave on D^I. In case $b_1 < \sum_{i=1}^k a_i$, then $f(y, z)$ is not quasiconcave on $G \subset D$.

(b) If $l > 1$, then $f(y, z)$ is not quasiconcave on $G \subset D$.

(c) Suppose $k = 1$. If $a_1 > \sum_{j=1}^l b_j$, then $f(y, z)$ is strictly pseudoconvex on D^I, and if $a_1 = \sum_{j=1}^l b_j$, then $f(y, z)$ is pseudoconvex on D^I. In case $a_1 < \sum_{j=1}^l b_j$, then $f(y, z)$ is not quasiconvex on $G \subset D$.

(d) If $k > 1$, then $f(y, z)$ is not quasiconvex on $G \subset D$.

Proof. (a) Consider

$$g(y, z) = \log f(y, z) = \sum_{i=1}^k a_i \log y_i - b_1 \log z_1 \tag{5.11}$$

on D^I. In view of Propositions 5.1 and 5.6, the function f is quasiconcave (pseudoconcave, strictly pseudoconcave) on D^I if and only if g is. By Proposition 3.45 g is strongly pseudoconcave, and therefore strictly pseudoconcave, on D^I if for every $(y, z) \in D^I$ and $v \in R^{k+1}$, $v \neq 0$

$$v^T \nabla g(y, z) = 0 \qquad \text{implies } v^T \nabla^2 g(y, z) v < 0. \tag{5.12}$$

We have

$$v^T \nabla g(y, z) = \sum_{i=1}^k \frac{a_i}{y_i} v_i - \frac{b_1}{z_1} v_{k+1} \tag{5.13}$$

and

$$v^T \nabla^2 g(y, z) v = \sum_{i=1}^k \left[-\frac{a_i}{(y_i)^2} \right] (v_i)^2 + \frac{b_1}{(z_1)^2} (v_{k+1})^2. \tag{5.14}$$

Let $v = (\tilde{v}, v_{k+1})$, where $\tilde{v} \in R^k$. We have to verify (5.12) for all $v \neq 0$. For $v = (\tilde{v}, 0)$, $\tilde{v} \neq 0$, we find $v^T \nabla^2 g(y, z) v < 0$ since all $a_i > 0$ in (5.14). Hence (5.12) is true for all $v \neq 0$ if and only if it is true for $(\tilde{v}, 1)$ where $\tilde{v} \in R^k$. Let

$$C(\tilde{v}) = \sum_{i=1}^{k} \frac{a_i}{y_i} \tilde{v}_i - \frac{b_1}{z_1} \tag{5.15}$$

and

$$A(\tilde{v}) = \sum_{i=1}^{k} \left[-\frac{a_i}{(y_i)^2} \right] (\tilde{v}_i)^2 + \frac{b_1}{(z_1)^2}. \tag{5.16}$$

Consider

$$M = \max \{A(\tilde{v}) \colon C(\tilde{v}) = 0\}. \tag{5.17}$$

If $M < 0$ for any $(y, z) \in D^I$, then g is strictly pseudoconcave on D^I. Now (5.17) is a quadratic program with a strictly concave objective function on R^k. Hence the unique maximizer of (5.17) is found by solving

$$\nabla_{\tilde{v}}(A(\tilde{v}) - \lambda C(\tilde{v})) = 0, \tag{5.18}$$

$$C(\tilde{v}) = 0 \tag{5.19}$$

for some $\lambda \in R$ (Avriel, 1976). We find

$$M = \frac{b_1}{(z_1)^2} \left(1 - b_1 \Big/ \sum_{i=1}^{k} a_i \right). \tag{5.20}$$

If $b_1 > \sum_{i=1}^{k} a_i$, then $M < 0$, implying strict pseudoconcavity of g. For $b_1 = \sum_{i=1}^{k} a_i$ we have $M = 0$ in (5.20), and therefore $v^T \nabla^2 g(y, z) v \leq 0$ in (5.12). Then g is still pseudoconcave on D^I (see Crouzeix and Ferland, 1982). On the other hand, for $b_1 < \sum_{i=1}^{k} a_i$, $M > 0$. Thus (5.12) no longer holds, and therefore g is not quasiconcave on $G \subset D$, as can be seen from Proposition 3.16.

(b) For $l > 1$, $f(y, z)$ is not quasiconcave on $G \subset D$ since for a given $(\bar{y}, \bar{z}) \in D^I$ the restriction of f to the subsets

$$\begin{aligned} &\{(y, z) \in D^I \colon y = \bar{y}\} && \text{if } l = 2, \\ &\{(y, z) \in D^I \colon y = \bar{y},\ z_j = \bar{z}_j, j = 3, \ldots, l\} && \text{if } l > 2 \end{aligned} \tag{5.21}$$

is not quasiconcave because of the nonconvexity of the upper-level sets. The results in (c) and (d) are obtained by applying the results in (a) and (b) to the reciprocal function $1/f(y, z)$ and using Corollaries 5.2 and 5.7. □

Pseudoconcavity (pseudoconvexity) of f on D^I implies semistrict quasiconcavity (semistrict quasiconvexity) there. Using Definition 3.11 one can easily verify that in this case the continuous function f in (5.8) is semistrictly quasiconcave (semistrictly quasiconvex) on all of D. Hence we have the following lemma.

Lemma 5.12. (a) If $l = 1$ and $b_1 \geq \sum_{i=1}^{k} a_i$, then $f(y, z)$ is semistrictly quasiconcave on D.

(b) If $k = 1$ and $a_1 \geq \sum_{j=1}^{l} b_j$, then $f(y, z)$ is semistrictly quasiconvex on D.

In case $k = 1$ we can show the following result, by applying the criterion in Theorem 2.14 for convex functions:

Lemma 5.13. If $k = 1$, then $f(y, z)$ is (a) convex on D for $a_1 \geq 1 + \sum_{j=1}^{l} b_j$, (b) strictly convex on D^I for $a_1 > 1 + \sum_{i=1}^{l} b_j$, and (c) not convex on G for $a_1 < 1 + \sum_{j=1}^{l} b_j$, where G is an open convex set in D.

In addition, we mention the following result that can be shown in the same way as Lemma 5.13.

Lemma 5.14. The function $h(y) = \prod_{i=1}^{k} (y_i)^{a_i}$, $a_i > 0$, $i = 1, \ldots, k$, is (a) concave on the nonnegative orthant of R^k if $\sum_{i=1}^{k} a_i \leq 1$, (b) strictly concave on the positive orthant of R^k if $\sum_{i=1}^{k} a_i < 1$, and (c) not concave on any open subset G of the positive orthant of R^k if $\sum_{i=1}^{k} a_i > 1$.

Lemmas 5.11–5.14 will now be used together with the results on composite functions appearing in Section 5.1 to derive a number of criteria for products and ratios of concave and/or convex functions.

In view of Lemma 5.12 and Proposition 5.4 we have the following theorem.

Theorem 5.15. (a) Let ϕ_0 be a positive convex function on $C \subset R^n$ and let ϕ_k be nonnegative and concave functions on C. Assume that $b \geq \sum_{i=1}^{k} a_i$ where $a_i > 0$, $i = 1, \ldots, k$. Then,

$$F_1(x) = \prod_{i=1}^{k} [\phi_i(x)]^{a_i} / [\phi_0(x)]^b \tag{5.22}$$

is semistrictly quasiconcave on C.

(b) Let ϕ_0 be a nonnegative convex function on C and let $\phi_1, \ldots, \phi_l$ be positive concave functions on C. Assume that $a \geq \sum_{j=1}^{l} b_j$, where $b_j > 0$, $j = 1, \ldots, l$. Then,

$$F_2(x) = [\phi_0(x)]^a \bigg/ \prod_{j=1}^{l} [\phi_j(x)]^{b_j} \tag{5.23}$$

is semistrictly quasiconvex on C.

In addition, we obtain from Lemma 5.13 together with Proposition 2.16, the following theorem.

Theorem 5.16. If in Theorem 5.15(b) $a \geq 1 + \sum_{j=1}^{l} b_j$, then F_2 is convex on C. Also, F_2 is strictly convex on C if Φ_0 is strictly convex and positive, the Φ_i are strictly concave, and $a > 1 + \sum_{j=1}^{l} b_j$.

For differentiable functions we can show the following result, using Lemma 5.11 and Proposition 5.9:

Theorem 5.17. (a) If in Theorem 5.15 (a) $\Phi_0, \ldots, \Phi_k$ are differentiable on C and $\Phi_1, \ldots, \Phi_k$ are positive, then F_1 is pseudoconcave on C. Moreover F_1 is strictly pseudoconcave on C if $b > \sum_{i=1}^{k} a_i$, and either Φ_0 is strictly convex or at least one of the Φ_i, $i = 1, \ldots, k$ is strictly concave on C.

(b) If in Theorem 5.15(b) $\Phi_0, \ldots, \Phi_l$ are differentiable on C and Φ_0 is positive, then F_2 is pseudoconvex on C. Moreover, F_2 is strictly pseudoconvex on C if $a > \sum_{j=1}^{l} b_j$, Φ_0 is strictly convex or at least one of the Φ_j, $j = 1, \ldots, l$ is strictly concave on C.

We now specialize some of the results above. Taking $\Phi_0(x) \equiv 1$ in Theorem 5.15(a) and Theorem 5.16 we obtain the following corollary.

Corollary 5.18. (a) For nonnegative concave functions Φ_i defined on C with $a_i > 0$, $i = 1, \ldots, k$, the composite function

$$F_1(x) = \prod_{i=1}^{k} [\Phi_i(x)]^{a_i} \tag{5.24}$$

is semistrictly quasiconcave on C.

(b) For positive concave functions Φ_j on C with $b_j > 0$, $j = 1, \ldots, l$, the composite function

$$F_2(x) = 1 \bigg/ \prod_{j=1}^{l} [\Phi_j(x)]^{b_j} \tag{5.25}$$

is convex on C.

For $a_i = 1$ semistrict quasiconcavity of F_1 was also shown in Karamardian (1971).

From Lemma 5.14 and Proposition 2.16 we obtain (see also Berge, 1959) the following corollary.

Corollary 5.19. If in Corollary 5.18 $\sum_{i=1}^{k} a_i \leq 1$, then F_1 is concave on C.

We now derive criteria for the generalized concavity of products and ratios of two functions, specializing some of the above results. For convenience we introduce the following abbreviations that are used in the tables below:

s.qcv (s.qcx): semistrictly quasiconcave (semistrictly quasiconvex);
cv (cx): concave (convex).

For the product $\Phi_1(x) \cdot \Phi_2(x)$ we obtain the results in Table 5.1. They are derived from Theorem 5.15 and Corollary 5.18 by also using the identity

$$\Phi_1 \cdot \Phi_2 = \frac{\Phi_1}{[1/\Phi_2]} = \frac{\Phi_2}{[1/\Phi_1]}. \tag{5.26}$$

Example 5.1. Let $\Phi_1(x) = \sqrt{x}$ and $\Phi_2(x) = \exp(-x)$, $C = \{x \in R: x \geq 0\}$. From Table 5.1 we see that the product of these two functions,

$$F(x) = \sqrt{x} \exp(-x), \tag{5.27}$$

is semistrictly quasiconcave on C since $1/\phi_2(x) = \exp(x)$ is convex.

Table 5.1. The Product $\Phi_1 \cdot \Phi_2$

	Φ_2	
Φ_1	≥ 0, concave	> 0, convex
≥ 0 concave	s. qcv	s. qcv if $1/\Phi_2$ cx
> 0 convex	s. qcv if $1/\Phi_1$ cx	s. qcx if $1/\Phi_1$ or $1/\Phi_2$ cv

Table 5.2. The Product $l \cdot \Phi$

	Φ			
	concave		convex	
l	≥ 0	< 0	> 0	≤ 0
≥ 0	s. qcv	s. qcv if $1/\Phi$ cx	s. qcx if $1/\Phi$ cv	s. qcx
≤ 0	s. qcx	s. qcx if $1/\Phi$ cx	s. qcv if $1/\Phi$ cv	s. qcv

Let now Φ_1 be affine—that is, $\Phi_1(x) = l(x) = c^T x + \alpha$ for some $c \in R^n$ and $\alpha \in R$. Since affine functions are both concave and convex, we obtain from Theorem 5.15 for $l(x) \cdot \Phi(x)$ by again making use of the identity (5.26) the values shown in Table 5.2. If both functions are affine, we obtain the results shown in Table 5.3 for $l_1 \cdot l_2$ from Table 5.2.

For differentiable functions semistrict quasiconcavity (semistrict quasiconvexity) in Tables 5.1–5.3 can essentially be replaced by pseudoconcavity (pseudoconvexity); see Theorem 5.17. In Table 5.3, for example, we only have to exclude points $x \in R^n$, where $l_1(x) = l_2(x) = 0$. Then $F = l_1 \cdot l_2$ is pseudoconcave (pseudoconvex) wherever it is semistrictly quasiconcave (semistrictly quasiconvex).

Furthermore, we see from Theorem 5.17 that $F = \Phi_1 \cdot \Phi_2$ is strictly pseudoconcave if $\Phi_1 > 0$ is strictly concave and $\Phi_2 > 0$ is concave.

Let us now present some results for ratios of two functions. From Theorem 5.15, Table 5.4 can be obtained for $\Phi_1(x)/\Phi_2(x)$.

If both functions are either concave or convex, then Φ_1/Φ_2 is generally neither quasiconcave nor quasiconvex.

In case of an affine denominator, the sign restriction on the numerator can be dropped:

Table 5.3. The Product $l_1 \cdot l_2$

	l_2	
l_1	≥ 0	≤ 0
≥ 0	s. qcv	s. qcx
≤ 0	s. qcx	s. qcv

Table 5.4. The Ratio Φ_1/Φ_2

Φ_1	Φ_2: > 0, concave	Φ_2: > 0, convex
≥ 0 concave	—	s. qcv
≥ 0 convex	s. qcx	—

Proposition 5.20. If Φ is a concave (convex) function on C and l is affine and positive on C, then Φ/l is semistrictly quasiconcave (semistrictly quasiconvex) on C.

Proof. The function $f(y, z) = y/z$ is pseudoconcave and pseudoconvex on all of $E = \{(y, z) \in R^2 | z > 0\}$. This follows from the fact that for all $v \in R^2$ and all $(y, z) \in E$, $v^T \nabla f(y, z) = 0$ implies $v^T \nabla^2 f(y, z) v = 0$ (Crouzeix and Ferland, 1982). Then f is semistrictly quasiconcave and semistrictly quasiconvex on E, and therefore Φ/l has the stated properties, applying Proposition 5.4.

In view of Proposition 5.20 we now have for the ratio $\Phi(x)/l(x)$ the values shown in Table 5.5. Furthermore, in case of an affine numerator we obtain from Theorem 5.15 for $l(x)/\Phi(x)$ the values in Table 5.6.

For the ratio of two affine functions we see from Table 5.5 that $F = l_1/l_2$ is both semistrictly quasiconcave and semistrictly quasiconvex on the two half spaces $\{x \in R^n: l_2(x) > 0\}$ and $\{x \in R^n: l_2(x) < 0\}$.

For differentiable functions, semistrict quasiconcavity (semistrict quasiconvexity) in Tables 5.4–5.6 can essentially be replaced by pseudoconcavity (pseudoconvexity), as seen from Theorem 5.17. In particular, $F = l_1/l_2$ is pseudoconcave (pseudoconvex) wherever it is semistrictly quasiconcave (semistrictly quasiconvex).

Table 5.5. The Ratio Φ/l

Φ	l: > 0	l: < 0
concave	s. qcv	s. qcx
convex	s. qcx	s. qcv

Table 5.6. The Ratio l/Φ

	Φ			
	concave		convex	
l	>0	<0	>0	<0
≥ 0	s. qcx	s. qcx	s. qcv	s. qcv
≤ 0	s. qcv	s. qcv	s. qcx	s. qcx

Moreover, one can easily show that Φ_1/Φ_2 is strictly pseudoconcave if $\Phi_1 > 0$ is strictly concave and $\Phi_2 > 0$ is convex, or if $\Phi_1 > 0$ is concave and $\Phi_2 > 0$ is strictly convex.

References

AVRIEL, M. (1976), *Nonlinear Programming: Analysis and Methods*, Prentice-Hall, Englewood Cliffs, New Jersey.

BEREANU, B. (1972), Quasiconvexity, strict quasiconvexity and pseudoconvexity of composite objective functions, *Rev. Fr. Automat. Inf. Rech. Oper.* **6**, 15-26.

BERGE, C. (1959), *Espaces Topologiques Fonctions Multivoques*, Dunod, Paris.

CROUZEIX, J. P., and FERLAND, J. A. (1982), Criteria for quasiconvexity and pseudoconvexity: Relationships and comparisons, *Math. Programming* **23**, 193-205.

KARAMARDIAN, S. (1971), A class of nonlinear programming problems, *Cah. Cent. d'Etud. Rech. Oper.* **13**, 41-48.

MANGASARIAN, O. L. (1970), Convexity, pseudo-convexity and quasi-convexity of composite functions, *Cah. Cent. d'Etud. Rech. Oper.* **12**, 114-122.

SCHAIBLE, S. (1971), Beiträge zur quasikonvexen Programmierung, Doctoral Dissertation, Universität Köln, West Germany.

SCHAIBLE, S. (1972), Quasiconvex optimization in general real linear spaces, *Z. Oper. Res.* **16**, 205-213.

SCHAIBLE, S. (1981a), Quasiconvex, pseudoconvex and strictly pseudoconvex quadratic functions, *J. Opt. Theory Appl.* **35**, 303-338.

SCHAIBLE, S. (1981b), Generalized convexity of quadratic functions, in *Generalized Concavity in Optimization and Economics*, Edited by S. Schaible and W. T. Ziemba, Academic Press, New York, 183-197.

6

Special Functional Forms II: Quadratic Functions

In this chapter we characterize quadratic functions that are generalized concave. Let

$$Q(x) = \tfrac{1}{2}x^T Ax + b^T x, \tag{6.1}$$

where A is a real symmetric $n \times n$ matrix and $b \in R^n$. We want to derive conditions on A and b under which functions of the form $Q(x)$ are quasiconcave, pseudoconcave, etc., on a given convex set C of R^n. We assume that C has a nonempty interior, i.e., C is a *solid* set of R^n. In the first part of this chapter C is an arbitrary (solid) convex set. In the latter part we will discuss the special case where C is the positive or nonnegative orthant of R^n.

For quadratic functions we do not have to distinguish between semistrict quasiconcavity and quasiconcavity (Martos, 1967). To see this, note that a quadratic function that is constant on a line segment $[x^1, x^2]$ is also constant on the entire line through x^1 and x^2. From this, it follows that (3.94) and (3.96) are equivalent. Therefore, a quadratic function that is quasiconcave is also semistrictly quasiconcave.

We first focus our attention on quasiconcave, pseudoconcave, and strictly pseudoconcave functions Q. Later, we shall discuss strictly quasiconcave and strongly pseudoconcave quadratic functions. We shall show that these latter two classes of generalized concave functions coincide with the class of strictly pseudoconcave functions in the case of quadratic functions, at least on open convex sets C.

In our first approach, we shall characterize generalized concave quadratic functions Q in terms of eigenvalues of A. Other criteria will then be derived from these criteria. Such an approach was suggested in Schaible (1981a, 1981b).

6.1. Main Characterization of Quasiconcave, Pseudoconcave, and Strictly Pseudoconcave Quadratic Functions

In order to obtain a first set of criteria we proceed in the same way as in Section 5.2, i.e., we employ generalized concavity properties of composite functions. The criteria below were obtained in this way in Schaible (1971, 1973a).

The fundamental observation underlying our analysis is that every quadratic function $Q(x) = \frac{1}{2}x^TAx + b^Tx$ can be reduced to a certain normal form given by (6.11) below. To see this, we begin with letting $\lambda_1, \ldots, \lambda_n$ denote the eigenvalues of A and $t^1, \ldots, t^n$ an orthonormal basis of eigenvectors associated with $\lambda_1, \ldots, \lambda_n$. We assume that $\lambda_i > 0$ for $i = 1, \ldots, p$, $\lambda_i < 0$ for $i = p + 1, \ldots, r$ and $\lambda_i = 0$ for $i = r + 1, \ldots, n$. Let

$$\bar{P} = (t^1, \ldots, t^n) \tag{6.2}$$

and

$$\Lambda = \operatorname{diag}(\lambda_1, \ldots, \lambda_n). \tag{6.3}$$

Here diag $(\lambda_1, \ldots, \lambda_n)$ denotes a diagonal matrix with entries $\lambda_1, \ldots, \lambda_n$ along its main diagonal. Since $t^1, \ldots, t^n$ is an orthonormal basis of eigenvectors of A we have (see Noble, 1969)

$$\bar{P}^TA\bar{P} = \Lambda. \tag{6.4}$$

Now we define $\bar{y}$ by

$$\bar{y} = \bar{P}^Tx \quad \text{or} \quad x = \bar{P}\bar{y}. \tag{6.5}$$

Then $Q(x)$ may be rewritten as

$$Q(x) = \frac{1}{2}\sum_{i=1}^{r} \lambda_i(\bar{y}_i)^2 + \sum_{i=1}^{n} (b^Tt^i)\bar{y}_i. \tag{6.6}$$

Applying the translation

$$\bar{y} = \bar{\bar{y}} + \bar{v}, \qquad \text{where } \bar{v} = (-b^T t^1/\lambda_1, \ldots, -b^T t^r/\lambda_r, 0, \ldots, 0)^T, \tag{6.7}$$

we obtain

$$Q(x) = \tfrac{1}{2} \sum_{i=1}^{r} \lambda_i (\bar{\bar{y}}_i)^2 + \sum_{i=r+1}^{n} (b^T t^i) \bar{\bar{y}}_i + \delta \tag{6.8}$$

where

$$\delta = -\tfrac{1}{2} \sum_{i=1}^{r} (b^T t^i)^2 / \lambda_i. \tag{6.9}$$

Finally, we rescale the variables by applying the transformation

$$\bar{\bar{y}} = \bar{\bar{P}} y, \tag{6.10}$$

where $\bar{\bar{P}} = \operatorname{diag}(\lambda_1^{-1/2}, \ldots, \lambda_p^{-1/2}, (-\lambda_{p+1})^{-1/2}, \ldots, (-\lambda_r)^{-1/2}, 1, \ldots, 1)$,

which yields the normal form

$$Q(x) = \tfrac{1}{2} \sum_{i=1}^{p} (y_i)^2 - \tfrac{1}{2} \sum_{i=p+1}^{r} (y_i)^2 + \sum_{i=r+1}^{n} (b^T t^i) y_i + \delta. \tag{6.11}$$

Summarizing, we can say that the normal form (6.11) was obtained by applying the affine transformation

$$x = Py + v, \qquad \text{where } P = \bar{P}\bar{\bar{P}} \quad \text{and} \quad v = \bar{P}\bar{v} \tag{6.12}$$

in which P is nonsingular.

We now distinguish between the following two cases:

$$Q(x) = Q_1(y) = \tfrac{1}{2} \sum_{i=1}^{p} (y_i)^2 - \tfrac{1}{2} \sum_{i=p+1}^{r} (y_i)^2 + \delta, \tag{6.13}$$

and the case

$$Q(x) = Q_2(y) = \tfrac{1}{2} \sum_{i=1}^{p} (y_i)^2 - \tfrac{1}{2} \sum_{i=p+1}^{r} (y_i)^2 + \sum_{i=r+1}^{n} (b^T t^i) y_i + \delta, \tag{6.14}$$

where not all terms $b^T t^i$ vanish. As we see from (6.14), $Q(x)$ reduces to the normal form $Q_1(y)$ if and only if b is orthogonal to all eigenvectors of A corresponding to zero eigenvalues which is the case, using (6.4), if and only if

$$\text{rank}\,(A, b) = \text{rank}\,(\bar{P}^T A, \bar{P}^T b) = \text{rank}\,(\Lambda \bar{P}^T, \bar{P}^T b) = r = \text{rank}\,(A). \qquad (6.15)$$

We have shown that $Q(x)$ can be considered as a composite function involving one of the normal forms $Q_i(y)$ and an affine transformation of variables. From Propositions 5.5 and 5.10, we see that $Q(x)$ is quasiconcave (pseudoconcave, strictly pseudoconcave) on C if and only if its normal form $Q_i(y)$ is quasiconcave (pseudoconcave, strictly pseudoconcave) on $D = \{y \in R^n: x = Py + v \in C\}$. Hence, it suffices to characterize normal forms that are generalized concave. The domain D can again be assumed to be solid and convex since it is the affine image of such a set.

Suppose $Q_i(y)$ is not concave on the solid convex set D. This is the case if and only if $p \geq 1$ since $Q_i(y)$ is concave on a solid subset if and only if it is concave on R^n (Cottle, 1967), and for those functions $p = 0$. So we restrict ourselves to the case $p \geq 1$ in the sequel. Since the upper-level sets of a quasiconcave function are convex (Definition 3.1), the following quadratic functions are not quasiconcave on D:

$$Q_1 \text{ with } p > 1 \quad \text{and } Q_2. \qquad (6.16)$$

To see this, consider Q_1 as a function of y_1 and y_2 and Q_2 as a function of y_1 and y_{i^0}, where i^0 is an index for which $b^T t^{i^0} \neq 0$. Such restrictions to two variables do not have convex upper level sets in D. Thus, Q_1, Q_2 in (6.16) are not quasiconcave.

Hence the only nonconcave normal form that may be quasiconcave on a solid convex set D is

$$Q_1(y) = G(y) + \delta, \qquad \text{where } G(y) = \tfrac{1}{2}(y_1)^2 - \tfrac{1}{2}\sum_{i=2}^{r} (y_i)^2. \qquad (6.17)$$

In order to investigate this function we first prove the following lemma.

Lemma 6.1. Let F be defined by

$$F(y) = \left[(y_1)^2 - \sum_{i=2}^{r} (y_i)^2 \right]^{1/2}. \qquad (6.18)$$

Then F is concave on the convex cones

$$D_1 = \left\{ y \in R^n \colon (y_1)^2 - \sum_{i=2}^{r} (y_i)^2 \geq 0,\ y_1 \geq 0 \right\} \tag{6.19}$$

and

$$D_2 = \left\{ y \in R^n \colon (y_1)^2 - \sum_{i=2}^{r} (y_i)^2 \geq 0,\ y_1 \leq 0 \right\}. \tag{6.20}$$

Proof. We shall prove the assertion by induction. For $k = 1, \ldots, r$ we define

$$F_k(y) = \left[(y_1)^2 - \sum_{i=2}^{k} (y_i)^2 \right]^{1/2} \tag{6.21}$$

$$D_{1,k} = \left\{ y \in R^n \colon (y_1)^2 - \sum_{i=2}^{k} (y_i)^2 \geq 0,\ y_1 \geq 0 \right\} \tag{6.22}$$

and

$$D_{2,k} = \left\{ y \in R^n \colon (y_1)^2 - \sum_{i=2}^{k} (y_i)^2 \geq 0,\ y_1 \leq 0 \right\}. \tag{6.23}$$

We now show that F_k is concave on $D_{1,k}$ and on $D_{2,k}$ for $k = 1, \ldots, r$. Certainly this is true for $k = 1$. Now suppose F_k is concave on $D_{1,k}$, $D_{2,k}$ for some $k < r$. The following identity holds for $y \in D_{1,k+1} \subset D_{1,k}$ and $y \in D_{2,k+1} \subset D_{2,k}$:

$$[F_{k+1}(y)]^2 = [F_k(y) - y_{k+1}][F_k(y) + y_{k+1}]. \tag{6.24}$$

Hence, since F_k is nonnegative,

$$D_{1,k+1} = \{ y \in D_{1,k} \colon F_k(y) - y_{k+1} \geq 0, \quad F_k(y) + y_{k+1} \geq 0, \quad y_1 \geq 0 \}. \tag{6.25}$$

Since F_k is concave on $D_{1,k}$ as assumed, $D_{1,k+1}$ is a convex set in view of (6.25), and so is $D_{2,k+1}$. Furthermore, F_{k+1} is concave on these sets, since F_{k+1} is the square root of the product of two nonnegative concave functions according to (6.24) and (6.25) (see Corollary 5.19). Hence $F(y) = F_r(y)$ is concave on the closed convex cones $D_1 = D_{1,r}$, $D_2 = D_{2,r}$. □

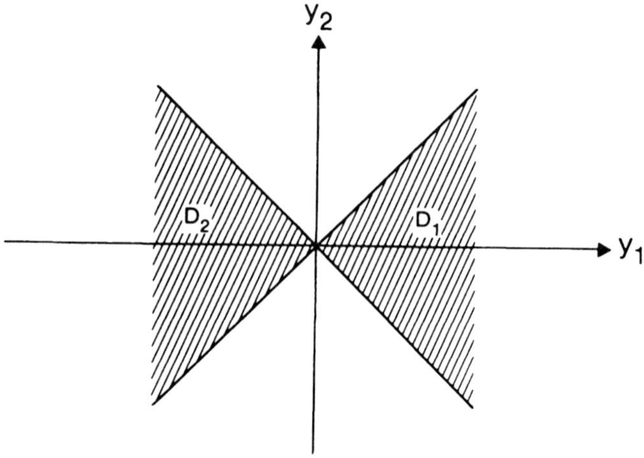

Figure 6.1. The convex cones D_1, D_2 for $r = n = 2$.

Illustrations of the convex cones D_1, D_2 are given in Figures 6.1 and 6.2. For $r > 2$ these cones are no longer polyhedral. Such cones were studied by Greub (1967). From Lemma 6.1 and Proposition 5.1 we conclude that

$$G(y) = \tfrac{1}{2}(y_1)^2 - \tfrac{1}{2} \sum_{i=2}^{r} (y_i)^2 = \tfrac{1}{2}[F(y)]^2 \tag{6.26}$$

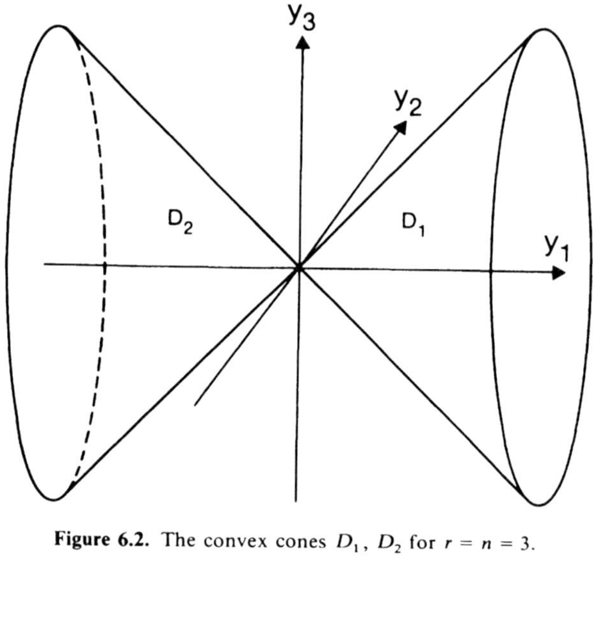

Figure 6.2. The convex cones D_1, D_2 for $r = n = 3$.

is quasiconcave on the closed convex cones

$$D_1 = \{y: G(y) \geq 0, y_1 \geq 0\} \tag{6.27}$$

and

$$D_2 = \{y: G(y) \geq 0, y_1 \leq 0\}. \tag{6.28}$$

Similarly, we see from Proposition 5.6 and Lemma 6.1 that G is pseudoconcave on

$$D_1^0 = \{y: G(y) \geq 0, y_1 > 0\} \tag{6.29}$$

and

$$D_2^0 = \{y: G(y) \geq 0, y_1 < 0\}. \tag{6.30}$$

It can also be proved that D_1, D_2 and D_1^0, D_2^0 are *maximal* domains of quasiconcavity and pseudoconcavity, respectively. For this, note that the upper-level sets of G in $D\backslash(D_1 \cup D_2)$ are not convex for any solid convex set $D \subset R^n$, where $D\backslash(D_1 \cup D_2) \neq \varnothing$, which can be seen by restricting G to the variables y_1, y_2. Likewise, D_1^0, D_2^0 are maximal domains of pseudoconcavity.

In the sequel we exclude the concave case by using the term *merely quasiconcave* (*merely pseudoconcave*) function for a nonconcave quasiconcave (pseudoconcave) function. Summarizing the results above, we have the following theorem.

Theorem 6.2. A quadratic function $Q(x) = \frac{1}{2}x^TAx + b^Tx$ is merely quasiconcave (merely pseudoconcave) on a solid convex set $C \subset R^n$ if and only if

(i) $\operatorname{rank}(A, b) = \operatorname{rank} A$, (6.31)

(ii) A has exactly one positive eigenvalue, (6.32)

(iii) $C \subset C_1$ or $C \subset C_2$ ($C \subset C_1^0$ or $C \subset C_2^0$) (6.33)

where

$$C_k = \{x \in R^n: x = Py + v, y \in D_k\} \tag{6.34}$$

and

$$C_k^0 = \{x \in R^n: x = Py + v, y \in D_k^0\}, \qquad k = 1, 2. \tag{6.35}$$

The maximal domains of quasiconcavity (pseudoconcavity) are the images of convex cones D_k (D_k^0) under the affine transformation $x = Py + v$. Such a transformation was used to reduce $Q(x)$ to its normal form $Q_1(y) = G(y) + \delta$. In view of (6.27) and (6.28) then

$$C_1 = \{x \in R^n: Q(x) \geq \delta, y_1 \geq 0\}, \qquad C_2 = \{x \in R^n: Q(x) \geq \delta, y_1 \leq 0\} \tag{6.36}$$

and

$$C_1^0 = \{x \in R^n: Q(x) \geq \delta, y_1 > 0\}, \qquad C_2^0 = \{x \in R^n: Q(x) \geq \delta, y_1 < 0\}. \tag{6.37}$$

In deriving Theorem 6.2 we observed the following interesting property of quasiconcave quadratic functions:

Theorem 6.3. If Q is merely quasiconcave on a solid convex set $C \subset R^n$ then there exists a $\delta \in R$ such that

$$Q(x) \geq \delta \text{ for all } x \in C, \tag{6.38}$$

and

$$K(x) = [Q(x) - \delta]^{1/2} \tag{6.39}$$

is concave on C.

Hence, merely quasiconcave quadratic functions $Q(x)$ are bounded from below on C in contrast to concave quadratic functions which are unbounded on R^n, the maximal domain of concavity. Furthermore, quasiconcave quadratic functions are concavifiable by scaling. (For further elaboration on concave transformable functions see Chapter 8.) It is this property that will enable us to derive many of the other criteria below.

The maximal domain of quasiconcavity in (6.36) can be fully characterized in terms of eigenvalues and eigenvectors of A together with b. To see this, note that $y_1 = \sqrt{\lambda_1}[(t^1)^T x + b^T t^1/\lambda_1]$ in view of (6.12), taking into consideration also (6.2), (6.7), and (6.10). Hence

$$C_1 = \{x \in R^n: Q(x) \geq \delta, (t^1)^T x \geq -b^T t^1/\lambda_1\} \tag{6.40}$$

and

$$C_2 = \{x \in R^n \colon Q(x) \geq \delta, (t^1)^T x \leq -b^T t^1/\lambda_1\}, \tag{6.41}$$

where δ is given by (6.9). C_1^0 and C_2^0 can be similarly described, with the second inequality on the right-hand sides of (6.40) and (6.41) to be replaced by a strict inequality.

We now derive another representation of C_1 and C_2. It involves a stationary point of Q—that is, a point $s \in R^n$ such that

$$\nabla Q(s) = As + b = 0. \tag{6.42}$$

For a merely quasiconcave function, condition (i) in Theorem 6.2 guarantees the existence of a stationary point s. In view of (6.42) the variable transformation

$$x = z + s \tag{6.43}$$

yields

$$Q(x) = \tfrac{1}{2} z^T A z + Q(s). \tag{6.44}$$

Then the affine transformation

$$z = ((\lambda_1)^{-1/2} t^1, (-\lambda_2)^{-1/2} t^2, \ldots, (-\lambda_r)^{-1/2} t^r, t^{r+1}, \ldots, t^n) y \tag{6.45}$$

reduces Q to its normal form $Q(x) = G(y) + \delta$, where

$$\delta = Q(s). \tag{6.46}$$

Furthermore, from (6.43) and (6.45) we see that $y_1 = \sqrt{\lambda_1}(t^1)^T(x - s)$. Thus we have

$$C_1 = \{x \in R^n \colon Q(x) \geq Q(s), (t^1)^T x \geq (t^1)^T s\} \tag{6.47}$$

and

$$C_2 = \{x \in R^n \colon Q(x) \geq Q(s), (t^1)^T x \leq (t^1)^T s\}. \tag{6.48}$$

This shows that

$$C_1 = s + \{z \in R^n \colon z^T A z \geq 0, (t^1)^T z \geq 0\} \tag{6.49}$$

and

$$C_2 = s + \{z \in R^n: z^T A z \geq 0, (t^1)^T z \leq 0\}. \tag{6.50}$$

Note that if A is a singular matrix, then a stationary point s as given by (6.42) is not unique. However, the characterization of C_1 and C_2 in (6.47)-(6.49) is independent of the particular stationary point used. To see this, let s^1, s^2 be two distinct stationary points—that is,

$$As^i = -b \qquad \text{for } i = 1, 2. \tag{6.51}$$

Multiplying this equation by $\frac{1}{2}s^i$ and adding $b^T s^i$ to both sides we obtain

$$Q(s^i) = \tfrac{1}{2}(s^i)^T A s^i + b^T s^i = \tfrac{1}{2} b^T s^i. \tag{6.52}$$

Then

$$Q(s^1) = \tfrac{1}{2} b^T s^1 = \tfrac{1}{2}(-As^2)^T s^1 \tag{6.53}$$

$$= \tfrac{1}{2}(-As^1)^T s^2 = \tfrac{1}{2} b^T s^2 = Q(s^2), \tag{6.54}$$

using (6.42) for s^1 and s^2. Hence $Q(s)$ in (6.47)-(6.48) does not depend on the particular stationary point s. The same is true for $(t^1)^T s$ in (6.47)-(6.48) since for the eigenvector t^1 associated with the eigenvalue λ_1 we have

$$At^1 = \lambda_1 t^1. \tag{6.55}$$

Then

$$(t^1)^T s_1 - (t^1)^T s_2 = \lambda_1^{-1}(At^1)^T(s^1 - s^2) = (\lambda_1)^{-1}(t^1)^T[A(s^1 - s^2)] = 0 \tag{6.56}$$

in view of (6.42). This completes the proof that the representation of C_1, C_2 in (6.47)-(6.48) is independent of the stationary point s used.

The characterization of maximal domains of quasiconcavity (and pseudoconcavity) in (6.49)-(6.50) was also obtained by Ferland (1971, 1972) via an approach different from the one used here. The above discussion shows the equivalence of Ferland's results to those obtained by Schaible (1971, 1973a). The latter approach was used here.

Crouzeix (1980) showed that Ferland's characterization of quasiconcave functions can be used to prove concavifiability of these functions. Using Schaible's approach, we obtained the concavifiability property (Lemma 6.1, Theorem 6.3) as a side product when proving the criteria for quasiconcave and pseudoconcave functions of Theorem 6.2.

We point out that δ can be calculated once a stationary point s of Q is known. In addition to (6.46) we have

$$\delta = \tfrac{1}{2}b^T s, \tag{6.57}$$

as can be seen from (6.52). If A is nonsingular, then $s = -A^{-1}b$ [see (6.42)]. Hence we have

$$\delta = -\tfrac{1}{2}b^T A^{-1} b. \tag{6.58}$$

Example 6.1. Let us consider $Q(x) = -5(x_1)^2 + 8x_1x_2 + (x_2)^2 - 6x_1 - 12x_2$ on R^2. Hence $A = \begin{pmatrix} -10 & 8 \\ 8 & 2 \end{pmatrix}$ and $b = \begin{pmatrix} -6 \\ -12 \end{pmatrix}$. We want to determine the domains C_1 and C_2. The eigenvalues of A are $\lambda_1 = 6$, $\lambda_2 = -14$ with the corresponding eigenvectors $t^1 = (1/\sqrt{5}, 2/\sqrt{5})^T$ and $t^2 = (-2/\sqrt{5}, 1/\sqrt{5})^T$. According to (6.12) the following transformation reduces Q to its normal form $Q_1(y) = \frac{1}{2}(y_1)^2 - \frac{1}{2}(y_2)^2 + \delta$:

$$x = \begin{pmatrix} 1/\sqrt{30} & -2/\sqrt{70} \\ 2/\sqrt{30} & 1/\sqrt{70} \end{pmatrix} y + \begin{pmatrix} 1 \\ 2 \end{pmatrix}. \tag{6.59}$$

We also use the inverse transformation

$$y = \begin{pmatrix} (6/5)^{1/2} & 2(6/5)^{1/2} \\ -2(14/5)^{1/2} & (14/5)^{1/2} \end{pmatrix} x - \begin{pmatrix} 5(6/5)^{1/2} \\ 0 \end{pmatrix}. \tag{6.60}$$

Furthermore, we obtain from (6.9) that $\delta = -15$. Hence in view of (6.40)-(6.41), we find

$$C_1 = \{x \in R^2\colon Q(x) \geq -15, x_1 + 2x_2 \geq 5\} \tag{6.61}$$

and

$$C_2 = \{x \in R^2\colon Q(x) \geq -15, x_1 + 2x_2 \leq 5\}. \tag{6.62}$$

As we saw before, C_1 and C_2 can also be determined using a stationary point s of $Q(x)$. Since A is nonsingular, s is unique. We find $s = -A^{-1}b = (1, 2)^T$. By (6.58) we have again $\delta = -15$. From (6.47)-(6.48) we obtain C_1 and C_2 as above.

In Figure 6.3 the convex cones C_1 and C_2 are illustrated. On the boundary of C_1 and C_2 $Q(x) = \delta$, and thus $\frac{1}{2}(y_1)^2 - \frac{1}{2}(y_2)^2 = 0$. The cones are therefore bounded by the lines $y_2 = y_1$ and $y_2 = -y_1$, or equivalently

$$(\sqrt{6} + 2\sqrt{14})x_1 + (2\sqrt{6} - \sqrt{14})x_2 = 5\sqrt{6} \tag{6.63}$$

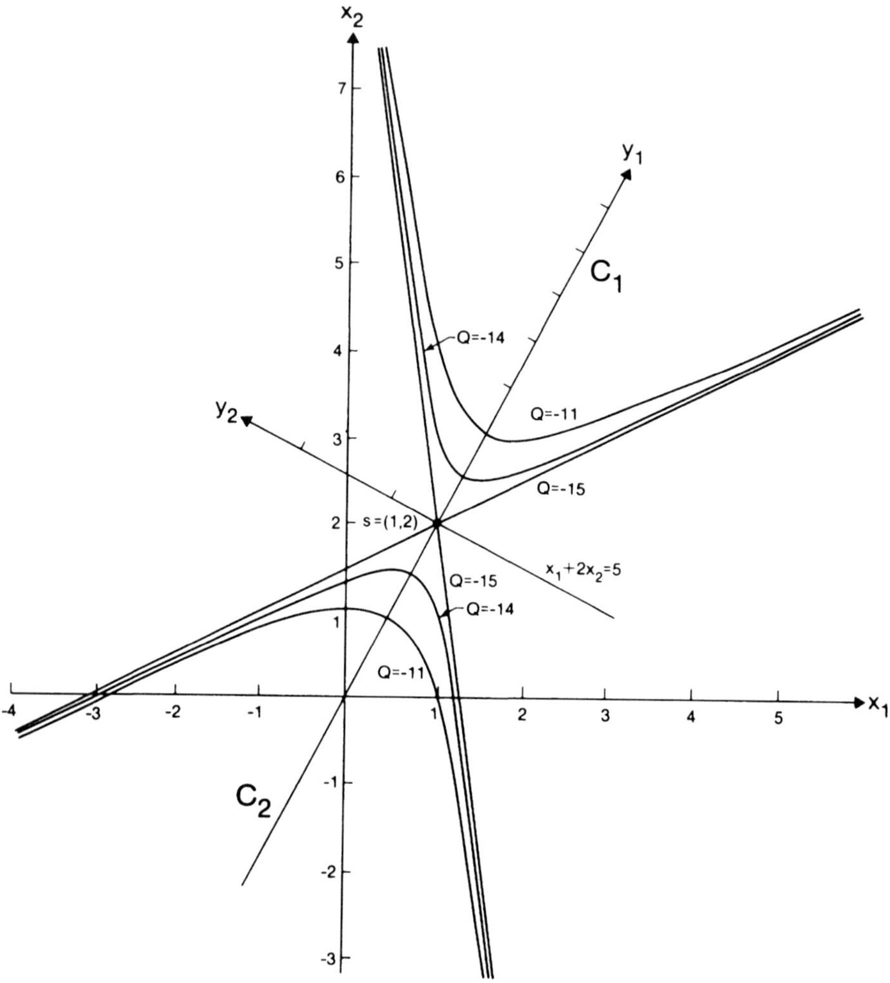

Figure 6.3. The quadratic function of Example 6.1 and its maximal domains of quasiconcavity.

and

$$(\sqrt{6} - 2\sqrt{14})x_1 + (2\sqrt{6} + \sqrt{14})x_2 = 5\sqrt{6}. \tag{6.64}$$

The stationary point $x = s = (1, 2)^T$ is the only point the cones C_1 and C_2 have in common.

In Theorem 6.2, quasiconcave and pseudoconcave quadratic functions were characterized. As we see from this result, the following property holds for quadratic functions:

Corollary 6.4. Let C be an open convex set in R^n and let $\bar{C}$ denote the closure of C. Then $Q(x) = \frac{1}{2}x^T Ax + b^T x$ is quasiconcave on $\bar{C}$ if and only if $Q(x)$ is pseudoconcave on C.

Hence any characterization of a pseudoconcave function Q on an open convex set can be viewed as a criterion for a quasiconcave function Q on a solid set, and vice versa.

The next theorem presents a criterion for strictly pseudoconcave functions Q that have not been characterized so far. As known, a concave quadratic function is strictly concave if and only if the matrix A is nonsingular. We now derive an analogous result for pseudoconcave quadratic functions.

Theorem 6.5 (Schaible, 1977). Let Q be a nonconcave quadratic function, on an open set C in R^n. Then Q *is* strictly pseudoconcave on C if and only if Q is pseudoconcave on C and A is nonsingular.

Proof. It suffices to show the result for the normal form $G(y) = \frac{1}{2}(y_1)^2 - \frac{1}{2}\sum_{i=2}^{r}(y_i)^2$ on an open convex set $D \in R^n$. For $r < n$ G is not strictly pseudoconcave on D, since G is constant with regard to y_{r+1}. (See Definition 3.13.) On the other hand, if $r = n$, then any restriction of G to a line segment in D satisfies the inequality for strictly pseudoconcave functions in Definition 3.13. Hence G is strictly pseudoconcave if and only if $r = n$, i.e., if A is nonsingular. □

In Theorem 6.2 we gave a complete characterization of quasiconcave quadratic functions. Since Q is quasiconvex if and only if $-Q$ is quasiconcave, Theorem 6.2 can be used to derive a similar characterization for quasiconvex quadratic functions. In Table 6.1 we summarize the criteria we have found for quasiconcave and quasiconvex quadratic functions. As before, C denotes a solid convex set in R^n. In addition, the following notation is used:

$$A_1 = \{x \in R^n: c^T x + \gamma \geq 0\} \quad \text{and} \quad A_2 = \{x \in R^n: c^T x + \gamma \leq 0\}, \tag{6.65}$$

for some $c \in R^n$ and $\gamma \in R$;

$$B_1 = \{x \in R^n: c^T x + \alpha \geq 0, d^T x + \beta \geq 0\},$$

$$B_2 = \{x \in R^n: c^T x + \alpha \leq 0, d^T x + \beta \leq 0\}, \tag{6.66}$$

$$B_3 = \{x \in R^n: c^T x + \alpha \geq 0, d^T x + \beta \leq 0\},$$

$$B_4 = \{x \in R^n: c^T x + \alpha \leq 0, d^T x + \beta \geq 0\} \tag{6.67}$$

Table 6.1. Quasiconcavity and Quasiconvexity of $Q(x)$

	$r-p=0$		$r-p=1$: rank (A, b) = rank A		$r-p=1$: rank $(A, b) >$ rank A		$r-p>1$	
$p=0$	cv	cx	cv	qcx if $C \subset A_1$ or $C \subset A_2$	cv	not qcx	cv	not qcx
	$Q(x) = b^T x$		$Q(x) = \mu(c^T x + \gamma)^2 + \sigma$, $\mu < 0$					
$p=1$ rank (A, b) = rank A	qcv if $C \subset A_1$ or $C \subset A_2$	cx	qcv if $C \subset B_1$ or $C \subset B_2$	qcx if $C \subset B_3$ or $C \subset B_4$			qcv if $C \subset C_1$ or $C \subset C_2$; $[Q(x) - \delta]^{1/2}$ is cv	not qcx
	$Q(x) = \mu(c^T x + \gamma)^2 + \sigma$, $\mu > 0$		$Q(x) = (c^T x + \alpha)(d^T x + \beta) + \sigma$ c, d linearly independent					
rank (A, b) $>$ rank A	not qcv	cx			not qcv	not qcx	not qcv	not qcx
$p>1$	not qcv	cx	not qcv	qcx if $C \subset C'_1$ or $C \subset C'_2$; $-[\delta - Q(x)]^{1/2}$ is cx	not qcv	not qcx	not qcv	not qcx

for c, $d \in R^n$ and $\alpha, \beta \in R$;

$$C'_1 = \{x \in R^n : x = Py + v, Q(x) \le \delta, y_1 \ge 0\},$$

$$C'_2 = \{x \in R^n : x = Py + v, Q(x) \le \delta, y_1 \le 0\}, \tag{6.68}$$

where P and v were derived in (6.12) and δ in (6.9).

Note that representations of C'_1, C'_2 similar to those of C_1, C_2 in (6.40) and (6.41), (6.47)-(6.50) can be derived for the quasiconvex case.

The sets $A_1, A_2, B_1, B_2, B_3, B_4, C_1, C_2, C'_1, C'_2$ are maximal domains of quasiconcavity or quasiconvexity. See Table 6.1 for details.

We also use in Table 6.1 the following abbreviations:

cv:	concave on C;
cx:	convex on C;
qcv:	merely quasiconcave on C;
qcx:	merely quasiconvex on C;
not qcv:	not quasiconcave on C;
not qcx:	not quasiconvex on C.

To illustrate the use of Table 6.1, take, for example, the case of $p = 0$, $r - p = 1$ and rank $(A, b) =$ rank A. Then the quadratic function Q is of the form $Q(x) = \mu(c^T x + \gamma)^2 + \sigma$ where $\mu < 0$. Q is concave on any solid convex set C, and it is merely quasiconvex on C if $C \subset A_1$ or $C \subset A_2$. In this case either $c^T x + \gamma \ge 0$ or $c^T x + \gamma \le 0$ for all $x \in C$. If C has points in both half-spaces then Q is not quasiconvex on C.

To take another example, let $p > 1$, $r - p = 1$ and rank$(A, b) =$ rank A. In this case Q is not quasiconcave on any solid convex set C. It is quasiconvex on C if $C \subset C'_1$ or $C \subset C'_2$, in which case Q is convex transformable, i.e., $-[\delta - Q(x)]^{1/2}$ is convex on C.

We see from Table 6.1 that the only (nonlinear) quadratic functions that decompose R^n completely into sets where they are either quasiconcave or quasiconvex are products of affine functions (apart from an additive constant), namely,

$$Q(x) = \mu(c^T x + \gamma)^2 + \sigma \tag{6.69}$$

and

$$Q(x) = (c^T x + \alpha)(d^T x + \beta) + \sigma, \tag{6.70}$$

where c and d are linearly independent. Generalized concavity of these functions was already studied in Section 5.2.

Furthermore, we observe in Table 6.1 that the only quadratic functions that are both quasiconcave and quasiconvex on the whole of R^n are affine functions.

Before we move on to other characterizations of generalized concave quadratic functions we briefly discuss an application of the boundedness property in (6.38). Consider the quadratic program

$$\inf\{Q(x): x \in S\}, \tag{6.71}$$

where Q is merely quasiconcave on the closed convex polyhedron $S \subset R^n$. From (6.38) we see that Q is bounded from below on S, hence $\inf\{Q(x): x \in S\} > -\infty$. Furthermore, the infimum is attained since Q is quadratic and S is a closed convex polyhedron. Therefore, the quadratic program above has an optimal solution. We also know that since Q is quasiconcave, there exists an optimal solution that is an extreme point of S provided S has one (Martos, 1975).

In the sequel, we will derive other characterizations of generalized concave quadratic functions. In view of Corollary 6.4, it suffices to obtain criteria for functions Q that are pseudoconcave on an open convex set since from these characterizations we can easily find criteria for quasiconcave functions Q on solid convex sets. Hence in the following we focus our attention on characterizing pseudoconcavity on open convex sets. In addition, we will derive criteria for strictly pseudoconcave quadratic functions.

6.2. Characterizations in Terms of an Augmented Hessian

The criteria in this section involve an augmented Hessian of Q.

Theorem 6.6 (Schaible, 1977). A quadratic function Q is pseudoconcave on an open convex set $C \subset R^n$ if and only if there exists a function $r: C \to R$ such that the augmented Hessian

$$H(x; r) = \nabla^2 Q(x) - r(x)\nabla Q(x)\nabla Q(x)^T \tag{6.72}$$

is negative semidefinite for all $x \in C$. If Q is not concave on C then for every $x \in C$ the smallest possible number r such that $H(x; r)$ is negative semidefinite on C is given by

$$r_0(x) = [2(Q(x) - \delta)]^{-1}. \tag{6.73}$$

Proof. If Q is concave on C, then $H(x; r)$ is negative semidefinite for $r(x) = 0$. If Q is merely pseudoconcave, then by Theorem 6.3 $[Q(x) - \delta]^{1/2}$ is concave on C. Hence the Hessian of $[Q(x) - \delta]^{1/2}$ is negative semidefinite (see Chapter 2). This implies that $H(x; r)$ is negative semidefinite for $r(x) = \{2[Q(x) - \delta]\}^{-1}$.

Conversely, suppose that $H(x; r)$ is negative semidefinite for some $r: C \to R$. We can assume that $r > 0$ since $H(x; r')$ is negative semidefinite for any $r' > r$. Let $x = Py + v$ be an affine transformation that reduces Q to its normal form, i.e., $Q(x) = Q_1(y)$ or $Q(x) = Q_2(y)$. Obviously, $K(y; s) = \nabla^2 Q_i(y) - s(y)\nabla Q_i(y)\nabla Q_i(y)^T$, $i = 1, 2$, where $s(y) = r(Py + v)$, is negative semidefinite. Let us denote the entries of $K(y; s)$ by K_{ij}. Suppose $p \geq 2$. Then

$$\det\begin{pmatrix} K_{11} & K_{12} \\ K_{21} & K_{22} \end{pmatrix} = \begin{pmatrix} 1 - s(y_1)^2 & -sy_1y_2 \\ -sy_1y_2 & 1 - s(y_2)^2 \end{pmatrix} \tag{6.74}$$

$$= 1 - s((y_1)^2 + (y_2)^2). \tag{6.75}$$

On the other hand, $K_{ii} = 1 - s(y_i)^2 \leq 0$ which implies $s(y_i)^2 \neq 0$, $i = 1, 2$. Hence $s(y_i)^2 > 0$, since $s > 0$, and therefore $1 - s((y_1)^2 + (y_2)^2) < 1 - s(y_1)^2 \leq 0$. Then

$$\det\begin{pmatrix} K_{11} & K_{12} \\ K_{21} & K_{22} \end{pmatrix} < 0, \tag{6.76}$$

which contradicts the negative semidefiniteness of $K(y; s)$. Therefore $p \leq 1$.

If $p = 0$, then Q is concave and thus pseudoconcave. Let now $p = 1$. Assume $Q(x) = Q_2(y)$. Let $R = b^T t^\rho \neq 0$ in (6.14) for $\rho > r$. Then

$$0 \leq \det\begin{pmatrix} K_{11} & K_{1,\rho} \\ K_{\rho,1} & K_{\rho,\rho} \end{pmatrix} = \det\begin{pmatrix} 1 - s(y_1)^2 & -sRy_1 \\ -sRy_1 & -s(R)^2 \end{pmatrix} = -s(R)^2 < 0 \tag{6.77}$$

since $s > 0$ and $R \neq 0$. Therefore, $K(y; s)$ cannot be negative semidefinite. Thus $Q(x) = Q_1(y) = \frac{1}{2}(y_1)^2 - \frac{1}{2}\sum_{j=2}^r (y_j)^2 + \delta$. In view of Theorem 6.2, Q is pseudoconcave on C, if the image D of C under the transformation $x = Py + v$ is contained in $\{y \in R^n: (y_1)^2 - \sum_{j=2}^r (y_j)^2 \geq 0\}$. By expansion we find

$$\det\begin{pmatrix} K_{11} & \cdots & K_{1r} \\ K_{r1} & \cdots & K_{rr} \end{pmatrix} = (-1)^r\left\{-1 + s\left[(y_1)^2 - \sum_{j=2}^r (y_j)^2\right]\right\}$$

$$\begin{cases} \geq 0 & \text{if } r \text{ is even} \\ \leq 0 & \text{if } r \text{ is odd.} \end{cases} \tag{6.78}$$

This implies $(y_1)^2 - \sum_{j=2}^{r}(y_j)^2 > 0$ on D. Hence Q is pseudoconcave on C. Because of the inequality (6.78), s has to satisfy $s(y) \geq [(y_1)^2 - \sum_{j=2}^{r}(y_j)^2]^{-1}$ for all $y \in D$. Thus $r(x) \geq \{2[Q(x) - \delta]\}^{-1}$ for $x \in C$. We already saw that $H(x; r)$ is negative semidefinite for $r(x) = \{2[Q(x) - \delta]\}^{-1}$. Therefore, for merely pseudoconcave Q we obtain that $r_0(x) = \{2[Q(x) - \delta]\}^{-1}$ is the smallest possible r such that $H(x; r)$ is negative semidefinite. □

In order to apply this criterion, r_0 and therefore δ, has to be determined. For this, (6.9), (6.57), or (6.58) may be used. Similarly to Theorem 6.6, we can show the following:

Theorem 6.7 (Schaible, 1977). A quadratic function Q is strictly pseudoconcave on an open convex set $C \subset R^n$ if and only if $H(x; r)$ in (6.72) is negative definite on C for sufficiently large $r(x)$. If Q is not concave on C, then any $r(x) > r_0(x)$ may be chosen.

Proof. Let Q be concave on C. Then $p = 0$. Applying Definition 3.13 to the normal form of Q, one can see that in this case Q is strictly pseudoconcave if and only if either $r = n$ or $r = n - 1$, where $Q(x) = Q_2(y)$. This is equivalent to

$$K(y; s) = \nabla^2 Q_i(y) - s(y)\nabla Q_i(y)\nabla Q_i(y)^T \tag{6.79}$$

being negative definite for some nonnegative $s(y)$. Thus $H(x; r)$ is negative definite for sufficiently large $r(x)$. Conversely, let Q be nonconcave on C. As seen from Theorem 6.5, in this case a pseudoconcave function Q is strictly pseudoconcave if and only if $r = n$. Let $r = n$. Then the leading principal minors of $K(y; s)$ of even (odd) order are positive (negative) for $s(y) > [(y_1)^2 - \sum_{j=2}^{n}(y_j)^2]^{-1}$ in view of (6.78). On the other hand, if $r < n$, then $\det K(y; s) = 0$. Hence $K(y; s)$ is negative definite if and only if $r = n$. □

Example 6.2. Let $Q(x) = x_1x_2$ and $C = \{x \in R^2: x_1 > 0, x_2 > 0\}$. From Table 5.3 it follows that Q is pseudoconcave on C. The same will now be shown with the help of Theorem 6.6. We have

$$H(x; r) = \begin{pmatrix} 0 & 1 \\ 1 & 0 \end{pmatrix} - r(x)\begin{pmatrix} (x_2)^2 & x_1x_2 \\ x_1x_2 & (x_1)^2 \end{pmatrix} = \begin{pmatrix} -r(x_2)^2 & 1 - rx_1x_2 \\ 1 - rx_1x_2 & -r(x_1)^2 \end{pmatrix}. \tag{6.80}$$

$H(x; r)$ is negative semidefinite for $r \geq 0$ if $\det H(x; r) = -1 + 2rx_1x_2 \geq 0$, i.e., $r(x) \geq 1/(2x_1x_2) = 1/[2Q(x)]$. By Theorem 6.6 Q is pseudoconcave on C since $H(x; r)$ is negative semidefinite on C for $r(x) \geq 1/(2x_1x_2)$. Note that $r_0(x) = 1/[2Q(x)]$ is the smallest possible $r(x)$ as also shown in Theorem 6.6. Actually, Q is even strictly pseudoconcave on C in view of Theorem 6.7.

Example 6.3. In Example 6.1 we considered the function $Q(x) = -5(x_1)^2 + 8x_1x_2 + (x_2)^2 - 6x_1 - 12x_2$. We saw that Q is pseudoconcave on

$$C_1^0 = \{x \in R^2\colon Q(x) \geq -15,\ x_1 + 2x_2 > 5\}, \tag{6.81}$$

$$C_2^0 = \{x \in R^2\colon Q(x) \geq -15,\ x_1 + 2x_2 < 5\}. \tag{6.82}$$

Since the matrix A in Q is nonsingular, Q is strictly pseudoconcave.

We now want to verify this result by applying Theorem 6.7. For $H(x; r)$ we find

$$\begin{pmatrix} -10 - r(-10x_1 + 8x_2 - 6)^2 & 8 - r(8x_1 + 2x_2 - 12)(-10x_1 + 8x_2 - 6) \\ 8 - r(8x_1 + 2x_2 - 12)(-10x_1 + 8x_2 - 6) & 2 - r(8x_1 + 2x_2 - 12)^2 \end{pmatrix}. \tag{6.83}$$

$H(x; r)$ is negative definite for $r \geq 0$ if $\det[H(x; r)] > 0$. This holds if

$$r(x) > \frac{1}{2[-5(x_1)^2 + 8x_1x_2 + (x_2)^2 - 6x_1 - 12x_2 + 15]} = \frac{1}{2[Q(x) + 15]}. \tag{6.84}$$

By Theorem 6.7, Q is strictly pseudoconcave on C_1^0, C_2^0. We also verified that the largest lower bound on $r(x)$ is $r_0(x) = \{2[Q(x) - \delta]\}^{-1}$.

In general, the criterion in Theorem 6.6 may be applied as follows: If $b = 0$, then check whether

$$H(x; r_0) = A - r_0(x)(Ax)(Ax)^T \qquad \text{with } r_0(x) = 1/x^TAx \tag{6.85}$$

is negative semidefinite on C. If $b \neq 0$, then find a stationary point s of Q by solving (6.42). If there does not exist one, then rank$(A, b) >$ rank A, and thus Q is not (merely) pseudoconcave on C. If there exists one, then check whether

$$H(x; r_0) = A - r_0(x)(Ax + b)(Ax + b)^T, \qquad r_0(x) = [2Q(x) - b^Ts]^{-1} \tag{6.86}$$

is negative semidefinite on C. If A is nonsingular, $r_0(x) = [2Q(x) + b^TA^{-1}b]^{-1}$ can be used [see (6.58) and (6.73)].

6.3. Characterizations in Terms of the Bordered Hessian

As already discussed in Chapter 3, Arrow and Enthoven (1961) used the leading principal minors of the bordered Hessian

$$D(x) = \begin{pmatrix} 0 & \nabla Q(x)^T \\ \nabla Q(x) & \nabla^2 Q(x) \end{pmatrix} \tag{6.87}$$

to derive necessary conditions and sufficient conditions for quasiconcavity of twice continuously differentiable functions (see also Ferland, 1971). There remains a gap between these conditions even in the case of quadratic functions, as pointed out in Ferland (1971). More recently, Avriel and Schaible (1978) introduced modified conditions involving *all* principal minors of $D(x)$ that turn out to be necessary *and* sufficient for pseudoconcave quadratic functions. These criteria will now be derived employing Theorems 6.6 and 6.7 above. Hence they are based again on the concavifiability of pseudoconcave quadratic functions shown in Theorem 6.3.

In order to formulate the conditions, we introduce the following notation. Let $A_{\gamma,k}$ be the principal submatrix of order k of $\nabla^2 Q(x) = A$ formed by the $(i_1, \ldots, i_k)$th row and column of A, $\gamma = \{i_1, \ldots, i_k\}$. For $k = 1, \ldots, n$ let

$$\Gamma_k = \{\gamma\colon \gamma = \{i_1, \ldots, i_k\}, 1 \le i_1 < \cdots < i_k \le n\}. \tag{6.88}$$

The leading principal submatrix of order k of A is denoted by A_k. Correspondingly, let $D_{\gamma,k}(x)$ be the principal submatrix of order $k + 1$ of $D(x)$ associated with $A_{\gamma,k}$, i.e.,

$$D_{\gamma,k}(x) = \begin{pmatrix} 0 & \dfrac{\partial Q(x)}{\partial x_{i_1}} & \cdots & \dfrac{\partial Q(x)}{\partial x_{i_k}} \\ \dfrac{\partial Q(x)}{\partial x_{i_1}} & \dfrac{\partial^2 Q(x)}{\partial x_{i_1}^2} & \cdots & \dfrac{\partial^2 Q(x)}{\partial x_{i_1} x_{i_k}} \\ \vdots & \vdots & & \vdots \\ \dfrac{\partial Q(x)}{\partial x_{i_k}} & \dfrac{\partial^2 Q(x)}{\partial x_{i_k} x_{i_1}} & \cdots & \dfrac{\partial^2 Q(x)}{\partial x_{i_k}^2} \end{pmatrix}. \tag{6.89}$$

Similarly, let $D_k(x)$ be the leading principal submatrix of order $k + 1$ of $D(x)$. From Theorem 6.6 we derive the following theorem.

Theorem 6.8 (Avriel and Schaible, 1978). A quadratic function Q is pseudoconcave on an open convex set $C \subset R^n$ if and only if for all $x \in C$, $\gamma \in \Gamma_k$, and $k = 1, \ldots, n$

$$\text{(i)} \quad (-1)^k \det D_{\gamma,k}(x) \geq 0 \tag{6.90}$$

and

$$\text{(ii)} \quad \det D_{\gamma,k}(x) = 0 \quad \Rightarrow \quad (-1)^k \det A_{\gamma,k} \geq 0. \tag{6.91}$$

Proof. In view of Theorem 6.6, we have to show that the conditions (i) and (ii) are equivalent to negative semidefiniteness of $H(x; r)$; i.e., for all $x \in C$, $\gamma \in \Gamma_k$, $k = 1, \ldots, n$

$$(-1)^k \det H_{\gamma,k} \geq 0. \tag{6.92}$$

Here $\det H_{\gamma,k}$ is the principal minor of order k of $H(x; r)$ formed by the $(i_1, \ldots, i_k)$th rows and columns of $H(x; r)$. In order to see that (6.90) and (6.91) are equivalent to (6.92) we prove

$$\det H_{\gamma,k} = \det[A_{\gamma,k} - r(x) g_{\gamma,k}(x) g_{\gamma,k}^T(x)] = \det A_{\gamma,k} + r(x) \det D_{\gamma,k}(x), \tag{6.93}$$

where

$$g_{\gamma,k}(x) = \left(\frac{\partial Q(x)}{\partial x_{i_1}}, \ldots, \frac{\partial Q(x)}{\partial x_{i_k}} \right)^T.$$

Suppose $r(x) \neq 0$. From Schur's formula (Gantmacher, 1959a), we obtain

$$\det \begin{pmatrix} 1/r(x) & g_{\gamma,k}^T(x) \\ g_{\gamma,k}(x) & A_{\gamma,k} \end{pmatrix} = \frac{1}{r(x)} \det [A_{\gamma,k} - r(x) g_{\gamma,k}(x) g_{\gamma,k}(x)^T]. \tag{6.94}$$

On the other hand, expanding the determinant we find

$$\det \begin{pmatrix} 1/r(x) & g_{\gamma,k}^T(x) \\ g_{\gamma,k}(x) & A_{\gamma,k} \end{pmatrix} = \frac{1}{r(x)} \det A_{\gamma,k} + \det D_{\gamma,k}(x). \tag{6.95}$$

Then (6.94) and (6.95) yield (6.93). From (6.93) we see that (6.92) and (i) and (ii) are equivalent. □

In a very similar way we can obtain from Theorem 6.7 the following theorem.

Theorem 6.9 (Avriel and Schaible, 1978). A quadratic function Q is strictly pseudoconcave on an open convex set $C \subset R^n$ if and only if for all $x \in C$ and $k = 1, \ldots, n$

$$\text{(i)} \quad (-1)^k \det D_k(x) \geq 0 \tag{6.96}$$

and

$$\text{(ii)} \quad \det D_k(x) = 0 \;\Rightarrow\; (-1)^k \det A_k > 0. \tag{6.97}$$

Ferland (1971), extending a result of Arrow and Enthoven (1961), proved that the condition $(-1)^k \det D_k(x) > 0$ for all $k = 1, \ldots, n$ and $x \in C$ is sufficient for pseudoconcavity. In the case of quadratic functions this result follows from Theorem 6.9, where it is shown that then Q is actually strictly pseudoconcave. Thus Ferland's criterion covers only strictly pseudoconcave functions Q. Also, it is not necessary for strict pseudoconcavity, as Example 6.6 below shows. On the other hand, conditions (i) and (ii) in Theorem 6.9 are necessary and sufficient for strict pseudoconcavity.

Ferland (1971), extending another result of Arrow and Enthoven (1961), proved that the condition $(-1)^k \det D_k(x) \geq 0$ for all $k = 1, \ldots, n$ and $x \in C$ is necessary for pseudoconcavity on C. In the case of quadratic functions this follows from Theorem 6.8 in our approach. The function $Q(x) = (x_1)^2$, $C = R$ shows that this is not a sufficient condition for pseudoconcavity. A necessary and sufficient condition is presented in Theorem 6.8 above. Note that it involves all principal minors, not just the leading ones. Also, condition (6.91) is needed in case one of the principal minors $\det D_{\gamma,k}(x)$ vanishes. This happens quite frequently, as Example 6.7 below shows.

Let us turn now to some examples.

Example 6.4. Let $Q(x) = x_1 x_2$ and $C = \{x \in R^2 \colon x_1 > 0, x_2 > 0\}$. We have

$$D(x) = \begin{pmatrix} 0 & x_2 & x_1 \\ x_2 & 0 & 1 \\ x_1 & 1 & 0 \end{pmatrix}, \tag{6.98}$$

and thus $\det D_1(x) = -(x_2)^2 < 0$ and $\det D_2(x) = 2x_1x_2 > 0$. By Theorem 6.9, Q is strictly pseudoconcave on C as we saw already in Example 6.2.

Example 6.5. Let $Q(x) = \frac{1}{2}(x_1)^2 + x_1x_2 + 2x_1 + 3x_2$ and $C = \{x \in R^2\colon x_1 > 0, x_2 > 0\}$. Then we have

$$D(x) = \begin{pmatrix} 0 & x_1 + x_2 + 2 & x_1 + 3 \\ x_1 + x_2 + 2 & 1 & 1 \\ x_1 + 3 & 1 & 0 \end{pmatrix}, \tag{6.99}$$

and thus $\det D_1(x) = -(x_1 + x_2 + 2)^2 < 0$ on C and $\det D_2(x) = (x_1 + 2x_2 + 1)(x_1 + 3) > 0$ on C. By Theorem 6.9, Q is strictly pseudoconcave on C.

Example 6.6. Let $Q(x) = -\frac{1}{2}(x_2)^2 + x_2$ and $C = R^2$. We have

$$D(x) = \begin{pmatrix} 0 & -x_1 & 1 \\ -x_1 & -1 & 0 \\ 1 & 0 & 0 \end{pmatrix}. \tag{6.100}$$

Then $\det D_1(x) = -(x_2)^2 \le 0$, $\det A_1 = -1 < 0$ and $\det D_2(x) = 1 > 0$ on C. By Theorem 6.9, Q is strictly pseudoconcave on C. Note that here we need both (i) and (ii) of Theorem 6.9.

Example 6.7. Let $Q(x) = x_1x_3$ and $C = \{x \in R^3\colon x_1 > 0, x_3 > 0\}$. We have

$$D(x) = \begin{pmatrix} 0 & x_3 & 0 & x_1 \\ x_3 & 0 & 0 & 1 \\ 0 & 0 & 0 & 0 \\ x_1 & 1 & 0 & 0 \end{pmatrix}. \tag{6.101}$$

Then

$$\det D_{\gamma,1}(x) = \begin{cases} -(x_3)^2 < 0, & \gamma = \{1\}, \\ 0, & \gamma = \{2\}, \\ -(x_1)^2 < 0, & \gamma = \{3\}, \end{cases} \tag{6.102}$$

and for $\gamma = \{2\}$ we have $\det A_{\gamma,1} = 0$. Furthermore

$$\det D_{\gamma,2}(x) = \begin{cases} 0, & \gamma = \{1, 2\}, \\ 2x_1x_3, & \gamma = \{1, 3\}, \\ 0, & \gamma = \{2, 3\}, \end{cases} \tag{6.103}$$

and for $\gamma = \{1, 2\}$ and $\gamma = \{2, 3\}$ we have $\det A_{\gamma,2} = 0$. Finally,

$$\det D(x) = 0 \quad \text{and} \quad \det A = 0. \tag{6.104}$$

By Theorem 6.8 Q is pseudoconcave on C as it was also seen earlier in Section 5.2. From Theorem 6.9 we see that Q is not strictly pseudoconcave on C since (6.97) is not satisfied.

We finally mention another characterization of generalized concave functions Q in terms of the bordered Hessian $D(x)$.

Theorem 6.10 (Ferland, 1981). A nonconcave quadratic function Q is pseudoconcave on an open convex set $C \subset R^n$ if and only if the bordered Hessian $D(x)$ has exactly one (simple) positive eigenvalue for all $x \in C$.

As shown in Schaible (1981a), this criterion can also be derived from our main characterization in Theorem 6.2. One can easily see that it suffices to prove Theorem 6.10 for normal forms. Theorem 6.10 can then be established by using Theorem 6.2 and a result by Jacobi on the number of positive eigenvalues (Gantmacher, 1959a). For more details, see Schaible (1981a).

In addition to Theorem 6.10 we have the following theorem.

Theorem 6.11 (Schaible, 1981a). A nonconcave quadratic function Q is strictly pseudoconcave on an open convex set C in R^n if and only if the bordered Hessian $D(x)$ has exactly one positive eigenvalue for all $x \in C$ and $D(x)$ is nonsingular for all $x \in C$.

6.4. Strictly Quasiconcave and Strongly Pseudoconcave Quadratic Functions

So far we characterized quasiconcave quadratic functions Q. We already pointed out that in case of quadratic functions there is no difference between quasiconcavity and semistrict quasiconcavity. In Chapter 3 two other concepts of generalized concavity are discussed: strict quasiconcavity and strong pseudoconcavity. We now want to characterize quadratic functions Q that belong to these families of generalized concave functions.

We first show the following.

Theorem 6.12 (Schaible, 1981a). A quadratic function Q is strictly quasiconcave on an open convex set C in R^n if and only if Q is strictly pseudoconcave on C.

Proof. A strictly pseudoconcave function is strictly quasiconcave, as a comparison of Theorems 3.26 and 3.41 shows. This is even true for nonquadratic functions. We now show that for quadratic functions the reverse is also true.

If Q is concave, then it follows from the definition in Chapter 3 that strict quasiconcavity implies strict pseudoconcavity. Let us assume that Q is not concave. A strictly quasiconcave function Q is quasiconcave, which, in turn, is pseudoconcave on an open convex set C (see Theorem 6.2). By Theorem 6.5, Q is strictly pseudoconcave provided $r = n$. Now for a nonconcave strictly quasiconcave function we necessarily have $r = n$, as can be seen from the normal form (6.13). □

We now turn to strongly pseudoconcave quadratic functions, for which we can show the following.

Theorem 6.13 (Schaible, 1981a). A quadratic function Q is strongly pseudoconcave on an open convex set C in R^n if and only if Q is strictly pseudoconcave on C.

Proof. Strong pseudoconcavity implies strict pseudoconcavity (Definition 3.15). Now assume Q is strictly pseudoconcave on C. Then from Theorem 6.7 we see that $H(x; r) = \nabla^2 Q(x) - r(x)\nabla Q(x)\nabla Q(x)^T$ is negative definite for some $r: C \to R$. Thus for every $x \in C$ and $v \in R^n$ such that $v^T v = 1$ and $v^T \nabla Q(x) = 0$ we have $v^T \nabla^2 Q(x) v < 0$. By Proposition 3.45 this implies strong pseudoconcavity. □

Summarizing the above results, we see that for quadratic functions on open convex sets the following relations hold:

$$\text{(I) quasiconcave} \Leftrightarrow \text{semistrictly quasiconcave} \Leftrightarrow \text{pseudoconcave,} \tag{6.105}$$

and

$$\text{(II) strictly quasiconcave} \Leftrightarrow \text{strictly pseudoconcave} \Leftrightarrow \text{strongly pseudoconcave.} \tag{6.106}$$

Hence there are only two distinct classes of generalized concave quadratic functions, if we restrict ourselves to open convex sets. The second class (II) is a proper subclass of the first one (I), and it is characterized by a nonsingular matrix A. Note, however, that for functions Q on convex sets that are not open not all the above relations hold. For instance, a quasiconcave function Q is not necessarily pseudoconcave, as seen from Theorem 6.2.

6.5. Quadratic Functions of Nonnegative Variables

We now specialize the above criteria to the case where C is the positive orthant $R^n_+ = \{x \in R^n: x_j > 0, j = 1, \ldots, n\}$. In view of Corollary 6.4, when characterizing pseudoconcave quadratic functions on the open set R^n_+ we simultaneously obtain criteria for quasiconcavity on the nonnegative orthant $\bar{R}^n_+ = \{x \in R^n: x_j \geq 0, j = 1, \ldots, n\}$. Later in this section we will see that often quadratic functions that are pseudoconcave on R^n_+ are also pseudoconcave on $\bar{R}^n_+\setminus\{0\}$ or $\bar{R}^n_+$.

Historically, criteria for generalized concave quadratic functions on $\bar{R}^n_+$ were discovered before the above criteria for general convex sets C were obtained. Martos (1969, 1971) established criteria for quasiconvex and pseudoconvex functions Q on $\bar{R}^n_+$ using the concept of positive subdefinite matrices (see Definition 3.7). Based on Martos' approach, Cottle and Ferland (1971, 1972) obtained several additional characterizations. All these criteria involve only finitely many tests in contrast to the definitions of quasiconcavity and pseudoconcavity.

In Schaible (1971, 1973a) it was shown that the criteria in Martos (1969) for quadratic forms on $\bar{R}^n_+$ can be obtained by specializing the characterizations in Theorem 6.2 that were derived for arbitrary solid convex sets. In this section we shall proceed along the same line. As in Schaible (1981b) we will show that the criteria of Cottle, Ferland, and Martos for the case $C = \bar{R}^n_+$ can be obtained by specializing Theorem 6.2 and some of its corollaries. Moreover, we will also derive criteria for strictly pseudoconcave quadratic functions Q on $\bar{R}^n_+$.

Apart from Theorems 6.2 and 6.5 we shall draw heavily on the results of Theorems 6.8 and 6.9. We recall that these follow from Theorem 6.3. Hence concavifiability of pseudoconcave quadratic functions is the underlying property that we use to derive the criteria below. We begin with proving the following lemma.

Lemma 6.14. *If* $Q(x) = \frac{1}{2}x^TAx + b^Tx$ *is merely pseudoconcave on* R^n_+, *then* $A \geq 0$ *and* $b \geq 0$.

Proof. According to Corollary 6.4, Q is quasiconcave on $\bar{R}^n_+$. Hence $Q(\alpha e^i + \beta e^j)$ is quasiconcave for $\alpha \geq 0$, $\beta \geq 0$, where e^i denotes the ith unit vector in R^n, $i \neq j$. As we see from Theorem 6.3, Q is bounded from below on $\bar{R}^n_+$. Therefore

$$g(\alpha) = Q(\alpha e^i) = \tfrac{1}{2}a_{ii}(\alpha)^2 + b_i\alpha \qquad (6.107)$$

is bounded from below for $\alpha \geq 0$. Hence $a_{ii} \geq 0$. We now consider

$$h(\beta) = Q(\bar{\alpha}e^i + \beta e^j) = \tfrac{1}{2}a_{ii}(\bar{\alpha})^2 + b_i\bar{\alpha} + (a_{ij}\bar{\alpha} + b_j)\beta + \tfrac{1}{2}a_{jj}(\beta)^2, \tag{6.108}$$

where $\bar{\alpha} \geq 0$ is fixed. If $a_{jj} = 0$, then $a_{ij}\bar{\alpha} + b_j \geq 0$, since $h(\beta)$ is bounded from below. On the other hand, if $a_{jj} > 0$ then also $a_{ij}\bar{\alpha} + b_j \geq 0$ must hold, since otherwise $h(\beta)$ would have a strict local minimum for some $\beta > 0$, contradicting, via Proposition 3.5, quasiconcavity of $h(\beta)$ for $\beta > 0$. Hence

$$a_{ij}\bar{\alpha} + b_j \geq 0 \qquad \text{for } \bar{\alpha} \geq 0. \tag{6.109}$$

Since this holds for $\bar{\alpha} = 0$ we have $b_j \geq 0$. Furthermore, since (6.109) is true for all $\bar{\alpha} > 0$, we conclude that $a_{ij} \geq 0$. □

Lemma 6.14 together with Theorem 6.2 will now enable us to derive criteria for pseudoconcave quadratic functions on R^n_+ that were first proved by Martos (1971). He used the concept of positive subdefinite matrices developed in Martos (1969). Our proof does not make use of this concept.

Theorem 6.15. Let $Q(x) = \frac{1}{2}x^TAx + b^Tx$ be nonconcave on R^n. Then Q is pseudoconcave on R^n_+ if and only if

$$A \geq 0, \tag{6.110}$$

$$b \geq 0, \tag{6.111}$$

$$p = 1, \tag{6.112}$$

and

$$\text{there exists an } s \in R^n \text{ such that } As + b = 0 \text{ and } b^Ts \leq 0. \tag{6.113}$$

Proof. Suppose that Q is merely pseudoconcave on R^n_+. Then (6.110) and (6.111) follow from Lemma 6.14. Theorem 6.2(i) and (ii) yield (6.112) and the existence of a stationary point s in (6.113). By Theorem 6.3 $Q(x) = \frac{1}{2}x^TAx + b^Tx \geq \delta$, where $\delta = \frac{1}{2}b^Ts$ in view of (6.57). Since $A \geq 0$ and $b \geq 0$ and the components of $x > 0$ can become arbitrarily small we conclude that $\delta = \frac{1}{2}b^Ts \leq 0$. Hence (6.113) holds.

Conversely, in view of (6.112) and (6.113) conditions (i), (ii) of Theorem 6.2 are fulfilled. We show that either $R^n_+ \subset C^0_1$ or $R^n_+ \subset C^0_2$. Then the assertion follows from Theorem 6.2. To see this, we use the following representation of C^0_1 [see (6.40) and (6.57)]:

$$C^0_1 = \{x \in R^n \colon Q(x) \geq \tfrac{1}{2}b^Ts, (t^1)^Tx > -(t^1)^Tb/\lambda_1\}. \tag{6.114}$$

Since $Q(x) \geq 0$ and $b^T s \leq 0$ because of (6.110), (6.111), and (6.113), we have $Q(x) \geq \frac{1}{2} b^T s$ on R^n_+. For a nonnegative matrix, by a theorem of Frobenius, there always exists an unisigned eigenvector t^1 associated with the positive eigenvalue λ_1 (Gantmacher, 1959b). Assume $t^1 \geq 0$. Hence $(t^1)^T x > 0$ on R^n_+ and $-(t^1)^T b/\lambda_1 \leq 0$. Thus $(t^1)^T x > -(t^1)^T b/\lambda_1$ holds on R^n_+. Therefore $R^n_+ \subset C^0_1$. Similarly, if $t^1 \leq 0$, we have $R^n_+ \subset C^0_2$. □

We remark here that Martos (1971) makes the following observations:

I: If A is nonsingular (i.e., Q is strictly pseudoconcave in view of Theorem 6.5), then (6.113) can be replaced by the condition $b^T A^{-1} b \geq 0$.

II: If there are two stationary points s^1, s^2 of Q, then $b^T s^1 = b^T s^2$. We saw this earlier in (6.54).

III: If $As + b = 0$ is solvable, then we have $b_i = 0$ if the ith row of A is zero.

Based on Martos' (1969) concept of positive subdefinite matrices, Cottle and Ferland (1972) proved the following result which is obtained here by specializing Theorem 6.15:

Corollary 6.16. Let $Q_0(x) = \frac{1}{2} x^T A x$ be nonconcave on R^n. Then $Q_0(x)$ is pseudoconcave on R^n_+ if and only if

$$A \geq 0 \tag{6.115}$$

and

$$p = 1. \tag{6.116}$$

From Theorem 6.15 and Corollary 6.16 we obtain the following.

Corollary 6.17. If $Q(x) = \frac{1}{2} x^T A x + b^T x$ is merely pseudoconcave on R^n_+, then $Q_0(x) = \frac{1}{2} x^T A x$ is merely pseudoconcave on R^n_+.

Note that the reverse implication of Corollary 6.17 does not necessarily hold for a quadratic form Q_0. Instead we have the following theorem.

Theorem 6.18 (Schaible, 1981b). Let $Q_0(x) = \frac{1}{2} x^T A x$ be merely pseudoconcave on R^n_+. Then $Q(x) = Q_0(x) + b^T x$ is pseudoconcave on R^n_+ for all $b \geq 0$ if and only if either $n = 1$ or $n = 2$ and $A = \begin{pmatrix} 0 & \alpha \\ \alpha & 0 \end{pmatrix}$ for some $\alpha > 0$.

The interested reader is referred to Schaible (1981b) for the proof. It turns out that for $n > 2$ the set of $b \in R^n$ such that $Q_0(x) + b^Tx$ is pseudoconcave on R^n_+ is an r-dimensional closed convex cone in the nonnegative orthant $\bar{R}^n_+$, where r is the rank of A. This cone can be characterized in terms of the eigenvalues and eigenvectors of A.

We now present criteria for pseudoconcavity in terms of principal minors rather than eigenvalues of A. Determinants of the related matrix $D(x)$ were already used in Theorems 6.8 and 6.9, where we characterized pseudoconcave functions Q on arbitrary open convex sets. As a corollary of Theorem 6.8 one can show the following.

Theorem 6.19. Let $Q_0(x) = \frac{1}{2}x^TAx$ be nonconcave on R^n. Then $Q_0(x)$ is pseudoconcave on R^n_+ if and only if

$$A \geq 0 \tag{6.117}$$

and

$$(-1)^k \det A_{\gamma,k} \leq 0 \qquad \text{for all } \gamma \in \Gamma_k, \qquad k = 1, \ldots, n. \tag{6.118}$$

This was first proved by Cottle and Ferland (1972) using positive subdefinite matrices. Schaible and Cottle (1980) derived the same result from Theorem 6.8. The proof is similar to the one of the next theorem and will therefore be omitted.

From Theorem 6.19 we see that for nonconcave quadratic forms on the positive orthant $C = R^n_+$, condition (ii) in Theorem 6.8 can be replaced by the stronger condition

$$\det D_{\gamma,k}(x) = 0 \quad \Rightarrow \quad \det A_{\gamma,k} = 0. \tag{6.119}$$

In addition to Theorem 6.19 we now derive a criterion for strictly pseudoconcave quadratic forms using Theorem 6.9.

Theorem 6.20 (Schaible, 1981b). Let $Q_0(x) = \frac{1}{2}x^TAx$ be nonconcave on R^n_+. Then $Q_0(x)$ is strictly pseudoconcave on R^n_+ if and only if

$$A \geq 0 \tag{6.120}$$

and

$$(-1)^k \det A_k < 0 \qquad \text{for all } k = 2, \ldots, n. \tag{6.121}$$

Proof. Suppose that Q_0 is merely strictly pseudoconcave on R^n_+. Then in view of Lemma 6.14, it remains to show that (6.121) holds. First we prove that

$$(-1)^k \det A_k \leq 0 \qquad \text{for } k = 2, \ldots, n. \tag{6.122}$$

Suppose to the contrary, that

$$(-1)^k \det A_k > 0 \qquad \text{for some } k \geq 2. \tag{6.123}$$

Let $x = \binom{x_k}{0} \in R^n$, where $x_k \in R^k_+$. We have

$$\det D_k(x) = \det \begin{pmatrix} 0 & (A_k x_k)^T \\ A_k x_k & A_k \end{pmatrix} \tag{6.124}$$

$$= -\det \begin{pmatrix} x_k^T A_k x_k & (A_k x_k)^T \\ 0 & A_k \end{pmatrix} \tag{6.125}$$

$$= -(x_k^T A_k x_k) \det A_k. \tag{6.126}$$

Because of (6.123), we have $A_k \neq 0$. Hence $A_k \geq 0$ implies

$$x_k^T A_k x_k > 0 \qquad \text{on } R^k_+. \tag{6.127}$$

Thus (6.123), (6.126), and (6.127) yield $(-1)^k \det D_k(x) < 0$. Then there exists an $\varepsilon \in R^{n-k}_+$ such that for $x_\varepsilon = \binom{x_k}{\varepsilon} \in R^n_+$ we have $(-1)^k \det D_k(x_\varepsilon) < 0$, contradicting (i) in Theorem 6.9. Therefore, (6.122) holds. It remains to show that $\det A_k \neq 0$ for $k = 2, \ldots, n$. Since $(-1)^k \det A_k \leq 0$, we see from Theorem 6.9 that $(-1)^k \det D_k(x) > 0$ on R^n_+. We partition A and x as follows:

$$A = \begin{pmatrix} A_k & B_k \\ B_k^T & C_k \end{pmatrix}, \qquad x = \begin{pmatrix} x_k \\ y_k \end{pmatrix}, \qquad \text{where } x_k \in R^k_+, \qquad y_k \in R^{n-k}_+. \tag{6.128}$$

Let $y_k = \bar{y}_k$ be fixed. We consider

$$f(x_k) = Q(x_k, \bar{y}_k) = \tfrac{1}{2} x_k^T A_k x_k + x_k^T B_k \bar{y}_k + \tfrac{1}{2} \bar{y}_k^T C_k \bar{y}_k. \tag{6.129}$$

f is strictly pseudoconcave on R_+^k. We now show that f is not concave there. Suppose, to the contrary, that f is concave on R_+^k. Then A_k is negative semidefinite, implying $a_{ii} \le 0$ for $i = 1, \dots, k$. Since $A \ge 0$, then $a_{ii} = 0$ for $i = 1, \dots, k$. For a symmetric negative semidefinite matrix this implies $A_k = 0$, yielding $\det D_k(x) = 0$ for $k \ge 2$. However, we already saw in Theorem 6.9 that $\det D_k(x) \ne 0$ on R_+^k. Therefore f cannot be concave on R_+^k. In view of Theorem 6.5 we conclude that $\det A_k \ne 0$. This completes the first part of the proof.

Now we show that $(-1)^k \det D_k(x) > 0$ on R_+^n for $k = 1, \dots, n$. Then strict pseudoconcavity of Q follows from Theorem 6.9. Denote the vector of the first k components of Ax by $(Ax)_k$. Then, $-\det D_1(x) = [(Ax)_1]^2 > 0$ since $A \ge 0$, $x > 0$ and the nonsingular matrix A does not contain a row of zeros. Now let $k \ge 2$. We partition A and x as in (6.128). Then,

$$\det D_k(x) = \det \begin{pmatrix} 0 & (A_k x_k + B_k y_k)^T \\ (A_k x_k + B_k y_k) & A_k \end{pmatrix} \tag{6.130}$$

$$= \det \begin{pmatrix} x_k^T A_k x_k + 2x_k^T B_k y_k + y_k^T C_k y_k & (A_k x_k + B_k y_k)^T \\ (A_k x_k + B_k y_k) & A_k \end{pmatrix}$$

$$- (x_k^T A_k x_k + 2x_k^T B_k y_k + y_k^T C_k y_k) \det A_k \tag{6.131}$$

$$= \det \begin{pmatrix} y_k^T C_k y_k & (B_k y_k)^T \\ (B_k y_k) & A_k \end{pmatrix} - (x^T A x) \det A_k. \tag{6.132}$$

For a nonconcave function $Q_0(x) = \frac{1}{2} x^T A x$, we have $A \ne 0$. Since $A \ge 0$, then

$$x^T A x > 0 \qquad \text{on } R_+^n. \tag{6.133}$$

In view of (6.121), (6.132), and (6.133), $(-1)^k \det D_k(x) > 0$ on R_+^n if

$$(-1)^k \det N_k \ge 0, \tag{6.134}$$

where

$$N_k = \begin{pmatrix} y_k^T C_k y_k & (B_k y_k)^T \\ (B_k y_k) & A_k \end{pmatrix}. \tag{6.135}$$

To see that (6.134) holds, consider the Schur complement of A_k in A (Gantmacher, 1959a):

$$S_k = C_k - B_k^T A_k^{-1} B_k. \tag{6.136}$$

According to Schur's formula

$$\det S_k = (\det A)/(\det A_k). \tag{6.137}$$

In view of (6.121), $(-1)^{n-k} \det S_k > 0$. In the same way, we can show that every leading principal minor of S_k of even (odd) order is positive (negative), considering the Schur complement of A_k in any leading principal submatrix of A containing A_k. Hence S_k is a negative definite matrix.

Now consider the Schur complement of A_k in N_k:

$$\bar{S}_k = y_k^T C_k y_k - (B_k y_k)^T A_k^{-1} (B_k y_k). \tag{6.138}$$

We see from (6.136) that $\bar{S}_k = y_k^T S_k y_k$. Since S_k is negative definite, $\bar{S}_k < 0$, for $y_k \neq 0$. In particular $\bar{S}_k < 0$ for $y_k > 0$. According to Schur's formula,

$$\det N_k = (\det \bar{S}_k) \cdot (\det A_k). \tag{6.139}$$

Hence (6.121) implies $(-1)^k \det N_k > 0$. Thus (6.134) holds, and the proof is complete. □

Theorem 6.20 shows that for nonconcave quadratic forms on the positive orthant condition (ii) in Theorem 6.9 is void. Hence, $Q_0(x) = \frac{1}{2} x^T A x$ is strictly pseudoconcave on R_+^n if and only if $(-1)^k \det D_k(x) > 0$ for all $x \in R_+^n$ and $k = 1, \ldots, n$.

In order to check strict pseudoconcavity of a quadratic form on R_+^n essentially (6.116) or (6.121) has to be verified. Both tests can be done by principal pivoting using simple pivots or (2×2)-block pivots. Cottle (1974) suggested an algorithm for computing the number of positive (and negative) eigenvalues as well as the sign of the leading principal minors of a symmetric matrix.

We see from Theorems 6.19 and 6.20 that checking (mere) pseudoconcavity and strict pseudoconcavity of a quadratic form on R_+^n is not more troublesome than checking concavity and strict concavity. In both cases the sign of all principal minors (or leading principal minors) has to be determined.

We now give two examples that illustrate the criteria in Theorems 6.19 and 6.20.

Example 6.8. Let $Q_0(x) = (x_1)^2 + 5x_1x_2 + 3(x_2)^2$. Hence $A = \begin{pmatrix} 2 & 5 \\ 5 & 6 \end{pmatrix}$. We have $A \geq 0$, and $(-1)^2 \det A_2 = \det A = -13 < 0$. By Theorem 6.20, $Q_0(x)$ is strictly pseudoconcave on R_+^2.

Example 6.9. Let $Q_0(x) = x_1x_2 + x_2x_3 + 2x_1x_3 + 2(x_3)^2$. Hence

$$A = \begin{pmatrix} 0 & 1 & 2 \\ 1 & 0 & 1 \\ 2 & 1 & 4 \end{pmatrix}.$$

To show that $Q_0(x)$ is pseudoconcave on R^3_+, we verify (6.117), (6.118). Obviously (6.117) holds. Furthermore,

$$(-1)\det A_{\gamma,1} = \begin{cases} 0, & \gamma = \{1\} \\ 0, & \gamma = \{2\} \\ -4, & \gamma = \{3\} \end{cases} \tag{6.140}$$

$$(-1)^2 \det A_{\gamma,2} = \begin{cases} -1, & \gamma = \{1, 2\} \\ -4, & \gamma = \{1, 3\} \\ -1, & \gamma = \{2, 3\} \end{cases} \tag{6.141}$$

$$(-1)^3 \det A_{\gamma,3} = -\det A = 0. \tag{6.142}$$

Hence (6.118) is also satisfied, and therefore Q_0 is pseudoconcave on R^3_+. From Theorem 6.20 we see that Q_0 is not strictly pseudoconcave there.

We now extend the pseudoconcavity criteria for quadratic forms in Theorem 6.19 and Theorem 6.20 to quadratic functions Q. Let

$$\tilde{Q}_0(x, x_{n+1}) = \tfrac{1}{2}(x \quad x_{n+1})\begin{pmatrix} A & b \\ b^T & 0 \end{pmatrix}\begin{pmatrix} x \\ x_{n+1} \end{pmatrix} \quad \text{and} \quad \tilde{A} = \begin{pmatrix} A & b \\ b^T & 0 \end{pmatrix}. \tag{6.143}$$

Obviously, $Q(x) = \tilde{Q}_0(x, 1)$. Hence, if the quadratic form $\tilde{Q}_0(x, x_{n+1})$ is pseudoconcave on R^{n+1}_+, then $Q(x)$ is pseudoconcave on R^n_+. The reverse is true too. To show this, we prove that conditions (6.117) and (6.118) in Theorem 6.19 are satisfied for $\tilde{Q}_0$, considered on R^{n+1}_+. For a merely pseudoconcave function Q in R^n_+, Lemma 6.14 implies $\tilde{A} \geq 0$. To see that also (6.118) in Theorem 6.19 holds, note that by Theorem 6.8

$$(-1)^k \det D_{\gamma,k}(x) \geq 0 \qquad \text{on } R^n_+ \tag{6.144}$$

for all principal submatrices $D_{\gamma,k}(x)$ of $D(x)$. Because of continuity of $D(x)$, (6.144) is still true for $x = 0$. Comparing $\det D_{\gamma,k}(0)$ with the principal minors of $\tilde{A}$, we see that (6.118) holds. Therefore, we have the following result, first proved by Cottle and Ferland (1972) using a somewhat different approach:

Lemma 6.21. Let $Q(x) = \frac{1}{2}x^TAx + b^Tx$ be nonconcave on R^n. Then Q is pseudoconcave on R^n_+ if and only if $\tilde{Q}_0(x, x_{n+1})$ is pseudoconcave on R^{n+1}_+.

From Theorem 6.19 and Lemma 6.21 we obtain the following theorem.

Theorem 6.22. Let $Q(x) = \frac{1}{2}x^TAx + b^Tx$ be nonconcave on R^n. Then Q is pseudoconcave on R^n_+ if and only if

$$A \geq 0, \qquad b \geq 0, \tag{6.145}$$

and for all principal minors $\tilde{A}_{\gamma,k}$, $k = 1, \ldots, n+1$, of A we have

$$(-1)^k \det \tilde{A}_{\gamma,k} \leq 0. \tag{6.146}$$

We now derive a criterion for strictly pseudoconcave quadratic functions, extending Theorem 6.20.

Theorem 6.23. Let Q be nonconcave on R^n. Then Q is strictly pseudoconcave on R^n_+ if and only if

$$A \geq 0, \qquad b \geq 0 \tag{6.147}$$

$$(-1)^k \det A_k < 0 \qquad \text{for } k = 2, \ldots, n, \tag{6.148}$$

and

$$(-1)^{n+1} \det \tilde{A} \leq 0. \tag{6.149}$$

Proof. Suppose that Q is merely strictly pseudoconcave on R^n_+. In view of Theorem 6.22 it remains to be shown that (6.148) holds. This follows from Theorem 6.20, knowing that $Q_0(x) = \frac{1}{2}x^TAx$ is merely strictly pseudoconcave by Corollary 6.17 and Theorem 6.5. Conversely, suppose that (6.147)-(6.149) hold. From Theorem 6.20 we see that $Q_0(x) = \frac{1}{2}x^TAx$ is strictly pseudoconcave on R^n_+ because of (6.147) and (6.148). In view of Corollary 6.16, then $p = 1$. Thus, according to Theorem 6.15, Q is pseudoconcave on R^n_+ if $b^TA^{-1}b \geq 0$ holds. The last assertion follows from Schur's formula applied to the Schur complement of A in $\tilde{A}$ (Gantmacher, 1959a):

$$\det \tilde{A} = (-b^TA^{-1}b) \det A, \tag{6.150}$$

taking into account (6.148) and (6.149). Applying Theorem 6.5, we see that Q is strictly pseudoconcave since according to (6.148) $\det A \neq 0$. □

From Theorems 6.22 and 6.23 we see that checking (mere) pseudoconcavity and strict pseudoconcavity of a proper ($b \neq 0$) quadratic function Q is not much more laborious than checking concavity and strict concavity.

Strict pseudoconcavity of a quadratic function can be verified by applying either Theorem 6.23 or 6.15 together with Theorem 6.5. The appropriate tests can be accomplished by principal pivoting. Cottle's (1974) algorithm mentioned before may be used for checking either (6.148), (6.149) or (6.112), (6.113). It seems to be somewhat more advantageous to check the former two conditions.

Let us illustrate the criterion in Theorem 6.23.

Example 6.10. We consider again the function in Example 6.5: $Q(x) = \frac{1}{2}(x_1)^2 + x_1x_2 + 2x_1 + 3x_2$. We verified earlier strict pseudoconcavity of Q on R^2_+. Now we want to show this by Theorem 6.23. We have

$$\tilde{A} = \begin{pmatrix} 1 & 1 & 2 \\ 1 & 0 & 3 \\ 2 & 3 & 0 \end{pmatrix}.$$

Hence $\tilde{A} \geq 0$. Furthermore,

$$(-1)^2 \det \tilde{A}_2 = \det A = -1 < 0 \text{ and } (-1)^3 \det \tilde{A} = -3 < 0. \tag{6.151}$$

Thus, the conditions (6.147)-(6.149) in Theorem 6.23 are satisfied, and therefore $Q(x)$ is strictly pseudoconcave on R^2_+.

For strictly pseudoconcave quadratic functions a result analogous to Lemma 6.21 does not hold. However, we do have, owing to the identity $Q(x) = \tilde{Q}_0(x, 1)$, the following lemma.

Lemma 6.24. If $\tilde{Q}_0(x, x_{n+1})$, defined by (6.143), is merely strictly pseudoconcave on R^{n+1}_+, then Q is merely strictly pseudoconcave on R^n_+.

To see that the reverse of Lemma 6.24 is not true, consider the following example.

Example 6.11. Let

$$Q(x_1, x_2) = \frac{1}{2}(x_1 \quad x_2)\begin{pmatrix} 0 & 1 \\ 1 & 0 \end{pmatrix}\begin{pmatrix} x_1 \\ x_2 \end{pmatrix} + (1 \quad 0)\begin{pmatrix} x_1 \\ x_2 \end{pmatrix}. \tag{6.152}$$

$Q(x)$ is strictly pseudoconcave on R^2_+ as seen from Theorem 6.23. However,

$$\tilde{Q}_0(x_1, x_2, x_3) = \frac{1}{2}(x_1 \quad x_2 \quad x_3)\begin{pmatrix} 0 & 1 & 1 \\ 1 & 0 & 0 \\ 1 & 0 & 0 \end{pmatrix}\begin{pmatrix} x_1 \\ x_2 \\ x_3 \end{pmatrix} \tag{6.153}$$

is not strictly pseudoconcave on R^3_+ since

$$\det\begin{pmatrix} 0 & 1 & 1 \\ 1 & 0 & 0 \\ 1 & 0 & 0 \end{pmatrix} = 0,$$

violating (6.121).

Lemma 6.24 yields in view of Theorem 6.20 the following theorem.

Theorem 6.25. Let $Q(x) = \frac{1}{2}x^TAx + b^Tx$ be nonconcave on R^n. Then, Q is strictly pseudoconcave on R^n_+ if

$$\tilde{A} \geq 0 \tag{6.154}$$

and

$$(-1)^k \det \tilde{A}_k < 0 \qquad \text{for } k = 2, \ldots, n+1. \tag{6.155}$$

In addition to Lemma 6.24 we have the following condition involving $\tilde{Q}_0$, which is both necessary and sufficient for strict pseudoconcavity:

Theorem 6.26. Let $Q(x) = \frac{1}{2}x^TAx + b^Tx$ be nonconcave on R^n. Then $Q(x)$ is strictly pseudoconcave on R^n_+ if and only if $\tilde{Q}_0(x, x_{n+1})$ is pseudoconcave on R^{n+1}_+ *and* $\det A \neq 0$.

The proof of the last theorem follows from Lemma 6.21 and Theorem 6.5. The result shows that instead of strict pseudoconcavity of $\tilde{Q}_0$ we need pseudoconcavity of $\tilde{Q}_0$ and the additional condition $\det A \neq 0$ in order to characterize strictly pseudoconcave functions Q. Note that the conditions of Theorem 6.26 are satisfied in Example 6.11, whereas $\tilde{Q}_0$ is not strictly pseudoconcave.

So far we characterized quadratic functions that are pseudoconcave on the open set R^n_+. As seen earlier in Corollary 6.4, pseudoconcave functions Q on R^n_+ are quasiconcave on the nonnegative orthant $\bar{R}^n_+$. We now show that many of these functions are also pseudoconcave on $\bar{R}^n_+$ or at least on $\bar{R}^n_+\backslash\{0\}$. In many mathematical programming applications, pseudoconcavity of the objective function is more desirable than just quasiconcavity since then a Karush–Kuhn–Tucker point is a maximizing point (see Chapter 3). Therefore, the following criteria are of special interest in quadratic programming where a quadratic function is maximized over a convex polyhedron contained in the nonnegative orthant of R^n. We first show the following.

Lemma 6.27. Let $Q(x) = \frac{1}{2}x^TAx + b^Tx$ be merely pseudoconcave on R^n_+. Let a_k denote the kth column of A and let t^1_k be the kth component of the eigenvector t^1. Then $t^1_k = 0$ if and only if $a_k = 0$.

Proof. Assume $a_k = 0$. Then $At^1 = \lambda_1 t^1$ implies $0 = a_k^T t^1 = \lambda_1 t^1_k$. Because of $\lambda_1 \neq 0$ this yields $t^1_k = 0$. Conversely, we have $A = \sum_{i=1}^r \lambda_i t^i (t^i)^T$. Hence,

$$a_k = \sum_{i=1}^{r} \lambda_i t^i_k t^i. \tag{6.156}$$

If $r = 1$, then $a_k \neq 0$ yields $t^1_k \neq 0$. Now let $r > 1$. Suppose to the contrary that $t^1_k = 0$. Then by (6.156) $a_{kk} = \sum_{i=2}^r \lambda_i (t^i_k)^2$. Since $a_{kk} \geq 0$ by (6.145) and $\lambda_i < 0$ for $i = 2, \ldots, r$, we conclude that $t^i_k = 0$. Then (6.156) yields $a_k = 0$. □

For quadratic forms we can show the following.

Theorem 6.28 (Schaible, 1981b). Let $Q_0(x) = \frac{1}{2}x^TAx$ be merely pseudoconcave on R^n_+. Then Q is pseudoconcave on $\bar{R}^n_+\backslash\{0\}$ if and only if A does not contain a row of zeros.

Proof. In view of Theorem 6.2 and the representation of C^0_1 via (6.40) Q_0 is pseudoconcave on $\bar{R}^n_+\backslash\{0\}$ if and only if $(t^1)^Tx > 0$ there. Equivalently this means that $t^1 > 0$. The result in Lemma 6.27 then proves the assertion. □

As we see from Theorem 6.28, a quadratic form is pseudoconcave on the semipositive orthant $\bar{R}^n_+\backslash\{0\}$ if it is pseudoconcave on the positive orthant and if it not constant in any one of its variables. Note, however, that such a quadratic form cannot be pseudoconcave on the nonnegative orthant. To see this, take $\bar{x} = 0$ and $\bar{\bar{x}} > 0$. Then we have $(\bar{\bar{x}} - \bar{x})^T \nabla Q_0(\bar{x}) = 0$, but $Q_0(\bar{\bar{x}}) = \frac{1}{2}\bar{\bar{x}}^TA\bar{\bar{x}} > 0 = \frac{1}{2}\bar{x}^TA\bar{x} = Q_0(\bar{x})$. Hence Q_0 is not pseudoconcave on $\bar{R}^n_+$ (see Definition 3.13).

Turning now to quadratic functions rather than quadratic forms, we can show that $Q(x) = \frac{1}{2}x^TAx + b^Tx$ ($b \neq 0$) is also pseudoconcave on $\bar{R}^n_+$ if it is pseudoconcave on R^n_+. This result was first proved by Cottle and Ferland (1972) using a different approach. We derive it here from our earlier results, essentially using our main criterion in Theorem 6.2.

Theorem 6.29. Let $Q(x) = \frac{1}{2}x^TAx + b^Tx$ be merely pseudoconcave on R^n_+, and suppose that $b \neq 0$. Then Q is pseudoconcave on $\bar{R}^n_+$.

Proof. From Theorem 6.2 together with (6.40) we see that

$$R^n_+ \subset C^0_1 = \{x \in R^n\colon Q(x) \geq \delta, (t^1)^Tx > -(t^1)^Tb/\lambda_1\} \tag{6.157}$$

where $t^1 \geq 0$ must hold. It remains to be shown that

$$(t^1)^T x > -(t^1)^T b/\lambda_1 \tag{6.158}$$

for all $x \geq 0$. This is true if $(t^1)^T b > 0$. Since $b \geq 0$ and $t^1 \geq 0$ we have $(t^1)^T b \geq 0$. Since $b \neq 0$ there exists one nonzero component, say b_k. From Martos' observation III above we see that $a_k \neq 0$ since otherwise we would have $b_k = 0$. In view of Lemma 6.27, $a_k \neq 0$ implies $t^1_k \neq 0$. Hence $(t^1)^T b \geq t^1_k b_k > 0$. □

Theorems 6.28 and 6.29 are useful in quadratic programming since the conditions there guarantee pseudoconcavity on the nonnegative orthant (possibly excluding zero), and not just on the positive orthant.

In this chapter, we have derived different characterizations of generalized concave quadratic functions. Some of these criteria can be extended to larger classes of functions like certain cubic functions (Schaible, 1971, 1973b). In Chapter 8, we will see in which way some results for quadratic functions carry over to twice-differentiable functions.

References

ARROW, K. J., and ENTHOVEN, A. D. (1961), Quasi-concave programming, *Econometrica* **29**, 779-800.

AVRIEL, M., and SCHAIBLE, S. (1978), Second order characterizations of pseudoconvex functions, *Math. Programming* **14**, 170-185.

COTTLE, R. W. (1967), On the convexity of quadratic forms over convex sets, *Oper. Res.* **15**, 170-172.

COTTLE, R. W. (1974), Manifestations of the Schur complement, *Linear Algebra Appl.* **8**, 189-211.

COTTLE, R. W., and FERLAND, J. A. (1971), On pseudoconvex functions of nonnegative variables, *Math. Programming* **1**, 95-101.

COTTLE, R. W., and FERLAND, J. A. (1972), Matrix-theoretic criteria for the quasiconvexity and pseudoconvexity of quadratic functions, *Linear Algebra Appl.* **5**, 123-136.

CROUZEIX, J. P. (1980), Conditions of convexity of quasiconvex functions, *Math. Oper. Res.* **5**, 120-125.

FERLAND, J. A. (1971), Quasi-convex and pseudo-convex functions on solid convex sets, Technical Report 71-4, Department of Operations Research, Stanford University.

FERLAND, J. A. (1972), Maximal domains of quasiconvexity and pseudoconvexity for quadratic functions, *Math. Programming* **3**, 178-192.

FERLAND, J. A. (1981), Matrix-theoretic criteria for the quasiconvexity of twice continuously differentiable functions, *Linear Algebra Appl.* **38**, 51-63.

GANTMACHER, F. R. (1959a), *The Theory of Matrices*, Vol. I, Chelsea, New York.

GANTMACHER, F. R. (1959b), *The Theory of Matrices*, Vol. II, Chelsea, New York.

GREUB, W. H. (1967), *Linear Algebra*, Springer-Verlag, New York.

MARTOS, B. (1967), Quasi-convexity and quasi-monotonicity in nonlinear programming, *Stud. Sci. Math. Hung.* **2**, 265-273.

MARTOS, B. (1969), Subdefinite matrices and quadratic forms, *SIAM J. Appl. Math.* **17**, 1215-1223.

MARTOS, B. (1971), Quadratic programming with a quasiconvex objective function, *Oper. Res.* **19**, 82-97.

MARTOS, B. (1975), *Nonlinear Programming, Theory and Methods*, North-Holland, Amsterdam.

NOBLE, B. (1969), *Applied Linear Algebra*, Prentice Hall, Englewood Cliffs, New Jersey.

SCHAIBLE, S. (1971), Beiträge zur quasikonvexen Programmierung, Doctoral Dissertation, Köln, Germany.

SCHAIBLE, S. (1973a), Quasiconcave, strictly quasiconcave and pseudoconcave functions, *Methods Oper. Res.* **17**, 308-316.

SCHAIBLE, S. (1973b), Quasiconcavity and pseudoconcavity of cubic functions, *Math. Programming* **5**, 243-247.

SCHAIBLE, S. (1977), Second order characterizations of pseudoconvex quadratic functions, *J. Optimization Theory Appl.* **21**, 15-26.

SCHAIBLE, S. (1981a), Generalized convexity of quadratic functions, in *Generalized Concavity in Optimization and Economics*, Edited by S. Schaible and W. T. Ziemba, Academic Press, New York, pp. 183-197.

SCHAIBLE, S. (1981b), Quasiconvex, pseudoconvex and strictly pseudoconvex quadratic functions, *J. Optimization Theory Appl.* **35**, 303-338.

SCHAIBLE, S., and COTTLE, R. W. (1980), On pseudoconvex quadratic forms, in *General Inequalities II*, Edited by E. F. Beckenbach, Birkhäuser-Verlag, Basel.

SCHAIBLE, S., and ZIEMBA, W. T. (Eds.) (1981), *Generalized Concavity in Optimization and Economics*, Academic Press, New York.

7

Fractional Programming

In Chapter 5 we saw that ratios of concave and/or convex functions are often generalized concave. Optimization problems involving one or several ratios in the objective function are called fractional programs.

This class of generalized concave programs is of interest in a considerable number of applications. Decision problems in management science as well as extremum problems in other areas give rise to various types of fractional programs.

The present chapter is devoted exclusively to these generalized concave programs. In addition to a survey of applications, we will present theoretical and algorithmic results for fractional programs.

In Section 7.1, we introduce notation and definitions in fractional programming. Then, in Section 7.2, we survey major areas of applications and indicate which types of ratios are of particular interest in these applications. Properties of certain fractional programs are discussed in Section 7.3. There the relationship of these problems to concave programs is analyzed and a dual fractional program is introduced. Both the single ratio and the multiratio cases are treated. Based on this analysis, we present several solution strategies for fractional programs in Section 7.4.

7.1. Notation and Definitions

In this section, we present the terminology needed in our discussion of fractional programming.

Let f, g, h_j, $j = 1, \ldots, m$, be real-valued functions defined on a set C of R^n. We consider the ratio

$$q(x) = \frac{f(x)}{g(x)} \tag{7.1}$$

over the set

$$S = \{x \in C: h_j(x) \le 0; j = 1, \ldots, m\}. \tag{7.2}$$

We assume that g is positive on S. If g is negative, then $q = (-f)/(-g)$ may be used instead. The nonlinear program

$$\text{(P)} \qquad \sup\{q(x): x \in S\} \tag{7.3}$$

is called a *fractional program.*

In some applications more than one ratio appears in the objective function. Examples of such problems are

$$\sup\left\{\sum_{i=1}^{p} q_i(x): x \in S\right\} \tag{7.4}$$

and

$$\text{(GP)} \qquad \sup\{\min_{1 \le i \le p} q_i(x): x \in S\}. \tag{7.5}$$

Here $q_i(x) = f_i(x)/g_i(x)$, $i = 1, \ldots, p$, where f_i, g_i are real-valued functions on C with $g_i(x) > 0$ on S.

Problem (GP) is sometimes referred to as a *generalized fractional program* (Crouzeix, Ferland, and Schaible, 1983; Jagannathan and Schaible, 1983). Both problems (7.4) and (7.5) are related to the *multiobjective fractional program*

$$\max_{x \in S} \{q_1(x), \ldots, q_p(x)\}. \tag{7.6}$$

In this introduction to fractional programming, the relationship to generalized concavity will be emphasized. Our main attention is given to the objective functions. With regard to the feasible region S we assume that it is a convex set. Convexity of S is ensured by convex (or quasiconvex) constraint functions h_j defined on a convex set C of R^n. We will make this assumption throughout this chapter.

The generalized concavity properties of ratios of two functions were studied in detail in Section 5.2. They are summarized in Tables 5.4, 5.5 and 5.6. In particular, we have the following condition on q:

(K) The function $q = f/g$ is semistrictly quasiconcave on S if f is nonnegative and concave and g is positive and convex on S; in the special case where g is affine, f need not be sign restricted.

We call a fractional program (7.3) that satisfies condition (K) a *concave fractional program.* Note that the ratio q itself is not a concave function in general. Instead it is composed of a concave and a convex function.

For differentiable functions f and g the objective function q in a concave fractional program is pseudoconcave (see Section 5.2). The ratio $q(x) > 0$ is strictly pseudoconcave if either f is strictly concave or g is strictly convex.

A special case of a concave fractional program is the *linear fractional program,* where all functions f, g, h_j, $j = 1, \ldots, m$, are affine and C is the nonnegative orthant $\bar{R}^n_+$ of R^n, that is, we have

$$\sup \left\{ \frac{c^T x + \alpha}{d^T x + \beta} : Ax \leq b, x \geq 0 \right\}. \tag{7.7}$$

Here $c \in R^n$, $d \in R^n$, $b \in R^m$, $\alpha \in R$, $\beta \in R$, and A denotes an $m \times n$ matrix. A fractional program is called a *quadratic fractional program* if f and g are quadratic functions and the h_j are affine.

In a generalized fractional program (GP) as given by (7.5), the objective function

$$q(x) = \min_{1 \leq i \leq p} \frac{f_i(x)}{g_i(x)} \tag{7.8}$$

is semistrictly quasiconcave on S if each function $q_i = f_i/g_i$ has this property. This is the case if each q_i satisfies condition (K). Such a problem (GP) is called a *concave generalized fractional program.* In the special case of affine functions f_i, g_i, h_j, and $C = \bar{R}^n_+$ we call (GP) a *linear generalized fractional program.*

In contrast to the generalized fractional program (7.5), problem (7.4) has a nonquasiconcave objective function in general. This is even the case if the objective function is the sum of a linear ratio and a linear function (Schaible, 1977). However, for a very limited class of problems (7.4), quasiconcavity (or quasiconvexity) still holds: The following weighted sum of an absolute and relative term

$$\lambda f(x) + \mu \frac{f(x)}{g(x)} \tag{7.9}$$

is still semistrictly quasiconcave or semistrictly quasiconvex under suitable concavity/convexity assumptions on f and g and for restricted ranges of λ and μ (Schaible, 1984).

Recently, multiobjective fractional programs (7.6) have been studied too (Choo, 1980; Kornbluth and Steuer, 1981a; Schaible, 1983; Warburton, 1981). There again, it is commonly assumed that the q_i satisfy condition (K).

7.2. Applications

Fractional programs arise in various circumstances in management science, as well as in other areas. In this section we outline some of the major areas of applications of these generalized concave programs. For a more detailed presentation see Schaible (1978), or Schaible and Ibaraki (1983).

7.2.1. Optimizing Efficiency

Ratios to be optimized often describe some kind of an efficiency measure of a system. We mention a few examples.

7.2.1.1. Maximization of Productivity (Gutenberg, 1975).

Gilmore and Gomory (1963) discuss a cutting stock problem in the paper industry. Under the given circumstances, it is more appropriate to minimize the ratio of wasted and used amount of raw material instead of just minimizing the amount of wasted material. This cutting stock problem is formulated as a linear fractional program.

7.2.1.2. Maximization of Return on Investment (Heinen, 1971)

In some resource allocation problems, the ratio of profit to capital or profit to revenue is to be maximized. A related objective is maximization of return relative to cost. Resource allocation problems with this objective are discussed in Mjelde (1978, 1983). In these models the term "cost," may, for example, be related to expenditures, or stand for the amount of pollution, or the probability of a disaster in nuclear energy production. Depending on the nature of functions describing return, profit, cost, or capital in the aforementioned examples, we obtain different types of fractional programs. If, for instance, the price per unit depends linearly on the output, and cost and capital are linear funtions, then maximizing the return on investment gives rise to a concave quadratic fractional program; see Pack (1962, 1965).

7.2.1.3. Maximization of Return to Risk

In a portfolio selection problem with risk-free borrowing and lending and normally distributed random returns, Ziemba, Parkan, and Brooks-Hill (1974) obtain as a major subproblem the constrained maximization of the ratio of expected return and risk. The numerator is an affine function and the denominator a convex function. Hence the portfolio selection problem gives rise to a concave fractional program. It is not a quadratic fractional program as some of the problems in Section 7.2.1.2 are. Ziemba (1974) generalizes the results from normal distributions to stable distributions of the return, obtaining other concave fractional programs (see also Kallberg and Ziemba, 1981). Similar fractional programs that are nonconcave can also be found in portfolio theory. An example is the following model by Ohlson and Ziemba (1976):

$$\max \left\{ q(x) = \frac{c^T x}{(x^T D x)^\gamma} : x \in S \right\}. \tag{7.10}$$

Here c is a positive vector in R^n, D a positive definite $n \times n$ matrix, and $\gamma \in (0, 1/2)$. We know from results in Section 5.2 that in the limiting cases of $\gamma = 0$ and $\gamma = 1/2$, the ratio q is semistrictly quasiconcave. An analysis of q shows that under some additional assumptions, often satisfied in portfolio theory, q in (7.10) is semistrictly quasiconcave on certain cones of R^n as long as $\gamma \in [0, \gamma_0]$ but not quasiconcave for $\gamma \in (\gamma_0, 1/2)$. For a detailed analysis of the generalized concavity properties of the ratio q in (7.10) see Schaible and Ziemba (1985).

7.2.1.4. Minimization of Cost to Time

Ratios involving time occur frequently in applications. Dantzig, Blattner, and Rao (1966) discuss a routing problem for ships or planes, where a cycle in the transportation network is to be determined, that minimizes the cost-to-time ratio. This ratio appears also in Fox (1969) and Lawler (1976), who deal with similar problems. Sometimes stochastic processes give rise to the minimization of cost per unit time as demonstrated in a paper by Derman (1962). An example of such a stochastic process is discussed by Klein (1963), who formulates a maintenance problem as a Markov decision process that leads to a linear fractional program. Here the ratio of the expected cost for inspection, maintenance, and replacement, and the expected time between two inspections, is to be minimized.

So far we have described fractional programs that arise when the efficiency of a system is to be optimized. An analysis by Eichhorn (1972, 1978) shows that the terms "technical effectiveness" or "economical effectiveness" generally will be ratios of two functions. Only under more restrictive assumptions can these terms be expressed as differences of two functions. Hence the optimization of ratios occurs naturally when the effectiveness (or efficiency) of a system is to be maximized.

There are a number of problems that give rise to a generalized fractional program (GP), i.e., the maximization of the smallest of several ratios. An early application is von Neumann's model of an expanding economy where the growth rate is to be maximized; see von Neumann (1945). The growth rate is the smallest of several output-input ratios. More recently, generalized fractional programs have been discussed in goal programming and multicriteria optimization. For example, such models arise in resource allocation problems (Charnes and Cooper, 1977) or financial planning (Ashton and Atkins, 1979). In the latter case, ratios to be simultaneously maximized may be liquidity, earnings per share, dividend per share, or return on capital. For more examples of simultaneous optimization of ratios in different planning situations, see Kornbluth and Steuer (1981b).

7.2.2. Noneconomic Applications

Meister and Oettli (1967) consider the capacity of a discrete, constant communication channel. This term is defined as the maximum value of the transmission rate, which is a ratio of a concave and a linear function. Hence a concave fractional program has to be solved in order to determine the capacity of a communication channel. It is not a quadratic fractional program. Instead it involves logarithms in the numerator.

Fractional programs also occur in numerical analysis. The eigenvalues of a matrix can be calculated as constrained maxima of the Rayleigh quotient, a ratio of two quadratic forms (see Noble, 1969). Hence the eigenvalue problem leads to quadratic fractional programs. Generalized fractional programs (GP) are met in discrete rational approximation where the Chebychev norm is used (Barrodale, 1973).

7.2.3. Indirect Applications

There are a number of operations research problems that indirectly give rise to a fractional program. In these applications, fractional programs occur as surrogate problems or subproblems.

For example, large-scale linear programs can be reduced to a finite number of smaller problems. These subproblems are linear fractional programs if a dual or primal-dual decomposition method is applied (Lasdon, 1970). The ratio in these fractional programs originates in the minimum-ratio rule of the simplex method (Dantzig, 1963). The feasible region in this sequence of linear fractional programs is the same. It is determined by the specially structured constraints of the large-scale linear program. Hence a sequence of linear fractional programs with specially structured constraints is obtained.

Fractional programs also occur indirectly in stochastic linear programming as first shown by Charnes and Cooper (1963) and Bereanu (1965). Consider the following example:

$$\max\{a^T x\colon x \in S\} \tag{7.11}$$

where the vector a is a realization of a jointly normally distributed random variable. Several deterministic surrogates for this stochastic program were suggested. One of them is the maximum probability model:

$$\max\{P(a^T x \geq k)\colon x \in S\} \tag{7.12}$$

where k is a given constant. An example of this model would be the maximization of the probability that a certain profit level k is achieved.

As shown in Bereanu (1965) and Charnes and Cooper (1963), the maximum probability model reduces to the fractional program

$$\max\left\{q(x) = \frac{e^T x - k}{(x^T V x)^{1/2}}\colon x \in S\right\} \tag{7.13}$$

where e is the vector of expected values of the components of a and the matrix V is the variance-covariance matrix of a. The objective function in (7.13) is the ratio of an affine and a convex function. Hence, for the stochastic linear program in (7.11), the maximum probability model (7.12) is a nonlinear concave fractional program. Note that (7.13) is a nonlinear fractional program that is not quadratic. The same fractional program is obtained when solving the portfolio selection problem by Ziemba, Parkan, and Brooks-Hill (1974) mentioned before.

As in the linear case, a stochastic nonlinear program can often be reduced to a fractional program. Examples are discussed in Schaible (1978). Since stochastic programming is used in various decision problems, fractional programs arise indirectly in a variety of contexts.

7.3. Concave Fractional Programs

In this section we analyze concave fractional programs and their generalizations (7.5). First we discuss the possibilities of reducing such quasiconcave programs to a concave program. Later on, we introduce a duality theory for such programs.

In order to relate an optimization problem to a concave program, one might try to find either a suitable range transformation or a variable transformation (see Chapter 8). We already saw in Chapter 6 that any quasiconcave quadratic program can be reduced to a concave program using a certain range transformation (see Theorem 6.3). Unfortunately, this is not always possible for such programs as concave fractional programs that have quasiconcave objective functions. In fact, one can show that in the case of the simple linear ratio $q(x) = -x_2/x_1$, there does not exist an increasing function H such that $H(q(x))$ is concave on an open convex set in the half-space $S = \{x \in R^2: x_1 > 0\}$. (For details see Example 8.7.)

However, a subclass of concave fractional programs can be reduced to an equivalent concave program.

Proposition 7.1 (Schaible, 1974a). If in a fractional program (P) f is positive and concave on S and $[g(x)]^{\varepsilon}$ is convex on S for some $\varepsilon \in (0, 1)$, then the equivalent problem

$$\sup\{-[q(x)]^{-t}: x \in S\} \qquad (7.14)$$

is a concave program for $t \geq \varepsilon/(1-\varepsilon)$.

Proof. By Theorem 5.16,

$$[q(x)]^{-\varepsilon/(1-\varepsilon)} = \frac{([g(x)]^{\varepsilon})^{1+\varepsilon/(1-\varepsilon)}}{[f(x)]^{\varepsilon/(1-\varepsilon)}} \qquad (7.15)$$

is convex. Then for $t \geq \varepsilon/(1-\varepsilon)$ it follows from Proposition 2.16 that $-[q(x)]^{-t}$ is concave. □

Example 7.1. Consider the fractional program

$$\sup\left\{q(x)=\frac{f(x)}{x^T D x}: x \in S\right\} \tag{7.16}$$

where f is a positive concave function and $g(x) = x^T Dx$ is a positive convex quadratic form on S. Then the equivalent problem

$$\sup\{-q(x)^{-1}: x \in S\} \tag{7.17}$$

has a concave objective function since $[g(x)]^{1/2}$ is also convex, and thus we can use $t \geq 1$, in applying Proposition 7.1.

There is a way to relate a larger class of concave fractional programs to a concave program. This can be achieved by separating numerator and denominator by a parameter μ (Dinkelbach, 1967; Jagannathan, 1966; for a similar approach see Cambini, 1981; Geoffrion, 1967; Ritter, 1967). Consider the parametric problem

$$(P_\mu) \qquad \sup\{f(x)-\mu g(x): x \in S\}. \tag{7.18}$$

We assume that f and g are continuous and S is compact. Then

$$F(\mu)=\max\{f(x)-\mu g(x): x \in S\} \tag{7.19}$$

is a continuous decreasing function on R. Let $\bar{\mu}$ denote the unique zero of $F(\mu)$. It can be shown that an optimal solution $\bar{x}$ of $(P_{\bar{\mu}})$ solves (P) and $\bar{\mu} = q(\bar{x})$. If (P) is a concave fractional program then (P_μ) is a parametric concave program for each μ in the range of q. Hence, a concave fractional program can be reduced to a parametric concave program. For further details see Dinkelbach (1967).

In the above transformations, variables were not changed. We will now show that by a suitable transformation of variables every concave fractional program can be reduced to a parameter-free concave program (see also Example 8.22). For the special case of a linear fractional program this transformation was suggested by Charnes and Cooper (1962). It was extended to nonlinear fractional programs by Schaible (1973, 1976a).

We differentiate between problems with affine and nonaffine denominators. First we have the following proposition.

Proposition 7.2. A concave fractional program (P) with an affine denominator can be reduced to the concave program

$$(\mathrm{P}'_{=})\sup\{tf(y/t)\colon th_j(y/t)\leq 0; j=1,\ldots,m,\ tg(y/t)=1,\ y/t\in C,\ t>0\} \tag{7.20}$$

by the transformation

$$y=\frac{x}{g(x)},\qquad t=\frac{1}{g(x)}. \tag{7.21}$$

Proof. The transformation (7.21) is a one-to-one mapping of S onto the feasible region of $(\mathrm{P}'_{=})$. Also, $q(x) = tf(y/t)$ for corresponding points x and (y, t). Hence, (P) has an optimal solution, if and only if $(\mathrm{P}'_{=})$ has one, and the two optimal solutions are related to each other by (7.21).

Using the definitions of convex sets and convex functions in Chapter 2, one can show that the set

$$C' = \{(y, t) \in R^{n+1}\colon y/t \in C, t > 0\} \tag{7.22}$$

is convex and $th_j(y/t)$ is convex on C'. Similarly, concavity of f on C implies concavity of $tf(y/t)$ on C'. Hence $(\mathrm{P}'_{=})$ is a concave program. □

In the special case of a linear fractional program as given by (7.7), $(\mathrm{P}'_{=})$ becomes the linear program

$$\sup\{c^Ty+\alpha t\colon Ay-bt\leq 0,\ d^Ty+\beta t=1,\ y\geq 0,\ t>0\} \tag{7.23}$$

(Charnes and Cooper, 1962). Note that the strict inequality $t > 0$ can be replaced by $t \geq 0$ if (P) has an optimal solution.

In the case of a concave fractional program with a nonaffine denominator, we relax the equality $tg(y/t) = 1$ in (7.20) to the inequality $tg(y/t) \leq 1$. Then the following is true:

Proposition 7.3. A concave fractional program (P) with a nonaffine denominator is equivalent to the concave program

$$(\mathrm{P}')\sup\{tf(y/t)\colon th_j(y/t)\leq 0; j=1,\ldots,m,\ tg(y/t)\leq 1,\ y/t\in C,\ t>0\} \tag{7.24}$$

applying the transformation (7.21).

Proof. As seen before, an optimal solution of (P) corresponds to an optimal solution of $(P'_=)$. Moreover, by the proof of Proposition 7.2, $tg(y/t)$ is convex. Hence, to complete the proof it is left to show that for an optimal solution (y', t') of (P') we shall always have $t'g(y'/t') = 1$. To see this, note that $(y', t') = \varepsilon(y'', t'')$, where $0 < \varepsilon \leq 1$ and (y'', t'') is a feasible solution of $(P'_=)$. For $\varepsilon < 1$, $t'f(y'/t') = \varepsilon t''f(y''/t'') < t''f(y''/t'')$ by nonnegativity of f, contradicting optimality of (y', t'). Hence $\varepsilon = 1$, and therefore $t'g(y'/t') = 1$. □

The additional constraint, $tg(y/t) \leq 1$ in (P'), is nonlinear. The other constraints, $th_j(y/t) \leq 0$, are linear if the fractional program (P) has linear constraints $h_j(x) \leq 0$. Such an additional nonlinear constraint in (P') can be avoided if the numerator $f(x)$ in (P) is affine. In this case, the above variable transformation may better be applied to the equivalent problem

$$\sup\{-g(x)/f(x): x \in S\}, \tag{7.25}$$

as illustrated in the following example.

Example 7.2. The maximum probability model

$$\max\left\{q(x) = \frac{e^Tx - k}{(x^TVx)^{1/2}}: Ax \leq b, x \geq 0\right\} \tag{7.26}$$

is equivalent to

$$\max\left\{-\frac{(x^TVx)^{1/2}}{e^Tx - k}: Ax \leq b, x \geq 0\right\} \tag{7.27}$$

assuming that $e^Tx - k > 0$ on the feasible region. If we apply the transformation (7.21) to (7.27) we obtain an equivalent concave program

$$\max\{-(y^TVy)^{1/2}: Ay - bt \leq 0, e^Ty - kt = 1, y \geq 0, t > 0\}, \tag{7.28}$$

which, in turn, can be reduced to the concave quadratic program

$$\max\{-y^TVy: Ay - bt \leq 0, e^Ty - kt = 1, y \geq 0, t > 0\}. \tag{7.29}$$

In concave programming, the concept of duality plays a crucial role in both theory and applications (see, e.g., Avriel, 1976; Geoffrion, 1971; Mangasarian, 1969a; Rockafellar, 1970; Stoer and Witzgall, 1970). For a nonconcave program such as the concave fractional program, standard concave programming duality concepts are not useful, since basic duality relations are no longer true. This is even the case in linear fractional programming (Schaible, 1976c).

In fractional programming, therefore, duality has to be defined in a new way. In the case of a concave fractional program, the equivalence to a concave program (Propositions 7.2 and 7.3) can be used to develop a duality theory. Any classical dual of the equivalent concave programs (P′) or ($P'_=$) can be used in order to obtain a meaningful dual of (P).

Consider the concave fractional program

$$\text{(P)} \qquad \sup\left\{q(x) = \frac{f(x)}{g(x)}: x \in S\right\} \tag{7.30}$$

where $S = \{x \in C: h_j(x) \leq 0; j = 1, \ldots, m\}$. We assume that $C \subset R^n$ is a convex set, f is concave on C and is nonnegative on S if G is not affine, g is positive and convex on C, and the h_j are convex on C.

The reader can verify that the Lagrangean dual (see Chapter 2) of the concave program (P′) that is equivalent to (P) becomes, after retransformation of variables,

$$\text{(D)} \qquad \inf_u\left\{\sup_{x \in C} \frac{f(x) - u^T h(x)}{g(x)}: u \geq 0\right\}. \tag{7.31}$$

This reduces to the classical Lagrangean dual of a concave program if $g(x) = 1$.

Because of the equivalence of (P) and (P′) as shown in Proposition 7.3, classical duality relations can be extended to the pair (P) and (D). In particular, we find for the optimal values in (P) and (D) that weak duality relations hold—that is,

$$\sup(\text{P}) \leq \inf(\text{D}). \tag{7.32}$$

If there is a feasible solution of (P), sup (P) is finite, and some constraint qualification (Avriel, 1976; Mangasarian, 1969a) holds, then strong duality can be verified too—that is,

$$\sup(\text{P}) = \min(\text{D}). \tag{7.33}$$

Moreover, other duality relations such as converse duality can be proved. For more details see Schaible (1976c). Thus, standard duality relationships, which are satisfied for an ordinary concave program (see Theorem 2.31), also hold for a concave fractional program.

If all functions, f, g, h in (7.30) are differentiable on C and C is an open set, then Wolfe's dual (Mangasarian, 1969a) may be applied to (P′) [or ($P'_=$) if g is affine]. This yields the following dual of (P):

$$\inf \lambda \tag{7.34}$$

subject to

$$-\nabla f(x) + u^T \nabla h(x) + \lambda \nabla g(x) = 0 \tag{7.35}$$

$$(\mathrm{D_w}) \qquad -f(x) + u^T h(x) + \lambda g(x) \geq 0 \tag{7.36}$$

$$x \in C, u \in R^m, \lambda \in R, u \geq 0, \lambda \geq 0$$

$$(\lambda \text{ is not restricted in sign if } g \text{ is affine}). \tag{7.37}$$

As in the nondifferentiable case, weak, strong, and converse duality relations can be established. In the special case of a linear fractional program (7.7), the dual program ($\mathrm{D_w}$) can be reduced to the linear program

$$\inf \lambda \tag{7.38}$$

subject to

$$A^T u + \lambda d \geq c \tag{7.39}$$

$$-b^T u + \lambda \beta \geq \alpha \tag{7.40}$$

$$u \geq 0, \qquad \lambda \in R. \tag{7.41}$$

In the literature, several approaches to introduce duality in linear or concave fractional programming have been suggested. It can be shown that most of the resulting duals can be obtained as classical duals of the equivalent program (P′) or ($P'_=$). For details see Schaible (1973, 1976a, 1976c, 1978).

As in concave programming, the dual variables can be used to determine the sensitivity of an optimal solution with regard to right-hand-side changes. Consider the differentiable fractional program

$$\max \left\{ q(x) = \frac{f(x)}{g(x)} : h_j(x) \leq b_j, j = 1, \ldots, m, x \in C \right\}. \tag{7.42}$$

Let $\bar{q}(b)$ denote the optimal value of q as a function of $b = (b_1, \ldots, b_m)^T$. The b_j's may, for instance, represent capacities and $\bar{q}(b)$ the maximal return on investment.

Now let $\bar{x}$ be an optimal solution of (7.42) for $b = b^0$, and $(\bar{x}, \bar{u}, \bar{\lambda})$ an optimal solution of the dual $(\mathrm{D_W})$. Then under mild additional assumptions we have (Schaible, 1978)

$$\left.\frac{\partial \bar{q}(b)}{\partial b_j}\right|_{b=b^0} = \frac{1}{g(\bar{x})}\bar{u}_j, \qquad j = 1, \ldots, m. \tag{7.43}$$

Hence the marginal increase of $\bar{q}(b)$, with respect to b_j, is proportional to the dual variable $\bar{u}_j$. The value of $\partial\bar{q}/\partial b_j$ can be calculated from (7.43) once a dual optimal solution is known.

Example 7.3. Consider the quadratic fractional program

$$\max q(x) = \frac{-(x_1)^2 - (x_2)^2 + x_2 - 4}{x_1} \tag{7.44}$$

subject to

$$x_1 \geq b. \tag{7.45}$$

We want to calculate $\partial\bar{q}(b)/\partial b$ at $b = b^0 = 2$. For this we determine an optimal solution of $(\mathrm{D_W})$. After some simplifications, $(\mathrm{D_W})$ becomes

$$\min \lambda \tag{7.46}$$

subject to

$$2x_1 - u + \lambda = 0 \tag{7.47}$$

$$2x_2 = 1 \tag{7.48}$$

$$(x_1)^2 + (x_2)^2 - 2u \leq 4 \tag{7.49}$$

$$x_1 > 0, \qquad x_2 \in R, \qquad u \geq 0, \qquad \lambda \in R. \tag{7.50}$$

An optimal solution of the primal program is $\bar{x} = (2, 1/2)$, which together with $\bar{u} = 1/8$, $\bar{\lambda} = -31/8$ solves the dual. Hence, in view of (7.43),

$$\left.\frac{\partial \bar{q}(b)}{\partial b}\right|_{b=2} = -\frac{\bar{u}}{\bar{x}_1} = -\frac{1}{16}. \tag{7.51}$$

Sensitivity analysis, for the special case of a linear fractional program, has been extensively studied by Bitran and Magnanti (1976).

We have seen that with the help of the variable transformation (7.21) duality relations can be established for concave fractional programs. We now want to discuss duality of generalized fractional programs (7.5). Unfortunately, no variable transformation has been reported that reduces a concave generalized fractional program (GP) to a concave program. Duality, therefore, cannot be introduced in the same straightforward way as in the single ratio case. However, it is still possible to obtain a duality theory through convex analysis as shown by Jagannathan and Schaible (1983). We present this approach here. Other approaches can be found in Crouzeix (1977, 1981), Crouzeix, Ferland, and Schaible (1983), Flachs (1982), Gol'stein (1967), Passy and Keslassy (1979), and von Neumann (1945).

Consider the generalized fractional program

$$\text{(GP)} \qquad \sup\left\{\min_{1\le i\le p} \frac{f_i(x)}{g_i(x)}\right\} \tag{7.52}$$

subject to

$$h_j(x) \le 0, \qquad j = 1, \ldots, m, \tag{7.53}$$

$$x \in C.$$

Assume that C is a convex set in R^n, the f_i are concave on C, the g_i are positive and convex on C, the h_j are convex on C, and each f_i is nonnegative if not all g_i are affine.

We will consider only fractional programs (GP) whose feasible region is nonempty.

For notational convenience let $F(x) = (f_1(x), \ldots, f_p(x))^T$, $G(x) = (g_1(x), \ldots, g_p(x))^T$, and $h(x) = (h_1(x), \ldots, h_m(x))^T$.

Assume also that C is compact and F, G, and h are continuous on C although these conditions can be weakened. We obtain for the optimal value of (GP)

$$v(\text{GP}) = \sup\{\tau\colon F(x) - \tau G(x) \ge 0, h(x) \le 0, x \in C\}. \tag{7.54}$$

For fixed $\tau \in R$ let

$$S_\tau = \{x \in R^n: -F(x) + \tau G(x) \leq 0, h(x) \leq 0, x \in C\}. \qquad (7.55)$$

Then

$$v(\text{GP}) = \begin{cases} \sup\{\tau: S_\tau \neq \emptyset, \tau \geq 0\} & \text{if } G(x) \text{ is not affine} \\ \sup\{\tau: S_\tau \neq \emptyset\} & \text{otherwise.} \end{cases} \qquad (7.56)$$

Since there is at least one x that satisfies (7.53), it follows that $S_\tau \neq \emptyset$ for sufficiently small τ. If each g_i is affine, then $-F(x) + \tau G(x)$ is convex on C for all $\tau \in R$. On the other hand, if at least one g_i is not affine, then $v(\text{GP}) \geq 0$. Hence we can restrict τ in (7.56) to $\tau > 0$ in this case. For such τ $-F(x) + \tau G(x)$ is convex on C. Therefore, in the affine as well as the nonaffine case, the relevant sets S_τ in the supremum problem (7.56) are defined by convex inequalities.

We now apply the following theorem of alternatives for convex inequalities (Bohnenblust, Karlin, and Shapley, 1950; Mangasarian, 1969a): Given a continuous vector-valued convex function $k: X \to R^r$ defined on the (nonempty) compact convex set X of R^n, then exactly one of the following alternatives is true:

I. $k(x) \leq 0$, $x \in X$ is consistent;
II. there exists $y \in R^r$, $y \geq 0$ such that $y^T k(x) > 0$ for all $x \in X$.

Applying this result to the system of convex inequalities defining S_τ,

$$-F(x) + \tau G(x) \leq 0, \qquad h(x) \leq 0, \qquad x \in C, \qquad (7.57)$$

we find $S_\tau \neq \emptyset$ if and only if there does not exist $u \in R^p$, $v \in R^m$, $u \geq 0$, $v \geq 0$ such that

$$u^T(-F(x) + \tau G(x)) + v^T h(x) > 0 \qquad \text{for all } x \in C. \qquad (7.58)$$

Hence

$$v(\text{GP}) = \sup\{\tau: S_\tau \neq \emptyset\} = \inf\{\tau: S_\tau = \emptyset\} \qquad (7.59)$$

$$= \inf\{\tau: \text{there exist } u \geq 0, u \neq 0, v \geq 0$$

$$\text{such that } u^T(-F(x) + \tau G(x)) + v^T h(x) > 0 \text{ for all } x \in C\}. \qquad (7.60)$$

The condition $u \neq 0$ in (7.60) follows from (7.53) applying the above theorem of alternatives to the system $h(x) \leq 0$, $x \in C$.

Since the g_i are positive, we can divide in (7.60) by $u^T G(x) > 0$. The resulting problem is

$$\text{(GD)} \qquad \inf\left\{\sup_{x \in C} \frac{u^T F(x) - v^T h(x)}{u^T G(x)} : u \geq 0, u \neq 0, v \geq 0\right\}. \qquad (7.61)$$

We consider (GD) as a dual program of (GP). For the optimal value of (GD), we have shown that

$$v(\text{GD}) = v(\text{GP}). \qquad (7.62)$$

The dual (GD) is a generalized fractional program involving infinitely many ratios. The dual objective function in (7.61) is semistrictly quasiconvex on the feasible region since it is the supremum of linear ratios indexed by x.

In the single-ratio case ($p = 1$), the dual (GD) reduces to (D) in (7.31). The latter was obtained via the variable transformation (7.21). Hence the two approaches lead to the same dual in the single ratio case.

We now consider the dual (GD) for the special case of a linear generalized fractional program (GP)

$$\sup\left\{\min_{1 \leq i \leq p} \frac{e_{i.}^T x + \alpha_i}{d_{i.}^T x + \beta_i} : Ax \leq b, x \geq 0\right\}. \qquad (7.63)$$

Here A is an $m \times n$ matrix, $b \in R^m$. Furthermore, $e_{i.}^T (d_{i.}^T)$ denotes row i of a $p \times n$ matrix $E(D)$. The column j of E (D, A) is denoted by $e_{.j}$ $(d_{.j}, a_{.j})$. Let

$$\alpha = (\alpha_1, \ldots, \alpha_p)^T \quad \text{and} \quad \beta = (\beta_1, \ldots, \beta_p)^T.$$

Our assumptions on the generalized fractional program as stated above are satisfied by (7.63) if $D \geq 0$, $\beta > 0$ and (7.63) is feasible. Under these conditions $v(\text{GP}) = v(\text{GD})$ as we saw before. Furthermore, applying Farkas' theorem of alternatives (Mangasarian, 1969a) to (7.57) instead of the one by Bohnenblust *et al.*, (GD) becomes

$$\inf\left\{\sup\left[\frac{\alpha^T u + b^T v}{\beta^T u}, \sup_{1 \leq j \leq n} \frac{e_{.j}^T u - a_{.j}^T v}{d_{.j}^T u}\right] : u \geq 0, u \neq 0, v \geq 0\right\} \qquad (7.64)$$

with the convention

$$\frac{\rho}{0} = \begin{cases} +\infty & \text{if } \rho > 0 \\ -\infty & \text{if } \rho \leq 0. \end{cases} \tag{7.65}$$

Thus the dual of a linear generalized fractional program is again a linear generalized fractional program involving finitely many ratios. For more details on duality relations for linear generalized fractional programs, see Crouzeix, Ferland, and Schaible (1983).

7.4. Algorithms in Concave Fractional Programming

Several procedures for calculating an optimal solution to a fractional program have been suggested. The majority of these methods solve the concave fractional programs (7.3), and can be classified as follows:

I: direct solution of the quasiconcave program (P),
II: solution of the concave program (P′),
III: solution of the dual program (D),
IV: solution of the parametric concave program (P_μ).

In the remainder of this chapter we outline these methods.

Strategy I: Direct Solution of the Quasiconcave Program (P)

As we mentioned above, the objective function in a concave fractional program (P) is semistrictly quasiconcave, and it is pseudoconcave in the differentiable case. For these problems a local maximum is a global one. It is unique if either the numerator is strictly concave or the denominator is strictly convex. Furthermore, a solution of the first-order optimality conditions by Karush, Kuhn, and Tucker is a global maximum (see Chapter 3). Therefore, concave fractional programs can be solved by some of the standard concave programming techniques, as presented in Avriel (1976), Craven (1978), Martos (1975) and Zangwill (1969). For example, in case all constraints are linear, the method of Frank and Wolfe (1956) can be applied. According to this method at each iteration the objective function is linearized. In the case of a fractional program, we can either linearize the ratio as a whole (Mangasarian, 1969b) or linearize the numerator and denominator separately (Meister and Oettli, 1967). Then a sequence of linear programs or linear fractional programs is to be solved. In either way the solutions to the subproblems converge to a global maximum of (P).

If, in a concave fractional program, the objective function is also quasiconvex and if the feasible region is a compact convex polyhedron, then an optimal solution is attained at an extreme point (Martos, 1975). Linear fractional programs with a bounded feasible region belong to this class of quasimonotone programs (see Section 5.2). A simplexlike procedure can be applied to calculate a global maximum. It generates a sequence of adjacent extreme points of S with nondecreasing values of $q(x)$. The method is finite under mild additional assumptions. For details, see Martos (1975).

Strategy II: Solution of the Concave Program (P′)

Some of the concave programming algorithms are not suitable for pseudoconcave programming (Martos, 1975). Concave fractional programs, however, can be reduced to an equivalent concave program by a transformation of variables as we have seen in Propositions 7.2 and 7.3. This gives us access to *any* concave programming algorithm for solving concave fractional programs (P). That is, the equivalent problem (P′) or $(P'_{=})$ can be solved by any concave programming procedure (Avriel, 1976; Gill, Murray, and Wright, 1981). In case there is a special algebraic structure in the numerator and/or the denominator, one may prefer to solve (P′) rather than (P). As an example, we mention the maximum probability model (7.13). We saw that it can be reduced to a concave quadratic program (7.29), if the variable transformation (7.21) is applied. In this way the fractional program can be solved by a standard quadratic programming algorithm (Gill, Murray, and Wright, 1981; van de Panne, 1975).

Similarly, the transformation of a linear fractional program (7.7) yields a linear program (7.23). This can be solved by the simplex method (Dantzig, 1963). Wagner and Yuan (1968) have shown that several other methods in linear fractional programming are algorithmically equivalent to solving the linear program (7.23) with the simplex method. Bitran (1979) has done a numerical comparison of different solution procedures. For this he used randomly generated linear fractional programs. He gives some indication that the algorithm in Martos (1964) is computationally superior to several of the other methods.

We have seen in Proposition 7.1 that a restricted class of concave fractional programs can be reduced to an equivalent concave program by a range transformation. In contrast to the variable transformation (7.21), such a range transformation does not change the feasible region. It may therefore be particularly useful for fractional programs where variables are restricted to be integers. For integer fractional programming algorithms, see Bitran and Magnanti (1976), Chandra and Chandramohan (1980), Granot and Granot (1977) and Schaible and Ibaraki (1983).

Strategy III: Solution of the Dual Program (D)

Sometimes it may be advantageous to solve a dual program of (P′) rather than (P′). For instance, for concave quadratic fractional programs with an affine denominator, the dual (D_w) becomes a linear program with one additional concave quadratic constraint (Schaible, 1974b). Example 7.1 illustrates this.

As in the single ratio case, the dual (DP) of a generalized fractional program (GP) may have computational advantages over the primal. For a linear problem (GP) the dual is again of this type [see (7.64)]. In contrast to the primal, it has only nonnegativity constraints, if $D > 0$ is assumed.

Strategy IV: Solution of the Parametric Concave Program (P_μ)

When problem (P_μ) in (7.18) is used, the zero $\bar{\mu}$ of the decreasing function

$$F(\mu) = \max\{f(x) - \mu g(x): x \in S\} \tag{7.66}$$

has to be calculated. The disadvantage of solving a parametric problem rather than the parameter-free (P′) may be outweighted by other benefits. For instance, for a quadratic fractional program, the structure of the model is not well exploited when (P′) is used, whereas (P_μ) is a concave quadratic program for each μ. This problem can be treated by standard techniques (Gill, Murray, and Wright, 1981; van de Panne, 1975).

The zero of $F(\mu)$ can be calculated either by a parametric programming technique or by a method that solves (P_μ) for discrete values $\mu = \mu_i$ converging to $\bar{\mu}$. Such an iterative procedure was suggested by Dinkelbach (1967). For convergence properties and modifications of Dinkelbach's algorithm, see Ibaraki, Ishii, Iwase, Hasegawa, and Mine (1976), Ibaraki (1981, 1983), Schaible (1976b), and Schaible and Ibaraki (1983). An extension of Dinkelbach's algorithm to generalized fractional programs (GP) was suggested by Crouzeix, Ferland, and Schaible (1985, 1986).

In the following we list a selected group of references for fractional programming. A comprehensive bibliography for this growing field of nonlinear programming was provided by Schaible (1982).

References

ASHTON, D. J., and ATKINS, D. R. (1979), Multi-criteria programming for financing planning, *J. Oper. Res. Soc.* **30**, 259–270.

AVRIEL, M. (1976), *Nonlinear Programming: Analysis and Methods*, Prentice-Hall, Englewood Cliffs, New Jersey.

BARRODALE, I. (1973), Best rational approximation and strict quasi-convexity, *SIAM J. Numerical Anal.* **10**, 8–12.

BEREANU, B. (1965), Decision regions and minimum risk solutions in linear programming, in *Colloquium on Applications of Mathematics to Economics*, Edited by A. Prekopa, Publication House of the Hungarian Academy of Sciences, Budapest, Hungary.

BITRAN, G. R. (1979), Experiments with linear fractional problems, *Naval Res. Logistics Q.* **26**, 689–693.

BITRAN, G. R., and MAGNANTI, T. L. (1976), Duality and sensitivity analysis for fractional programs, *Oper. Res.* **24**, 675–699.

BOHNENBLUST, H. F., KARLIN, S., and SHAPLEY, L. S. (1950), Solutions of discrete two-person games, in *Contributions to the Theory of Games*, Vol. I, Annals of Mathematical Studies No. 24, pp. 51–72.

CAMBINI, A. (1981), An algorithm for a special class of fractional programs, in *Generalized Concavity in Optimization and Economics*, Edited by S. Schaible and W. T. Ziemba, Academic Press, New York.

CHANDRA, S., and CHANDRAMOHAN, M. (1980), A branch and bound method for integer nonlinear fractional programs, *Z. Angew. Math. Mech.* **60**, 735–737.

CHARNES, A., and COOPER, W. W. (1962), Programming with linear fractional functionals, *Naval Res. Logistics Q.* **9**, 181–186.

CHARNES, A., and COOPER, W. W. (1963), Deterministic equivalents for optimizing and satisficing under chance constraints, *Oper. Res.* **11**, 18–39.

CHARNES, A., and COOPER, W. W. (1977), Goal programming and multiobjective optimization, Part I, *Eur. J. Oper. Res.* **1**, 39–54.

CHOO, E. U. (1980), Multicriteria linear fractional programming. Ph.D. thesis, University of British Columbia, Vancouver, B.C.

CRAVEN, B. D. (1978), *Mathematical Programming and Control Theory*, Chapman & Hall, London, England.

CROUZEIX, J. P. (1977), Contributions à l'étude des fonctions quasi-convexes, Doctoral Thesis, Université de Clermont, France.

CROUZEIX, J. P. (1981), A duality framework in quasiconvex programming, in *Generalized Concavity in Optimization and Economics*, Edited by S. Schaible and W. T. Ziemba, Academic Press, New York.

CROUZEIX, J. P., FERLAND, J. A., and SCHAIBLE, S. (1983), Duality in generalized linear fractional programming, *Math. Programming* **27**, 342–354.

CROUZEIX, J. P., FERLAND, J. A., and SCHAIBLE, S. (1985), An algorithm for generalized fractional programs, *J. Optimization Theory Appl.* **47**, 35–49.

CROUZEIX, J. P., FERLAND, J. A., and SCHAIBLE, S. (1986), A note on an algorithm for generalized fractional programs, *J. Optimization Theory Appl.* **50**, 183–187.

DANTZIG, G. B. (1963), *Linear Programming and Extensions*, Princeton University Press, Princeton, New Jersey.

DANTZIG, G. B., BLATTNER, W., and RAO, M. R. (1966), Finding a cycle in a graph with minimum cost to time ratio with applications to a ship routing problem, in *Theory of Graphs*, Dunod, Paris, and Gordon and Breach, New York.

DERMAN, C. (1962), On sequential decisions and Markov chains, *Management Sci.* **9**, 16–24.

DINKELBACH, W. (1967), On nonlinear fractional programming, *Management Sci.* **13**, 492–498.

EICHHORN, W. (1972), Effektivität von Produktionsverfahren, *Oper. Res. Verfahren* **12**, 98–115.

EICHHORN, W. (1978), *Functional Equations in Economics*, Addison-Wesley, Reading, Massachusetts.

FLACHS, J. (1982), Global saddle-point duality for quasi-concave programs II, *Math. Programming* **24**, 326–345.

FOX, B. (1969), Finding minimum cost-time ratio circuits, *Oper. Res.* **17**, 546–550.

FRANK, M., and WOLFE, P. (1956), An algorithm for quadratic programming, *Naval Res. Logistics Q.* **3**, 95–110.

GEOFFRION, A. M. (1967), Solving bi-criterion mathematical programs, *Oper. Res.* **15**, 39–54.

GEOFFRION, A. M. (1971), Duality in nonlinear programming: A simplified applications-oriented development, *SIAM Rev.* **13**, 1–37.

GILL, P. E., MURRAY, A., and WRIGHT, M. H. (1981), *Practical Optimization*, Academic Press, New York.

GILMORE, P. C., and GOMORY, R. E. (1963), A linear programming approach to the cutting stock problem, Part II, *Oper. Res.* **11**, 863–888.

GOL'STEIN, E. G. (1967), Dual problems of convex and fractionally-convex programming in functional spaces, *Sov. Math. Dokl.* **8**, 212–216.

GRANOT, D., and GRANOT, F. (1977), On integer and mixed integer fractional programming problems, *Ann. Discrete Math.* **1**, 221–231.

GUTENBERG, E. (1975), *Einführung in die Betriebswirtschaftslehre*, Gabler, Wiesbaden, Germany.

HEINEN, E. (1971), *Grundlagen betriebswirtschaftlicher Entscheidungen, das Zielsystem der Unternehmung*, Gabler, Wiesbaden, Germany.

IBARAKI, T. (1981), Solving mathematical programming problems with fractional objective functions, in *Generalized Concavity in Optimization and Economics*, Edited by S. Schaible and W. T. Ziemba, Academic Press, New York.

IBARAKI, T. (1983), Parametric approaches to fractional programs, *Math. Programming* **26**, 345–362.

IBARAKI, T., ISHII, H., IWASE, J., HASEGAWA, T., and MINE, H. (1976), Algorithms for quadratic fractional programming problems, *J. Oper. Res. Soc. Jpn.* **19**, 174–191.

JAGANNATHAN, R. (1966), On some properties of programming problems in parametric form pertaining to fractional programming, *Management Sci.* **12**, 609–615.

JAGANNATHAN, R., and SCHAIBLE, S. (1983), Duality in generalized fractional programming via Farkas' lemma, *J. Optimization Theory Appl.* **41**, 417–424.

KALLBERG, J. D., and ZIEMBA, W. T. (1981), Generalized concave functions in stochastic programming and portfolio theory, in *Generalized Concavity in Optimization and Economics*, Edited by S. Schaible and W. T. Ziemba, Academic Press, New York.

KLEIN, M. (1963), Inspection-maintenance-replacement schedule under Markovian deterioration, *Management Sci.* **9**, 25–32.

KORNBLUTH, J. S. H., and STEUER, R. E. (1981a), Multiple objective linear fractional programming, *Management Sci.* **27**, 1024–1039.

KORNBLUTH, J. S. H., and STEUER, R. E. (1981b), Goal programming with linear fractional criteria, *Eur. J. Oper. Res.* **8**, 58-65.

LASDON, L. S. (1970), *Optimization Theory for Large Systems.* McMillan, New York.

LAWLER, E. L. (1976), *Combinatorial Optimization: Networks and Matroids.* Holt, Rinehart & Winston, New York.

MANGASARIAN, O. L. (1969a), *Nonlinear Programming.* McGraw-Hill, New York.

MANGASARIAN, O. L. (1969b), Nonlinear fractional programming, *J. Oper. Res. Soc. Jpn.* **12**, 1-10.

MARTOS, B. (1964), Hyperbolic programming, *Naval Res. Logistics Q.* **11**, 135-155 [originally published in *Math. Institute of Hungarian Acad. Sci.* **5**, 383-406 (1960) (Hungarian)].

MARTOS, B. (1975), *Nonlinear Programming: Theory and Methods.* North-Holland, Amsterdam.

MEISTER, B., and OETTLI, W. (1967), On the capacity of a discrete, constant channel, *Inf. Control* **11**, 341-351.

MJELDE, K. M. (1978), Allocation of resources according to a fractional objective, *Eur. J. Oper. Res.* **2**, 116-124.

MJELDE, K. M. (1983), *Methods on the Allocation of Limited Resources*, J. Wiley & Sons, Chichester, England.

NOBLE, B. (1969), *Applied Linear Algebra*, Prentice-Hall, Englewood Cliffs, New Jersey.

OHLSON, J. A., and ZIEMBA, W. T. (1976), Portfolio selection in a log-normal market when the investor has a power utility function, *J. Financial Quant. Anal.* **11**, 57-71.

PACK, L. (1962), Maximierung der Rentabilitaet als preispolitisches Ziel, in *Zur Theorie der Unternehmung*, Edited by H. Koch, Festschrift für E. Gutenberg, Gabler, Wiesbaden, Germany.

PACK, L. (1965), Rationalprinzip, Gewinnprinzip und Rentabilität, *Z. Betriebswirtsch.* **35**, 525-551.

PASSY, U., and KESLASSY, A. (1979), Pseudo duality and duality for explicitly quasiconvex functions, Mimeograph Series No. 249, Faculty of Industrial Engineering and Management, Technion, Haifa, Israel.

RITTER, K. (1967), A parameteric method for solving certain nonconcave maximization problems, *J. Comput. Syst. Sci.* **1**, 44-54.

ROCKAFELLAR, R. T. (1970), *Convex Analysis*, Princeton University Press, Princeton, New Jersey.

SCHAIBLE, S. (1973), Fractional programming: Transformations, duality and algorithmic aspects, Technical Report 73-9, Department of Operations Research, Stanford University, Stanford, California.

SCHAIBLE, S. (1974a), Maximization of quasiconcave quotients and products of finitely many functionals, *Cah. Cent. Etud. Rech. Opér.* **16**, 45-53.

SCHAIBLE, S. (1974b), Parameter-free convex equivalent and dual programs of fractional programs, *Z. Oper. Res.* **18**, 187-196.

SCHAIBLE, S. (1976a), Fractional programming: I. Duality, *Management Sci.* **22**, 858-867.

SCHAIBLE, S. (1976b), Fractional programming: II, On Dinkelbach's algorithm, *Management Sci.* **22**, 868-872.

SCHAIBLE, S. (1976c), Duality in fractional programming: A unified approach, *Oper. Res.* **24**, 452-461.

SCHAIBLE, S. (1977), On the sum of a linear and linear-fractional function, *Naval Res. Logistics Q.* **24**, 691-693.

SCHAIBLE, S. (1978), *Analyse und Anwendungen von Quotientenprogrammen*, Mathematical Systems in Economics No. 42, Hain-Verlag, Meisenheim, Germany.

SCHAIBLE, S. (1982), Bibliography in fractional programming, *Z. Oper. Res.* **26**, 211-241.

SCHAIBLE, S. (1983), Bicriteria quasiconcave programs, *Cah. Cent. Etud. Rech. Opér.* **25**, 93-101.

SCHAIBLE, S. (1984), Simultaneous optimization of absolute and relative terms, *Z. Angew. Math. Mech.* **64**, 363–364.

SCHAIBLE, S., and IBARAKI, T. (1983), Fractional programming (Invited Review), *Eur. J. Oper. Res.* **12**, 325–338.

SCHAIBLE, S., and ZIEMBA, W. T. (Editors) (1981), *Generalized Concavity in Optimization and Economics*, Academic Press, New York.

SCHAIBLE, S., and ZIEMBA, W. T. (1985), Generalized concavity of a function in portfolio theory, *Z. f. Oper. Res.* **29**, 161–186.

STOER, J., and WITZGALL, CH. (1970), *Convexity and Optimization in Finite Dimensions I.* Springer, Berlin.

VAN DE PANNE, C. (1975), *Methods for Linear and Quadratic Programming*, North-Holland, Amsterdam.

VON NEUMANN, J. (1945), A model of general economic equilibrium, *Rev. Econ. Stud.* **13**, 1–9.

WAGNER, H. M., and YUAN, J. S. C. (1968), Algorithmic equivalence in linear fractional programming, *Management Sci.* **14**, 301–306.

WARBURTON, A. R. (1981), Topics of multiple criteria optimization, Ph.D. thesis, University of British Columbia, Vancouver, B.C.

ZANGWILL, W. I. (1969), *Nonlinear Programming: A Unified Approach*, Prentice-Hall, Englewood Cliffs, New Jersey.

ZIEMBA, W. T. (1974), Choosing investment portfolios when the returns have a stable distribution, in *Mathematical Programming in Theory and Practice*, Edited by P. L. Hammer and G. Zoutendijk, North-Holland, Amsterdam.

ZIEMBA, W. T., PARKAN, C., and BROOKS-HILL, R. (1974), Calculation of investment portfolios with risk-free borrowing and lending, *Management Sci.* **21**, 209–222.

8

Concave Transformable Functions

In this chapter we address an important class of generalized concave functions that have been the subject of extensive recent research. This is the family of concave transformable (or transconcave) functions. As we have already pointed out in Chapter 4, there are several economic applications where it is important to be able to express a convex preference ordering by means of a concave utility function or, in other words, to be able to transform a certain quasiconcave function into a concave function having the same upper-level set mapping. In the mathematical programming context, it may be of value to know whether a certain nonconcave optimization problem can be transformed, via a one-to-one transformation, into a concave program. This, for example, may facilitate the use of concave programming duality theory in cases where such a transformation exists. Thus, in the first section of this chapter, we shall investigate properties of functions that can be transformed into concave functions by means of an increasing transformation. Our attention will be devoted mainly to classifications of these functions. In Section 8.2 we shall discuss necessary and sufficient conditions for the concavifiability of a twice continuously differentiable function by means of an increasing transformation. Several seemingly different approaches for deriving such conditions will be presented within a consistent unified framework.

The notion of concavifiability, as discussed in Sections 8.1 and 8.2, can be extended by considering a wider family of functions that are concave transformable by means of domain and range transformations. Therefore, in Section 8.3 we discuss functions that can be transformed into quasiconcave functions using a one-to-one transformation of their domain, and then into concave functions by an increasing function.

8.1. Concave Transformable Functions: Range Transformation

In this section we shall study functions that are concave transformable by a continuous increasing function. Let f be a real-valued continuous function defined on the convex set $C \subset R^n$, and denote by $I_f(C)$ the range of f; that is, the image of C under f. We have the following definition.

Definition 8.1. A function f is said to be G-concave if there exists a continuous real-valued increasing function G defined on $I_f(C)$, such that $G(f(x))$ is concave over C. Alternatively, letting G^{-1} denote the inverse of G, if

$$f(\lambda x^1 + (1-\lambda)x^2) \geq G^{-1}(\lambda G(f(x^1)) + (1-\lambda)G(f(x^2))) \tag{8.1}$$

holds for every $x^1 \in C$, $x^2 \in C$ and $0 \leq \lambda \leq 1$. Similarly, f is said to be G-convex if $G(f(x))$ is convex over C, or alternatively, if the inequality in (8.1) is reversed.

Note that G^{-1} is increasing if and only if G is increasing. Thus, it is easy to establish the equivalence of (8.1) and the concavity of Gf. In the sequel we shall discuss only properties of G-concave functions; the analogous results for G-convex functions can be established in the same manner. It should be noted, however, that the symmetry between (generalized) concavity and (generalized) convexity, which holds for all the families of functions introduced in Chapters 2 and 3, does not hold for G-concave functions. That is, if f is G-concave then $(-f)$ is not necessarily G-convex; it is, however, $\bar{G}$-convex with $\bar{G}(t) = -G(-t)$.

Concave (or convex) transformable functions were first considered by de Finetti (1949), and Fenchel (1951, 1956). A thorough discussion of this topic can be found in Kannai (1977, 1981). See also, Avriel (1972), Avriel and Zang (1974), Crouzeix (1977), Horst (1971a, b), and Lindberg (1981). These references and others such as, Aumann (1975), Avriel (1976), Avriel and Schaible (1978), Diewert (1978), Gerencsér (1973), Kannai and Mantel (1978), and Schaible and Zang (1980) discuss also conditions that are necessary and sufficient for concavifiability. Properties of individual demand functions, induced by concavifiable utility functions, were discussed by Hurwicz, Jordan, and Kannai (1984) and Kannai (1985).

In order to establish a very basic property of G-concave functions we note that in case a G-concave function f is not semistrictly quasiconcave on C, then by Definition 3.11 there exist points $x^1 \in C$, $x^2 \in C$, $f(x^1) \neq f(x^2)$, and a real number $0 < \bar{\lambda} < 1$ such that

$$f(\bar{\lambda}x^1 + (1 - \bar{\lambda})x^2) \le \min\{f(x^1), f(x^2)\} \tag{8.2}$$

holds. But since the inequality

$$G^{-1}(\min\{G(f(x^1)), G(f(x^2))\}) < G^{-1}(\bar{\lambda}G(f(x^1)) + (1 - \bar{\lambda})G(f(x^2))) \tag{8.3}$$

is satisfied for any increasing G, we have from the above two inequalities a contradiction to (8.1). We have just proved a necessary condition for G-concavity:

Proposition 8.1. Every G-concave function on a convex set C is also semistrictly quasiconcave on C.

Note that by Proposition 3.31, the continuity of f implies that f is also quasiconcave. The fact that a concavifiable function must be quasiconcave or, in other words, that G-concave functions must have convex upper-level sets is a well-known property (see Fenchel, 1951). Unfortunately, a converse result to the above proposition does not hold, and a semistrictly quasiconcave function is not necessarily G-concave, as the following example shows.

Example 8.1. Let $C = \{x \in R: 0 \le x \le 5\}$ and

$$f(x) = \begin{cases} 4x & \text{if } 0 \le x \le 1 \\ 5 - x & \text{if } 1 \le x \le 2 \\ [1 - (x-2)^2]^{1/2} + 2 & \text{if } 2 \le x \le 3 \\ 5 - x & \text{if } 3 \le x \le 5. \end{cases} \tag{8.4}$$

The function f is semistrictly quasiconcave on C, and it is shown in Figure 8.1. It can be easily seen that f is concave on [0, 2], [2, 3], and [3, 5]. If we consider the set $C_1 = [1, 5]$ then f is not concave on C_1. It is, however, G-concave on C_1 with

$$G(t) = \begin{cases} t & \text{if } 0 \le t \le 2 \quad \text{or} \quad 3 \le t \le 4 \\ 3 - [-(t-2)^2 + 1]^{1/2} & \text{if } 2 \le t \le 3 \end{cases} \tag{8.5}$$

and $Gf(x) = 5 - x$ on C_1. If we consider, however, the whole of C, then f is neither concave nor G-concave. Suppose, to the contrary, that such a function $G(t)$ exists. Note that the slope of f once the point $x = 2$ is

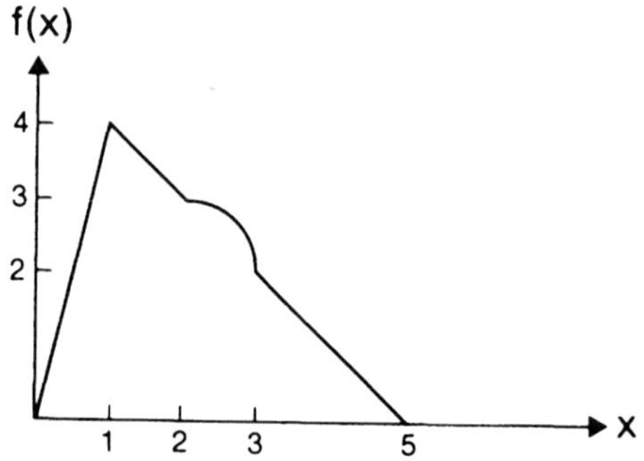

Figure 8.1. The function given by (8.4).

approached from above is zero and since $x = 2$ is not a maximum point, the slope of Gf as x approaches 2 from above cannot be zero. This implies that the slope of G as t approaches 3 from below must be unbounded, but then the slope of Gf as x approaches 3/4 from below will be unbounded. However, this cannot happen in the interior of the domain of a concave Gf. In a similar way, since the slope of f at $x = 3$ once it is approached from below is unbounded it turns out that the slope of G as t approaches 2 from above goes to zero, implying that the slope of Gf as x approaches 1/2 from above goes to zero too. But this contradicts the assumed concavity of Gf since $x = 1/2$ is not a maximum.

One of the main questions to be addressed in Section 8.2 is to find conditions under which a semistrictly quasiconcave function is G-concave. This problem is posed differently in the economics literature (Debreu, 1976; Kannai, 1974, 1977, 1980, 1981): Find conditions under which a continuous, complete, and convex preference ordering can be the upper-level set mapping of a concave function. The second problem is a more general one, in the sense that the knowledge of the *function* inducing the preference ordering is not required in the statement of the problem. However, once a quasiconcave function $f_1(x)$ inducing a specific continuous, complete, and convex preference ordering is known, then any concave function $f_2(x)$ that induces the same preference ordering can be shown to be an increasing transformation of $f_1(x)$. This will imply that, in this case, the second problem is equivalent to the first one. To show this we let f_1 and f_2 be *any* two functions inducing the same preference ordering on C; that is, for every $x^1 \in C, x^2 \in C$

$$f_1(x^1) > f_1(x^2) \quad \Leftrightarrow \quad f_2(x^1) > f_2(x^2) \tag{8.6}$$

holds. Then $f_2(x) = G(f_1(x))$, where G is the increasing function given by

$$G(t) = \{f_2(x): f_1(x) = t\}, \tag{8.7}$$

for every $t \in I_{f_1}(C)$. Note that G is continuous in case both f_1 and f_2 are so, and that we have established the correspondence $f_2(x) = G(f_1(x))$ without imposing concavity or quasiconcavity assumptions.

Before we turn to some properties of G-concave functions, we exhibit some examples for such functions.

Example 8.2. Let

$$f(x) = xe^{-x} \tag{8.8}$$

for $x \in C$, where

$$C = \{x \in R: x > 0\}. \tag{8.9}$$

Then f is strictly pseudoconcave on C. It is also G-concave with

$$G(t) = \log(t), \qquad t > 0 \tag{8.10}$$

since $G(f(x)) = \log x - x$ is a concave function on C.

Example 8.3. Let

$$f(x) = -(x_1)^{3/2}/x_2, \tag{8.11}$$

and let

$$C = \{x \in R^2: x_1 > 0, x_2 > 0\}. \tag{8.12}$$

Then f can be expressed as $f(x) = \bar{f}(\phi_1(x), \phi_2(x))$, where $\bar{f}(y) = y_1/y_2$ for $y_1 < 0$, $y_2 > 0$, and $\phi_1(x) = -(x_1)^{3/2}$, $\phi_2(x) = x_2$, for $x_1 > 0$, $x_2 > 0$. It was shown by Avriel and Schaible (1978) that $\bar{f}$ is pseudoconcave for $y_1 < 0$. Moreover, all other assumptions of Proposition 5.8 hold. Consequently, f is pseudoconcave. It is also G-concave over C, where

$$G(t) = -(t)^2 \tag{8.13}$$

for $t < 0$, since the function

$$G(f(x)) = -(x_1)^3/(x_2)^2 \tag{8.14}$$

is concave over C, owing to Proposition 3.2 in Schaible (1972).

Example 8.4. Let

$$Q(x) = \tfrac{1}{2}x^TAx + b^Tx, \tag{8.15}$$

where A is a symmetric $n \times n$ matrix and $b \in R^n$. Assume that $Q(x)$ is quasiconcave but not concave on a solid convex set C of R^n. Then, by Theorem 6.3, we have that $Q(x)$ is G-concave with

$$G(t) = (t - \delta)^{1/2}, \tag{8.16}$$

where $\delta \le Q(x)$ for all $x \in C$ must hold. We shall later on analyze this function in a more detailed manner. (See Example 8.19.)

Example 8.5. Let

$$f(x) = (x)^3 + x \tag{8.17}$$

and

$$C = \{x \in R: -1 \le x \le 1\}. \tag{8.18}$$

Then f is G-concave over C with

$$G(t) = -e^{-(9/8)t}. \tag{8.19}$$

We shall return to this example in the sequel when we discuss the family of r-concave functions, a subfamily of G-concave functions with $G(t) = -e^{-rt}$ for some positive r. There we shall show how the value of $r = 9/8$ was obtained. (See Example 8.16.)

The reader is also referred to Proposition 7.1 for a family of functions, consisting of the ratio of a positive concave and a convex function, which is, under certain conditions, G-concave.

We now turn to discuss some properties of G-concave functions. First, we establish a hierarchy among G-concave functions.

Proposition 8.2 (Avriel and Zang, 1974). Let f be G_1-concave on the convex set $C \subset R^n$, and let G_2 be a continuous increasing function on $I_f(C)$. If the function $\tau(t) = G_2G_1^{-1}(t)$ is concave on the image under G_1 of the range of f, then f is also G_2-concave over C.

Proof. By (8.1) for every $x^1 \in C$, $x^2 \in C$ and $0 \le \lambda \le 1$

$$f(\lambda x^1 + (1 - \lambda)x^2) \ge G_1^{-1}(\lambda G_1 f(x^1) + (1 - \lambda)G_1 f(x^2)) \tag{8.20}$$

holds, and consequently

$$\begin{aligned} G_2 f(\lambda x^1 + (1 - \lambda)x^2) &\ge G_2G_1^{-1}(\lambda G_1 f(x^1) + (1 - \lambda)G_1 f(x^2)) \\ &\ge \lambda G_2 f(x^1) + (1 - \lambda)G_2 f(x^2) \end{aligned} \tag{8.21}$$

is satisfied where the first inequality in (8.21) follows the monotonicity of G_2, and the second inequality from the concavity of τ. The first and last terms in (8.21) imply the G_2-concavity of f. □

A simple example shows that the converse to the above proposition is not true. Let $f(x) = -(x)^2$. Then f is G-concave both with $G_1(t) = -e^{-2t}$ and $G_2(t) = -e^{-t}$. However, $\tau(t) = G_2G_1^{-1}(t) = -(-t)^{1/2}$ is a convex function for $t \leq 0$.

Thus, if f is both G_1 and G_2-concave then $G_2G_1^{-1}(t)$ is not necessarily concave. If, however, f is G_1-concave and for every G_2 for which f is G_2-concave, $G_2G_1^{-1}(t)$ is concave, then G_1 can be considered as a *least concavifying function* for f. Note that this concept is not uniquely defined since $\alpha G_1(t) + \beta$ will also be a least concavifying function for f. Hence $G_1 f$ can be regarded as a *least concave representation* of the preference ordering induced by f, that is, $G_2 f$ is "more concave" than $G_1 f$. The above concepts were introduced by de Finetti (1949) and discussed further by Debreu (1976) and Kannai (1977, 1980, 1981).

As a result of Proposition 8.2 we have the following corollary.

Corollary 8.3. Let f be G-concave on the convex set $C \subset R^n$. If G is convex on $I_f(C)$, then f is a concave function over C.

Proof. Let t^1 and t^2 be any two points in $I_f(C)$. Then

$$G(\lambda t^1 + (1-\lambda)t^2) \leq \lambda G(t^1) + (1-\lambda)G(t^2) \tag{8.22}$$

holds for every $0 \leq \lambda \leq 1$. Let $\bar{t}^1 = G(t^1)$, $\bar{t}^2 = G(t^2)$. Then for each pair of points $\bar{t}^1$ and $\bar{t}^2$ in the image under G of $I_f(C)$

$$G(\lambda G^{-1}(\bar{t}^1) + (1-\lambda)G^{-1}(\bar{t}^2)) \leq \lambda \bar{t}^1 + (1-\lambda)\bar{t}^2 \tag{8.23}$$

holds for every $0 \leq \lambda \leq 1$. Using both sides of (8.22) as arguments for G^{-1}, we can see that G^{-1} is concave. Letting $G_1 = G$, $G_2 = t$, we obtain that $\tau(t) = G_2G^{-1}(t)$ is concave. By Proposition 8.2, f is G_2-concave. But G_2 is the identity function; hence f is concave. □

As an illustration of this corollary, note that every G-concave function with $G(t) = -\log(-t)$, $t < 0$, is also concave. A family of functions related to this particular G was discussed by Klinger and Mangasarian (1968) under the term of log-convex functions. [G-convexity with $G(t) = \log(t)$.] It was shown there that a log-convex function is also convex.

Unfortunately, some elementary algebraic properties of concave functions and their generalizations, discussed in Chapters 2 and 3, do not have their analogs for G-concave functions. Thus, if f is G-concave then $\alpha f + \beta$ is not necessarily G-concave for $\alpha > 0$; it is, however, $\bar{G}$-concave, where $\bar{G}(t) = G([t - \beta]/\alpha)$. It was pointed out in Chapter 3 that, unlike concave functions, families of generalized concave functions are usually not closed under addition. This is also the case with G-concave functions, and the sum of two G-concave functions is not necessarily G-concave. We return to this question in Section 8.3, where functions that are concave transformable by some one-to-one domain and range transformations are discussed. There we show that every family of G-concave functions is closed under some general addition operation. Obviously, one may wonder whether the sum of a G_1-concave and a G_2-concave function is G_3-concave? This conjecture is false too. Take $f_1(x_1, x_2) = -(x_1)^3$, $f_2(x_1, x_2) = -(x_2)^2$, and $C = \{x \in R^2: x_1 < 0, x_2 < 0\}$. Then f_1 is G_1-concave with $G_1(t) = t^{1/3}$ and f_2 is G_2-concave with $G_2(t) = t$. However, their sum $f(x) = -(x_1)^3 - (x_2)^2$ is not G-concave on C, as we shall see in Example 8.12.

There are, however, properties of concave functions and their generalizations, discussed in Chapter 3, that do hold for G-concave functions:

Proposition 8.4 (Avriel and Zang, 1974). Let $f_1, f_2, \ldots,$ be a finite or infinite collection of G-concave functions on a convex set $C \subset R^n$. Then the function

$$f(x) = \inf\{f_1(x), f_2(x), \ldots\} \tag{8.24}$$

is G-concave over C.

Proposition 8.5 (Avriel and Zang, 1974). The function f defined on the convex set $C \subset R^n$ is G-concave if and only if for every $x^1 \in C$, $x^2 \in C$ the single-variable function

$$F(\lambda) = f(\lambda x^1 + (1 - \lambda)x^2) \tag{8.25}$$

is G-concave for $0 \leq \lambda \leq 1$.

The proofs of the last two propositions are straightforward and are left for the reader.

We now turn to characterize differentiable G-concave functions. First we have the following proposition.

Proposition 8.6. Let f be differentiable on the convex set $C \subset R^n$ and let G be a differentiable increasing function on $I_f(C)$. Then f is G-concave if and only if for every $x^1 \in C$, $x^2 \in C$

$$G(f(x^2)) \leq G(f(x^1)) + G'(f(x^1))(x^2 - x^1)^T \nabla f(x^1) \tag{8.26}$$

holds. In case f and G are twice continuously differentiable on the above domains, then f is G-concave if and only if the matrix

$$\nabla^2 G(f(x)) = G'(f(x))\nabla^2 f(x) + G''(f(x))\nabla f(x)\nabla f(x)^T \tag{8.27}$$

is negative semidefinite for every $x \in C$.

Proof. From Theorem 2.12 and Definition 8.1 we obtain (8.26), and from Theorem 2.14 and Definition 8.1 we obtain the negative semidefiniteness of (8.27). □

In Proposition 8.1 we showed that G-concave functions are semistrictly quasiconcave. We now show that if differentiable G-concave functions are considered, then an inclusion in a more restricted family is obtained.

Proposition 8.7. Let f be a differentiable G-concave function on the convex set $C \subset R^n$ and let G be differentiable on $I_f(C)$. Then f is pseudoconcave.

Proof. Let $x^1 \in C, x^2 \in C$ and suppose that $(x^2 - x^1)^T \nabla f(x^1) \leq 0$ holds. Since $G'(f(x^1)) \geq 0$, we have from (8.26) that $G(f(x^2)) \leq G(f(x^1))$. It follows that $f(x^2) \leq f(x^1)$ and f is pseudoconcave. □

To conclude the discussion of G-concave functions, we obtain from Theorems 3.37 and 3.40 and Propositions 8.1 and 8.7 the following theorem.

Theorem 8.8. Let f be a G-concave function on the convex set $C \subset R^n$. Then, every local maximum of f is a global one. If, in addition, f and G are differentiable, then every stationary point of f is a global maximum over C. In both cases, the set of global maximizers is convex.

We now turn to a particular subfamily of G-concave functions. This family includes the functions which can be concavified using some exponential transformation.

Definition 8.2. Let f be a real-valued function defined on the convex set $C \subset R^n$. It is said to be r-concave if there exists a nonnegative number r, such that

$$f(\lambda x^1 + (1-\lambda)x^2) \geq \begin{cases} -\log\{\lambda\, e^{-rf(x^1)} + (1-\lambda)\, e^{-rf(x^2)}\}^{1/r} & \text{if } r \neq 0 \\ \lambda f(x^1) + (1-\lambda) f(x^2) & \text{if } r = 0 \end{cases} \tag{8.28}$$

holds for every $x^1 \in C$, $x^2 \in C$, and $0 \leq \lambda \leq 1$. Similarly, f is said to be r-convex if

$$f(\lambda x^1 + (1-\lambda)x^2) \leq \begin{cases} \log\{\lambda\, e^{rf(x^1)} + (1-\lambda)\, e^{rf(x^2)}\}^{1/r} & \text{if } r \neq 0 \\ \lambda f(x^1) + (1-\lambda) f(x^2) & \text{if } r = 0 \end{cases} \tag{8.29}$$

holds for every $x^1 \in C$, $x^2 \in C$ and $0 \leq \lambda \leq 1$.

Note that the terms corresponding to $r = 0$ in (8.28) and (8.29) can be obtained as the limit as $r \to 0$ of the terms corresponding to $r \neq 0$, respectively, and that a 0-concave (convex) function is concave (convex). The concept of r-convexity was introduced independently by Avriel (1972), Horst (1971a), and Martos (1966) (see also Avriel, 1976, and Martos, 1975). The definition of r-concave functions we use here differs from Avriel's by a trivial change of sign. Also, note that r-concave (utility) functions were considered in Aumann (1975), and that algorithms for problems involving such functions were considered by Avriel (1973).

The family of r-concave functions is an important subfamily of G-concave functions, since the form of G is well specified, and, as we shall later show, in certain cases it is possible to determine a suitable value of r, in order to concavify a given function. Moreover, we shall show that it is possible, in many cases, to characterize G-concave functions in terms of r-concave functions, and that under some mild conditions, every G-concave function, defined on a compact convex subset of R^n, is also r-concave. Thus in these cases, G-concavity can be characterized in terms of r-concavity, and to determine the exact form of G, only the value of r needs to be specified.

The following is a straightforward result of (8.28) and (8.29) that clarifies the concept of r-concavity (convexity) and relates it to G-concavity (convexity).

Proposition 8.9. Let f be a function defined on the convex set $C \subset R^n$. It is r-concave (convex) with $r > 0$ if and only if it is G-concave (convex) with $G(t) = -\exp(-rt)$ $[G(t) = \exp(rt)]$.

As a direct consequence to the above proposition, we have that f is r-concave if and only if the function g, given by $g = -f$, is r-convex. Note that in Example 8.5 we had a function that is (9/8)-concave.

All the properties that we have established so far for G-concave functions also hold for r-concave functions. Suppose that f is r-concave, and let $s > r$. Letting $G_1(t) = -\exp(-rt)$, $G_2(t) = -\exp(-st)$, we have that $\tau(t) = G_2G_1^{-1}(t) = -(-t)^{s/r}$ is concave for $t < 0$. Thus we obtain from Proposition 8.2 the following ranking relationship:

Proposition 8.10. Let f be r-concave on the convex set $C \subset R^n$. Then f is s-concave for every $s > r$.

As a result of Definition 8.1 and the above proposition, a concave function (0-concave) is r-concave for every $r > 0$. Moreover, some simple calculations show that (see Hardy, Littlewood, and Polya, 1952):

$$\begin{aligned}\lim_{r\to\infty} [-\log\{\lambda\, e^{-rf(x^1)} + (1-\lambda)\, e^{-rf(x^2)}\}^{1/r}] &= -\max\{-f(x^1), -f(x^2)\} \\ &= \min\{f(x^1), f(x^2)\} \qquad (8.30)\end{aligned}$$

holds. Consequently, (8.28) coincides with the definition of quasiconcavity as $r \to \infty$. Since by Proposition 8.1 every r-concave function is semistrictly quasiconcave, we have by Proposition 8.10 and the above discussion that r-concavity induces, as r increases, a gradual transition between concavity, semistrict quasiconcavity, and quasiconcavity.

Although r-concave functions are G-concave, there are some closedness properties that do not hold for G-concave functions, but are satisfied in the r-concave case. Let f be r-concave and let α be a real number. Then

$$-\exp[-r(f+\alpha)] = \exp(-r\alpha)[-\exp(-rf)] \qquad (8.31)$$

and by Proposition 2.15 and Proposition 8.9, $f + \alpha$ is r-concave. Moreover, in case f is r-concave then for any positive number k, kf is $\bar{r}$-concave with $\bar{r} = r/k$.

We observe that Proposition 8.6 holds for once or twice continuously differentiable r-concave functions. In the latter case, we have the following corollary (Avriel, 1972).

Corollary 8.11. Let f be twice continuously differentiable on the convex set $C \subset R^n$. Then f is r-concave on C with $r \geq 0$ if and only if the matrix $H(x; r)$ given by

$$H(x; r) = \nabla^2 f(x) - r\nabla f(x)\nabla f(x)^T \qquad (8.32)$$

is negative semidefinite for every $x \in C$.

Proof. In case $r = 0$ then f is concave and $H(x; 0)$ is negative semidefinite. Let $r > 0$; then substituting $G(t) = -\exp(-rt)$ in (8.27) we have

$$\nabla^2 G(f(x)) = r\, e^{-rf(x)} \nabla^2 f(x) - r^2\, e^{-rf(x)} \nabla f(x) \nabla f(x)^T. \qquad (8.33)$$

Since $re^{-rf(x)}$ is positive we have that $\nabla^2 Gf(x)$ is negative semidefinite if and only if $H(x; r)$ is so. □

We conclude this section showing that, under some mild assumptions, a G-concave function is also r-concave over compact convex subsets of its domain, which actually implies that the family of G-concave functions with $G'(t) > 0$ for $t \in I_f(C)$ is a subfamily of r-concave functions if restricted to compact subsets of C.

Proposition 8.12. Let f be a twice continuously differentiable G-concave function on the compact convex set $D \subset R^n$, where G is twice continuously differentiable and satisfying $G'(t) > 0$ on $I_f(D)$. Then f is r-concave on D.

Proof. By Proposition 8.6 and the hypothesis on G', the matrix

$$H(x; \bar{r}(x)) = \nabla^2 f(x) - \bar{r}(x) \nabla f(x) \nabla f(x)^T, \qquad (8.34)$$

where $\bar{r}(x) = -G''(f(x))/G'(f(x))$, is negative semidefinite. Moreover, it is negative semidefinite if $\bar{r}(x)$ is replaced by $r(x)$, where $r(x) \geq \bar{r}(x)$ holds for every $x \in D$. Let

$$r(x) \equiv r^* = \max\{0, \hat{r}\}, \qquad (8.35)$$

where

$$\hat{r} = \max_{x \in D} [-G''(f(x))/G'(f(x))]. \qquad (8.36)$$

Clearly, by the hypotheses, $\hat{r}$ exists. We have that f is r^*-concave on D. □

8.2. Conditions for Concavifiability of Twice Continuously Differentiable (C^2) Functions

In Section 8.1 we discussed properties of concavifiable functions. Some of these properties were actually necessary conditions for G or r-concavity. The most distinct ones are Propositions 8.1 and 8.7, which state that a G-concave function must be semistrictly quasiconcave, in general, or even pseudoconcave in case both f and G are differentiable.

The subject of this section will be necessary and sufficient conditions under which a general function (which must be at least semistrictly quasiconcave) is G-concave for some G. This question has been treated throughout the economic and mathematical programming literature. Here we are going to review only a small portion of this literature. In particular, we shall only consider functions f that are twice continuously differentiable on open convex subsets of R^n. For additional results, referring to functions that are not necessarily in C^2, see Crouzeix (1977), Debreu (1976), de Finetti (1949), Fenchel (1951, 1956), and Kannai (1974, 1977, 1982).

We have already shown in Theorem 6.3 (see also Example 8.4) that a merely quasiconcave quadratic function is G-concave on a solid convex set with $G(t)$ given by (8.16). In fact, as mentioned in Chapter 6, most of the characterizations of generalized concave quadratic functions are derived from this property. Consequently, in this section we shall not consider the quadratic case.

A first complete set of necessary and sufficient conditions for the concavifiability of C^2 functions was derived by Fenchel (1951, 1956). Later on, Kannai (1977, 1981) dwelt on these results and gave some different formulations. Other conditions were derived by Avriel (1972), Avriel and Schaible (1978), Diewert (1978), Gerencsér (1973), Katzner (1970), and Schaible and Zang (1980). Here we shall present all these results within a unified framework.

The problem we treat is to find necessary and sufficient conditions for the G-concavity of a twice continuously differentiable function f over the open convex set $C \subset R^n$. It can be formulated via an hierarchy of four conditions in two steps. In the first step, we discuss in Sections 8.2.1 and 8.2.2 two local conditions, that is, conditions that must be satisfied at every $x \in C$. The second step consists of global conditions that must hold, taking into account the entire domain C. This will be the subject of Section 8.2.4. Between them, Section 8.2.3 will discuss various possible representations and some technical aspects of the local conditions. Finally, in Section 8.2.5 we will discuss G-concave functions that are also r-concave. In this context we shall consider, for example, G-concave functions where $G'(t) > 0$ holds for every $t \in I_f(C)$. Functions in this class, if restricted to a compact convex subset of C, are, by Proposition 8.12, r-concave over this subset. For such functions (over compact sets) it is easier to establish G-concavity, since the functional form of G is known and only the value of r should be determined.

If we let

$$\alpha = \inf_{x \in C} f(x),$$

$$\beta = \sup_{x \in C} f(x), \tag{8.37}$$

where $-\infty \le \alpha, \beta \le \infty$ hold, then we require the concavifying function $G(t)$ to be increasing and twice continuously differentiable over the image of C under f denoted by $I_f(C)$. Note that a concave function has no minimum in an open domain. Thus, $I_f(C) = (\alpha, \beta]$ in case f has a maximum in C and $I_f(C) = (\alpha, \beta)$ otherwise. We also rule out the possibility that f is constant over C and thus $\alpha < \beta$ holds.

8.2.1. The Local Conditions

If f is G-concave and $\nabla f(\bar{x}) = 0$ for some nonmaximal $\bar{x} \in C$, then $\nabla G(f(\bar{x})) = G'(f(\bar{x}))\nabla f(\bar{x}) = 0$, and since $\bar{x}$ is also nonmaximal for $G(f(x))$ we have a contradiction to the G-concavity assumption. Thus, we obtain the following basic necessary condition for concavifiability expressed by Fenchel (1951):

(A) $\nabla f(x) \neq 0$ for all $x \in C$, except for points in C satisfying $f(x) = \beta$ if such points exist.

This condition, as well as some others to be derived below, is implied by the following necessary condition for G-concavity already obtained in Proposition 8.7.

(A′) f is pseudoconcave over C.

We also recall Proposition 8.6, which states that, under the above hypotheses, a necessary and sufficient condition for the concavifiability of f under G is the negative semidefiniteness of the matrix

$$\nabla^2 G(f(x)) = G'(f(x))\nabla^2 f(x) + G''(f(x))\nabla f(x)\nabla f(x)^T \qquad (8.38)$$

for every $x \in C$. In case there exists a point $\tilde{x} \in C$ satisfying $f(\tilde{x}) = \beta$, then the well-known necessary optimality criteria imply that $\nabla f(\tilde{x}) = 0$ and that $\nabla^2 f(\tilde{x})$ is negative semidefinite; thus $\nabla^2 G(f(\tilde{x}))$ is also negative semidefinite for every G satisfying the hypotheses. Moreover, if f is G-concave and $G'(\bar{t}) = 0$ for some $\alpha < \bar{t} < \beta$, then $\nabla G(f(x)) = 0$ for all $x \in C$ satisfying $f(x) = \bar{t}$, contradicting the G-concavity assumption since $\bar{t}$ is nonmaximal. Thus, $G'(t) > 0$ for all $\alpha < t < \beta$ and consequently a necessary and sufficient condition for the G-concavity of f is that the matrix $\nabla^2 f(x) + [G''(f(x))/G'(f(x))]\nabla f(x)\nabla f(x)^T$ be negative semidefinite for all nonmaximal points $x \in C$.

In the sequel we shall denote by $H(x; \rho)$ the augmented Hessian of f, introduced in Corollary 8.11, that is,

$$H(x; \rho) = \nabla^2 f(x) - \rho \nabla f(x) \nabla f(x)^T. \tag{8.39}$$

The above discussion implies that a necessary condition for the concavifiability of f over C is the existence of a function $\rho(x)$ defined on C such that $H(x; \rho(x))$ is negative semidefinite for all $x \in C$. Alternatively, if the function $\rho_0(x)$ defined by

$$\rho_0(x) \equiv \sup \{z^T \nabla^2 f(x) z / [z^T \nabla f(x)]^2 : \|z\| = 1, z^T \nabla f(x) \neq 0\} \tag{8.40}$$

satisfies $\rho_0(x) < \infty$ for every $x \in C$, then $H(x; \rho(x))$ is negative semidefinite for every function $\rho(x)$ satisfying $\rho(x) \geq \rho_0(x)$ and for every $x \in C$. Note that we may consider all points x in C, since points where $\nabla f(x) = 0$ holds must, in view of condition (A), be maximal in C and thus have a negative semidefinite Hessian. Therefore, for these points $\rho_0(x)$ can take on any value. In particular, since the maximization in (8.40) is taken then over an empty set, we may let $\rho_0(x) = -\infty$ in these cases.

Following Avriel and Schaible (1978), we let H be the family of functions for which a negative semidefinite augmented Hessian exists at every $x \in C$. Thus the second condition for concavifiability is

(B) $f \in H$.

Example 8.6. Let us again consider the function treated in Example 6.7

$$f(x) = x_1 x_3 \tag{8.41}$$

and

$$C = \{x \in R^3 : x_1 > 0, x_3 > 0\}. \tag{8.42}$$

For this pseudoconcave function we have

$$\nabla f(x) = \begin{pmatrix} x_3 \\ 0 \\ x_1 \end{pmatrix}, \qquad \nabla^2 f(x) = \begin{pmatrix} 0 & 0 & 1 \\ 0 & 0 & 0 \\ 1 & 0 & 0 \end{pmatrix}, \tag{8.43}$$

and

$$H(x; \rho) = \begin{pmatrix} -\rho(x_3)^2 & 0 & 1 - \rho x_1 x_3 \\ 0 & 0 & 0 \\ 1 - \rho x_1 x_3 & 0 & -\rho(x_1)^2 \end{pmatrix}. \tag{8.44}$$

This matrix is negative semidefinite provided $\rho \geq 0$ and $(\rho x_1 x_3)^2 - (1 - \rho x_1 x_3)^2 \geq 0$ hold; that is, $\rho \geq 1/(2x_1 x_3)$, implying that $f \in H$ and $\rho_0(x) = 1/(2x_1 x_3)$.

In Theorem 6.6, we showed that pseudoconcave quadratic functions belong to the family H. However, in general, pseudoconcave functions are not necessarily in H. We now exhibit an example for such a function where for no point $x \in C$, $H(x; \rho(x))$ is negative semidefinite for some $\rho(x)$.

Example 8.7 (Avriel and Schaible, 1978). Let $f(x) = -x_2/x_1$ on $C = \{x \in R^2: x_1 > 0\}$. It can be verified (see Mangasarian, 1969, Problem 3, p. 96, or the discussion following Table 5.6) that f is pseudoconcave. We have

$$\nabla f(x) = \begin{pmatrix} x_2/(x_1)^2 \\ -1/x_1 \end{pmatrix}, \qquad \nabla^2 f(x) = \begin{pmatrix} -2x_2/(x_1)^3 & 1/(x_1)^2 \\ 1/(x_1)^2 & 0 \end{pmatrix}, \tag{8.45}$$

and

$$H(x; \rho) = \begin{pmatrix} -2x_2/(x_1)^3 & 1/(x_1)^2 \\ 1/(x_1)^2 & 0 \end{pmatrix} - \rho \begin{pmatrix} (x_2)^2/(x_1)^4 & -x_2/(x_1)^3 \\ -x_2/(x_1)^3 & 1/(x_1)^2 \end{pmatrix}$$

$$= \frac{1}{(x_1)^2} \begin{pmatrix} -\dfrac{2x_2}{x_1} - \rho\left(\dfrac{x_2}{x_1}\right)^2 & 1 + \rho\dfrac{x_2}{x_1} \\ 1 + \rho\dfrac{x_2}{x_1} & -\rho \end{pmatrix}. \tag{8.46}$$

Hence if $H(x; \rho)$ is negative semidefinite then its determinant is nonnegative, that is,

$$\rho\left(\frac{2x_2}{x_1} + \rho\left(\frac{x_2}{x_1}\right)^2\right) - \left(1 + \rho\frac{x_2}{x_1}\right)^2 \geq 0 \tag{8.47}$$

holds. The above expression reduces, however, to $-1 \geq 0$ (independent of ρ) and $H(x; \rho)$ cannot be negative semidefinite.

Next, in Section 8.2.2, we will present a detailed characterization of functions in H via several equivalent sets of necessary and sufficient conditions. Then, in Section 8.2.3 we will develop three equivalent mathematical expressions for $\rho_0(x)$. These expressions are due to Diewert (1978), Fenchel (1951), and Schaible and Zang (1980). The reader who wishes to avoid these technicalities may proceed directly to Section 8.2.4, where we go up the hierarchy and establish two global conditions that imply that for a function in H

$$\rho_0(x) \leq \rho(x) \equiv [-G''(f(x))/G'(f(x))] \tag{8.48}$$

holds for some function G satisfying the above hypotheses and for all nonmaximal points $x \in C$.

8.2.2. Characterizations of Functions in H

To characterize functions in H, we specify for every $x \in C$, necessary and sufficient conditions for the existence of a number $\rho = \rho(x)$ such that $H(x; \rho)$ is negative semidefinite. Since for a given $x \in C$, $\nabla^2 f(x)$ and $\nabla f(x)$ are constants, we can state this problem in a somewhat more general framework: Given a symmetric $n \times n$ matrix A and a vector $b \in R^n$, find necessary and sufficient conditions for the existence of a number $\bar{r}$ such that the matrix $H(r)$, given by

$$H(r) = A - rbb^T, \tag{8.49}$$

is negative semidefinite for all $r \geq \bar{r}$.

Before we state and prove a set of conditions for this property to hold, we recall that a vector $v \in R^n$ is an eigenvector of A restricted to the subspace orthogonal to b, denoted by $b^\perp$, and a scalar λ is the corresponding restricted eigenvalue of A, if v is a stationary vector and λ a stationary value (that is, satisfying the first-order necessary conditions for an extremum) for the problem

$$\max \{v^T A v: v^T b = 0, v^T v = 1\}. \tag{8.50}$$

These conditions are equivalent to the existence of scalars β and λ and a vector v such that v, λ, and β satisfy

$$Av - \beta b - \lambda v = 0, \tag{8.51}$$

$$v^T b = 0, \tag{8.52}$$

$$v^T v = 1. \tag{8.53}$$

Note that if $b \neq 0$ then it is possible to find $n - 1$ orthogonal vectors satisfying (8.51)-(8.53). In the sequel, we denote these restricted eigenvectors and their corresponding restricted eigenvalues and β coefficients by v^j, λ_j, and β_j, $j = 1, \ldots, n - 1$, respectively. Also, β will denote the vector in R^{n-1} whose components are β_j, V will denote the $n \times (n - 1)$ matrix whose jth column is v^j, $j = 1, \ldots, n - 1$, and Λ will denote the $(n - 1) \times (n - 1)$ diagonal matrix whose diagonal entries are the restricted eigenvalues. Note that (8.51)-(8.53) imply that

$$(v^j)^T A v^j = \lambda_j \tag{8.54}$$

and

$$b^TAv^j/b^Tb = \beta_j \tag{8.55}$$

must hold. The number of nonzero restricted eigenvalues corresponding to these restricted eigenvectors is the rank of A restricted to $b^\perp$. In the sequel, we denote by B the matrix A bordered by the vector b, that is

$$B = \begin{pmatrix} 0 & b^T \\ b & A \end{pmatrix}. \tag{8.56}$$

We may now state and prove the following proposition.

Proposition 8.13. The matrix $H(r)$ is negative semidefinite for all $r \geq \bar{r}$ if and only if a pair of conditions (i.l) and (ii.j) (l = 1 or 2 or 3, j = 1 or 2 or 3 or 4) stated below hold:

(i.1) $v^Tb = 0 \Rightarrow v^TAv \leq 0$.

(i.2) The eigenvalues of A restricted to $b^\perp$ are all nonpositive.

(i.3) A is negative semidefinite whenever $b = 0$, and B has exactly one positive eigenvalue whenever $b \neq 0$.

(ii.1) $v^Tb = 0,\ v^TAv = 0 \Rightarrow Av = 0$.

(ii.2) If $b \neq 0$ then $\lambda_i = 0 \Rightarrow \beta_i = 0$ in (8.51).

(ii.3) If $b \neq 0$ and the rank of B is $q + 1$, then the rank of A is at most q.

(ii.4) If the rank of A restricted to $b^\perp$ is $q - 1$, then the rank of A is at most q.

Proof. The proof is constructed as follows: First we show that (i.1) and (ii.1) are equivalent to negative semidefiniteness of $H(r)$. Then we show that (i.1) $\Rightarrow$ (i.2) $\Rightarrow$ (i.3) $\Rightarrow$ (i.1). Finally, we show that under (i.1) we have (ii.1) $\Rightarrow$ (ii.2) $\Rightarrow$ (ii.3) $\Rightarrow$ (ii.4) $\Rightarrow$ (ii.1).

Negative semidefiniteness of $H(r)$ implies (i.1) and (ii.1): Let $H(r)$ be negative semidefinite. Then condition (i.1) obviously holds. To show (ii.1), let $v^Tb = 0$, $v^TAv = 0$. Then $v^TH(r)v = 0$. Since $H(r)$ is negative semidefinite, it follows that $H(r)v = 0$, and since $b^Tv = 0$ we have $Av = 0$.

(i.1) and (ii.1) implies negative semidefiniteness of $H(r)$: Let (i.1) and (ii.1) hold. If $v^Tb = 0$, then $v^TH(r)v = v^TAv \leqq 0$ in view of (i.1) for all r. Let $v^Tb \neq 0$. Then $b \neq 0$ and there exist $\alpha \in R^m$, $\alpha \in R$ such that

$$v = w + \alpha b, \qquad w^Tb = 0, \qquad \alpha \neq 0. \tag{8.57}$$

We must show that $\bar{r} = \{\sup v^TAv/(v^Tb)^2 : v^Tb \neq 0\}$ is finite, to ensure that $v^TH(r)v \leq 0$ for all $v \in R^n$ and $r \geq \bar{r}$. In view of (8.57)

$$\bar{r} = \sup\{[w^TAw + 2\alpha w^TAb + (\alpha)^2 b^TAb]/(w^Tb + \alpha b^Tb)^2 : w^Tb = 0, \alpha \neq 0\}$$

$$= \sup\{[(w/\alpha)^TA(w/\alpha) + 2(w/\alpha)^TAb]/(b^Tb)^2 + b^TAb/(b^Tb)^2 :$$

$$w^Tb = 0, \alpha \neq 0\}; \tag{8.58}$$

dividing numerator and denominator by $(\alpha)^2 > 0$ we get

$$\bar{r} = \sup\{(u^TAu + 2u^TAb + b^TAb)/(b^Tb)^2 : u^Tb = 0\}, \tag{8.59}$$

where $u = w/\alpha$. The above supremum is finite, since $u^TAu \leqq 0$ for all $u^Tb = 0$, $\alpha \neq 0$ in view of (i.1), and if $u^TAu = 0$, then $Au = 0$ in view of (ii.1).

(i.1) ⇒ *(i.2)*: This obviously holds in case $b = 0$. If $b \neq 0$ then multiplying (8.51) by $(v^j)^T$ we obtain $(v^j)^TA(v^j) = \lambda_j$ for $j = 1, \ldots, n-1$; (i.2) then follows from (i.1) immediately.

(i.2) ⇒ *(i.3)*: Again, the implication is obvious if $b = 0$. Let $b \neq 0$. Then it follows that the $(n+1) \times (n+1)$ matrix P given by

$$P = \begin{pmatrix} 1 & 0_{n-1}^T & 0 \\ 0_n & V & b/b^Tb \end{pmatrix}, \tag{8.60}$$

where 0_n denotes the origin in R^n $(0 = 0_1)$, is nonsingular. Recall that two $n \times n$ matrices Q and R are said to be *congruent* if there exists an $n \times n$ nonsingular matrix W such that $Q = W^TRW$. Consequently, B and $Q_1 = P^TBP$ are congruent, and using (8.51)-(8.55) we obtain

$$Q_1 = \begin{pmatrix} 0 & 0_{n-1}^T & 1 \\ 0_{n-1} & \Lambda & \beta \\ 1 & \beta^T & b^TAb/b^Tb \end{pmatrix}. \tag{8.61}$$

By adding multiples of the first row and column of Q_1, we obtain another congruent matrix

$$Q_2 = \begin{pmatrix} 0 & 0_{n-1}^T & 1 \\ 0_{n-1} & \Lambda & 0_{n-1} \\ 1 & 0_{n-1}^T & 0 \end{pmatrix}. \tag{8.62}$$

Thus, Q_2 and B are congruent. However, Q_2 possesses a very simple eigenstructure. Denoting by e^j the jth unit vector in R^{n-1}, it follows that the eigenvectors of Q_2 are $(1, 0_{n-1}^T, 1)$, $(0, e_j^T, 0)$, $j = 1, \ldots, n-1$, and $(1, 0_{n-1}^T, -1)$ with the corresponding eigenvalues $1, \lambda_j, j = 1, \ldots, n-1$, and -1. According to Sylvester's law of inertia (Marcus and Minc, 1964), the numbers of positive, negative, and zero eigenvalues of two congruent matrices are, respectively, the same. It follows from (i.2) that B has exactly one positive eigenvalue.

(i.3) $\Rightarrow$ *(i.1)*: Again, we have to consider the case where $b \neq 0$. Since B and Q_2, given by (8.62), are congruent, it follows from the above discussion that (i.3) implies $\lambda_j \leq 0$, $j = 1, \ldots, n-1$. This in turn implies through (8.54) that $(v^j)^T A v^j \leq 0$ for all eigenvectors of A restricted to $b^\perp$. Hence (i.1) follows.

Throughout the rest of the proof we suppose that (i.1) holds.

(ii.1) $\Rightarrow$ *(ii.2)*: Following (8.54) we have that to every restricted eigenvalue λ_j whose value is zero there corresponds a restricted eigenvector v^j satisfying $(v^j)^T A v^j = 0$. It follows from (ii.1) that $Av^j = 0$, and from (8.51) and $b \neq 0$ that $\beta_j = 0$.

(ii.2) $\Rightarrow$ *(ii.3)*: In case $q = n$, we are done. Let $q < n$ and suppose that (ii.2) holds, and that the rank of B is $q + 1$. Then there exists $n - q$ orthogonal vectors in R^{n+1} (v_0^i, v^i), $i = 1, \ldots, n-q$, where $v_0^i \in R$, $v^i \in R^n$, which are the eigenvectors of B corresponding to its zero eigenvalues. Hence

$$\begin{pmatrix} 0 & b^T \\ b & A \end{pmatrix} \begin{pmatrix} v_0^i \\ v^i \end{pmatrix} = 0, \qquad i = 1, \ldots, n-q, \tag{8.63}$$

implying that

$$b^T v^i = 0 \tag{8.64}$$

and

$$v_0^i b + Av^i = 0 \tag{8.65}$$

for $i = 1, \ldots, n-q$. Note that if $v^i = 0$ for some i, then (8.65) implies $v_0^i = 0$, hence $(v_0^i, v^i) = 0$. Therefore, without loss of generality we suppose that $\|v^i\| = 1$ for $i = 1, \ldots, n-q$. Comparing now (8.64) and (8.65) with (8.51) and (8.52), we have that v^i, $i = 1, \ldots, n-q$, are eigenvectors of A restricted to $b^\perp$ that correspond to zero eigenvalues $\lambda_i = 0$, with $\beta_i = v_0^i$. However, (ii.2) implies $\beta_i = v_0^i = 0$ for $i = 1, \ldots, n-q$. Since (v_0^i, v^i) are orthogonal $i = 1, \ldots, n-q$, it follows then that v^i, $i = 1, \ldots, n-q$ are orthogonal, and from (8.65) we have $Av^i = 0$, $i = 1, \ldots, n-q$. Hence, v^i, $i = 1, \ldots, n-q$ are eigenvectors of A with zero eigenvalues. It follows that the rank of A cannot exceed q.

$(ii.3) \Rightarrow (ii.4)$: First note that if $b = 0$ then rank A equals the rank of A restricted to $b^\perp$. Let $b \neq 0$. We show then that the difference between the rank of B and the rank of A restricted to $b^\perp$ is 2 or, in other words, that the number of zero eigenvalues of B equals the number of zero eigenvalues of A restricted to $b^\perp$. A vector (v_0^i, v^i), $v_0^i \in R$, $v^i \in R^n$ is an eigenvector of B with zero eigenvalue if and only if (8.64) and (8.65) hold and $v^i \neq 0$ [see discussion following (8.65)]. Comparing (8.64) and (8.65) with (8.51) and (8.52), we have that (v_0^i, v^i) is such an eigenvector of B if and only if v^i is an eigenvector of A restricted to $b^\perp$ with zero eigenvalue. Thus B and A restricted to $b^\perp$ have the same number of zero eigenvalues.

$(ii.4) \Rightarrow (ii.1)$: In case $q = n$, then A restricted to $b^\perp$ has full rank. By (i.1) it must be negative definite on $b^\perp$, and $v \neq 0$, $v^T b = 0$, $v^T A v = 0$ cannot hold.

Let $q < n$. By (i.1) and Corollary 3.18, A has at most one positive eigenvalue. However, in case A is negative semidefinite then by symmetry $v^T A v = 0 \Rightarrow Av = 0$. Therefore, it is only left to consider cases where $q < n$ and A has exactly one positive eigenvalue. We shall establish that if q is the rank of A and (ii.1) does not hold then there exist $n - q + 1$ linearly independent nonzero vectors v satisfying (8.51)-(8.53) with $\lambda = 0$. Thus $\lambda = 0$ for an $(n - q + 1)$-dimensional subspace of $b^\perp$, which implies that the rank of A restricted to $b^\perp$ is at most $q - 2$, contradicting (ii.4).

Let $A = UDU^T$, where U is an $n \times n$ orthogonal matrix whose columns are the eigenvectors of A and D is a diagonal matrix whose diagonal elements are the eigenvalues of A. Clearly, by a one-to-one transformation of v and b, it is sufficient to show that if

$$\text{(ii.1}')\quad z^T a = 0, \qquad z^T D z = 0 \Rightarrow Dz = 0$$

does not hold (where actually $z = U^T v$ and $a = U^T b$), then the rank of D restricted to $a^\perp$ is at most $q - 2$. Without loss of generality assume that

$$D = \begin{pmatrix} d_1 & & & & & \\ & d_2 & & & 0 & \\ & & \ddots & & & \\ & & & d_q & & \\ & & & & 0 & \\ & 0 & & & & \vdots \\ & & & & & 0 \end{pmatrix}, \tag{8.66}$$

where $d_1 < 0$, $d_2 < 0, \ldots, d_{q-1} < 0$ and $d_q > 0$. If (ii.1′) does not hold then there exists a vector $\bar{z} \in R^n$ satisfying $\|\bar{z}\| = 1$, $\bar{z}^T a = 0$, $\bar{z}^T D \bar{z} = 0$ and

$D\bar{z} \neq 0$, or equivalently

$$\sum_{i=1}^{n} \bar{z}_i a_i = 0, \tag{8.67}$$

$$\sum_{i=1}^{q} d_i(\bar{z}_i)^2 = 0, \tag{8.68}$$

$$\hat{z} = (\bar{z}_1, \bar{z}_2, \dots, \bar{z}_q)^T \neq 0, \tag{8.69}$$

where the signs of the d_i, $i = 1, \dots, q$ and (8.68) imply that $\bar{z}_q \neq 0$. However, (i.1) is equivalent to

(i.1') $z^T a = 0 \Rightarrow z^T D z \leq 0$.

By (8.67) and (8.68) it follows that $\bar{z}$ is a solution to

$$\max \{z^T D z \colon z^T a = 0, \|z\| = 1\}. \tag{8.70}$$

Therefore,

$$D\bar{z} - \beta a - \bar{\lambda}\bar{z} = 0 \tag{8.71}$$

holds, which together with (8.67) and (8.68) imply that $\bar{\lambda} = 0$ and

$$D\bar{z} = \beta a \neq 0 \tag{8.72}$$

holds. However, (8.66) and (8.69) imply that $a_{q+1} = a_{q+2} = \cdots a_n = 0$, and (8.67) becomes

$$\sum_{i=1}^{q} \bar{z}_i a_i = 0. \tag{8.73}$$

This equality implies that the linearly independent vectors in R^n given by

$$z^j = \begin{pmatrix} \hat{z} \\ e^j \end{pmatrix}, \qquad j = 1, \dots, n - q, \tag{8.74}$$

where e^j are the unit vectors in $R^{n-q}, j = 1, \dots, n - q$, are in $a^\perp$. Moreover, these vectors satisfy in view of (8.72)

$$Dz^j - \frac{\beta}{2} a - \lambda z^j = 0 \tag{8.75}$$

for $j = 1, \ldots, n-q$ with $\lambda = 0$. Since $\bar{z}_q \neq 0$ the vector $\tilde{z}$, whose ith component is zero for $i = 1, \ldots, q$ and one otherwise, is independent of $z^1, \ldots, z^{n-q}$. It is also in $a^\perp$ and it satisfies

$$D\tilde{z} - \tilde{\beta}a - \tilde{\lambda}\tilde{z} = 0 \tag{8.76}$$

with $\tilde{\beta} = \tilde{\lambda} = 0$.

Thus, comparing (8.75), (8.76), and (8.73) with (8.62), we find that the vectors $z^1, \ldots, z^{n-q}$ and $\tilde{z}$ are linearly independent eigenvectors of D restricted to $a^\perp$ with zero eigenvalues. □

Condition (i.1) of the above proposition is an obvious one: Since the matrix bb^T does not augment A on the subspace orthogonal to b, it is necessary for the negative semidefiniteness of $H(r)$ that A will be negative semidefinite on $b^\perp$. However, this condition alone is not sufficient for the negative semidefiniteness of $H(r)$, since it may well happen that there exists a sequence $\{v^k\} \to \bar{v}$ such that $(v^k)^T b > 0$, $\{(v^k)^T b\} \to 0$, $(v^k)^T A v^k > 0$ and $\{(v^k\}^T A v^k\} \to 0$ such that $\lim (v^k)^T A v^k / [(v^k)^T b]^2 \to \infty$. Along such a sequence it may be necessary to augment A asymptotically infinitely to retain negative semidefiniteness of $H(r)$. Under condition (ii.1) such a situation cannot occur. The following example demonstrates the fact that (i.1) alone is not a sufficient condition for the negative semidefiniteness of $H(r)$.

Example 8.8. Let

$$A = \begin{pmatrix} 1 & 0 \\ 0 & -1 \end{pmatrix}, \qquad b = \begin{pmatrix} 1 \\ -1 \end{pmatrix}. \tag{8.77}$$

Condition $v^T b = 0$ implies that v is a multiple of $(1, 1)^T$. Thus, $v^T A v = 0$ for all $v \in b^\perp$ and (i.1) holds. Consider now the vectors $v(\varepsilon) = (1, 1-\varepsilon)^T$. Then $v(\varepsilon)^T b = \varepsilon$ and $v(\varepsilon)^T A v(\varepsilon) = 2\varepsilon - (\varepsilon)^2$. Consequently $v(\varepsilon)^T H(r) v(\varepsilon) = -(r+1)(\varepsilon)^2 + 2\varepsilon$, and in order to be nonpositive for positive values of ε, $r \geq (2/\varepsilon) - 1$ must hold. As ε approaches zero from above, r becomes unbounded. We note, however, that $v^T b = 0$ and $v^T A v = 0$ imply that $Av \neq 0$ for nonzero multiples of $(1, 1)^T$. Thus (ii.1) does not hold. Since $v^T A v = 0$ for all $v \in b^\perp$, then the rank of A restricted to $b^\perp$ is zero whereas rank $(A) = 2$. Thus (ii.4) does not hold. Finally,

$$B = \begin{pmatrix} 0 & 1 & -1 \\ 1 & 1 & 0 \\ -1 & 0 & -1 \end{pmatrix} \tag{8.78}$$

and it is immediately observed that the last column of B is a combination of the first two. Thus, rank (B) = rank $(A) = 2$, contradicting (ii.3).

Condition (ii.4) is the well-known rank condition due to Fenchel (1951). However, the equivalence of this condition together with (i.1) to the negative semidefiniteness of $H(r)$ as well as the other rank condition (ii.2) is due to Zang (1981). The equivalence of conditions (ii.1) and (i.1) to the negative semidefiniteness of $H(r)$ was established in Schaible and Zang (1980). This equivalence is actually an extension of the well known Finsler theorem (1937), which states that $H(r)$ is negative definite for all $r > \bar{r}$ if and only if the matrix A restricted to $b^{\perp}$ is negative definite. Condition (ii.1) was used earlier by Diewert (1978) in the context of sufficient conditions for concavifiability, where it was conjectured that it is equivalent to condition (ii.2). Another version of condition (ii.1) is

$$\text{(ii.1}''\text{)} \quad v^T b = 0, \qquad v^T A v = 0 \qquad \text{implies } b^T A v = 0.$$

The equivalence of this condition to Fenchel's rank condition was shown by Kannai (1977). Condition (i.3) is due to Ferland (1981).

Next we establish another condition, which is necessary and sufficient for the negative semidefiniteness of $H(r)$ for all $r \geq \bar{r}$. This condition is due to Avriel and Schaible (1978). To prove it we need the following lemmas.

Lemma 8.14 (Schur's formula; see Gantmacher, 1959). Let

$$S = \begin{pmatrix} T_1 & T_2 \\ T_3 & T_4 \end{pmatrix} \tag{8.79}$$

be a square matrix, where T_1, T_4 are also square and T_1 is nonsingular. Then

$$\det(S) = \det(T_1)\det(T_4 - T_3T_1^{-1}T_2) \tag{8.80}$$

holds.

The next lemma is due to Avriel and Schaible (1978).

Lemma 8.15.

$$\det H(r) = \det(A - rbb^T) = \det A + r \det B \tag{8.81}$$

holds.

Proof. In case $r = 0$ we are clearly done. Let $r \neq 0$. Then, Lemma 8.14 implies

$$\det\begin{pmatrix} 1/r & b^T \\ b & A \end{pmatrix} = \frac{1}{r}(\det A - rbb^T). \tag{8.82}$$

It is easy to show that

$$\det\begin{pmatrix} 1/r & b^T \\ b & A \end{pmatrix} = \frac{1}{r}\det A + \det\begin{pmatrix} 0 & b^T \\ b & A \end{pmatrix}, \tag{8.83}$$

and (8.81) follows (8.82) and (8.83). □

To prove our next result, we recall and expand on some notation introduced in Chapter 6. Let A and B be defined as above, and Γ_k be the set consisting of monotone increasing sequences of k numbers from $\{1, \ldots, n\}$; that is, for $k = 1, \ldots, n$ let

$$\Gamma_k = \{\gamma\colon \gamma = (i_1, \ldots, i_k), 1 \leq i_1 < \cdots < i_k \leq n\}. \tag{8.84}$$

Let $A_{\gamma,k}$, $\gamma \in \Gamma_k$ denote the principal submatrix of order $k \times k$ of A formed by the $(i_1, \ldots, i_k)$th rows and columns of A, and let $b_{\gamma,k}$ be the corresponding subvector of b. Also, let A_k, $k = 1, \ldots, n$ denote the leading principal submatrix of order $k \times k$ of A, and let b_k be the corresponding subvector of b. We also associate with $A_{\gamma,k}$ the principal submatrix

$$D_{\gamma,k} = \begin{pmatrix} 0 & b_{\gamma,k}^T \\ b_{\gamma,k} & A_{\gamma,k} \end{pmatrix}, \qquad \gamma \in \Gamma_k \tag{8.85}$$

of B. Similarly, D_k, $k = 1, \ldots, n$, will denote the leading principal submatrix of order $(k + 1) \times (k + 1)$ of B. We now have the following proposition.

Proposition 8.16. The matrix $H(r)$ is negative semidefinite for all $r \geq \bar{r}$ if and only if for every $\gamma \in \Gamma_k$, $k = 1, \ldots, n$, $(-1)^k \det D_{\gamma,k} \geq 0$ and if $\det D_{\gamma,k} = 0$ then $(-1)^k \det A_{\gamma,k} \geq 0$.

The proof of this proposition follows easily from the proof of Theorem 6.8 and is therefore omitted. It is also similar to the proof of Finsler's theorem (1937) given in Bellman (1974).

Following Debreu (1952), we have that the following condition is equivalent to (i.1):

(i.4) A is negative semidefinite whenever $b = 0$, and if $b \neq 0$ then for every $\gamma \in \Gamma_k$, $k = 1, \ldots, n$, $(-1)^k \det D_{\gamma,k} \geq 0$ holds.

Thus, in view of the above proposition, the following condition is equivalent to (ii.1), (ii.2), (ii.3), or (ii.4), provided (i.1) holds:

(ii.5) For every $\gamma \in \Gamma_k$, $k = 1, \ldots,, n$, $\det D_{\gamma,k} = 0$ implies $(-1)^k \det A_{\gamma,k} \geq 0$.

Other conditions, which are equivalent to (i.1) and, in case (i.1) holds, to (ii.1), were derived by Crouzeix and Ferland (1982). We state these conditions without a proof.

(i.5) A has at most one positive eigenvalue, and whenever A has a positive eigenvalue then there exists a vector $v \in R^n$ such that $Av = b$ and $v^T b \geq 0$, and whenever $b = 0$ then A is negative semidefinite.

Crouzeix and Ferland also show that condition (i.5) together with

(ii.6) If $b \neq 0$ and A has a positive eigenvalue, and if $v \in R^n$ satisfies $Av = b$, then $v^T b > 0$

are equivalent to the negative semidefiniteness of $H(r)$. Thus (ii.6) is equivalent to (ii.1), (ii.2), (ii.3), (ii.4), or (ii.5). For some other conditions, which are equivalent to (i.1), see Chabrillac and Crouzeix (1982).

Before we discuss the straightforward implications of all the above conditions to the characterization of functions in H, we refer to the numerical aspects of the above criteria. In general, it is impossible to check conditions (i.1) and (ii.1). It is possible, however, to check (i.3) and (ii.3). This can be done by computing the inertia (a triplet including the number of negative, the number of positive, and the number of zero eigenvalues) of the matrices A and B. An algorithm that can carry out these calculations was suggested by Cottle (1974). It is also possible to check the validity of conditions (i.4) and (ii.5), as given in Proposition 8.16, in a finite number of calculations. However, it is argued by Cottle (1974) that the number of multiplications required to compute the inertia of a matrix is of the same order of magnitude as the number of multiplications required to compute its determinant. Thus, it is better to test (i.3) and (ii.3). In addition, Crouzeix and Ferland (1982) also show that it is more efficient to test (i.3) and (ii.3) than to test (i.5) and (ii.6). Moreover, in the case where $b = 0$, a simple testing of (i.3) may be applied utilizing the Cholesky factorization (see, for example, Stewart, 1973), of A, which exists whenever A is negative semidefinite. This factorization involves expressing A as $A = LDL^T$, where D is diagonal and L is a unit lower triangular matrix. Since A and D are congruent their inertias are equal.

A different method for testing (i.1) emanates from the next condition. In case $b \neq 0$, let Y be an $n \times (n-1)$ matrix whose columns form a basis for $b^\perp$:

(i.6) The $(n-1) \times (n-1)$ symmetric matrix Y^TAY is negative definite if $b \neq 0$ and A is negative semidefinite if $b = 0$.

It is simple to show that this condition is equivalent to (i.1). Moreover, it can be tested in the following way: If $b \neq 0$ assume, without loss of generality, that $b_n \neq 0$ and let $y^i = e^i - (b^Te^i/b^Tb)b$, $i = 1, \ldots, n-1$, where e^i is the ith unit vector. Then $y^i \in b^\perp$ for $i = 1, \ldots, n-1$ and the vectors $y^1, \ldots, y^{n-1}$ are linearly independent. Consequently, they can form the matrix Y of condition (i.6). Testing negative semidefiniteness of A if $b = 0$ and of Y^TAY if $b \neq 0$ can be carried out by applying the Cholesky factorization as mentioned above. There is some disadvantage in using this test, since there seems to be no way to extend this procedure to check (ii.1) or its equivalents, while testing (i.3) seems to be compatible with testing (ii.3).

Utilizing conditions (i.1)-(i.4) and (ii.1)-(ii.5), we can easily derive conditions that are necessary and sufficient for f to be in class H. If we compare (8.39) with (8.49), then all we should do is to substitute $\nabla^2 f(x)$ for A and $\nabla f(x)$ for b in the above conditions to obtain the following theorem.

Theorem 8.17. The function $f \in H$ if and only if for every $x \in C$, a pair of conditions (I.i) and (II.j) ($i = 1$ or 2 or 3 or 4, $j = 1$ or 2 or 3 or 4 or 5) stated below hold:

(I.1) $v^T\nabla f(x) = 0 \Rightarrow v^T\nabla^2 f(x)v \leq 0$.

(I.2) The eigenvalues of $\nabla^2 f(x)$ restricted to $\nabla f(x)^\perp$ are all non-positive.

(I.3) $\nabla^2 f(x)$ is negative semidefinite whenever $\nabla f(x) = 0$, and $B(x)$ has exactly one positive eigenvalue whenever $\nabla f(x) \neq 0$.

(I.4) $\nabla^2 f(x)$ is negative semidefinite whenever $\nabla f(x) = 0$, and if $\nabla f(x) \neq 0$, then for every $\gamma \in \Gamma_k$, $k = 1, \ldots, n$, $(-1)^k \det D_{\gamma,k}(x) \geqq 0$ holds.

(II.1) $v^T\nabla f(x) = 0$, $v^T\nabla^2 f(x)v = 0 \Rightarrow \nabla^2 f(x)v = 0$.

(II.2) If $\nabla f(x) \neq 0$, then $\lambda_j(x) = 0 \Rightarrow \beta_j(x) = 0$ in (8.51).

(II.3) If $\nabla f(x) \neq 0$ and the rank of $B(x)$ is $q + 1$, then the rank of $\nabla^2 f(x)$ is at most q.

(II.4) If the rank of $\nabla^2 f(x)$ restricted to $\nabla f(x)^\perp$ is $q - 1$, then the rank of $\nabla^2 f(x)$ is at most q.

(II.5) For all $\gamma \in \Gamma_k$, $k = 1, \ldots, n$ if $\det D_{\gamma,k}(x) = 0$, then $(-1)^k \det A_{\gamma,k}(x) \geq 0$ holds.

In the above conditions $B(x)$, $D_{\gamma,k}(x)$, $\lambda_j(x)$,and $\beta_j(x)$ stand for B, $D_{\gamma,k}$, λ_j, and β_j as introduced above, where $A = \nabla^2 f(x)$, $b = \nabla f(x)$ is substituted. To establish these conditions it is better, by the preceding discussion, to try to verify conditions (I.3) and (II.3). Thus, it is necessary to compute the inertia of either $\nabla^2 f(x)$ or $B(x)$ for *every* $x \in C$, which is often an impossible task. Note, however, that the same sort of complexity appears when trying to verify concavity of a general twice continuously differentiable function, by testing whether its Hessian is negative semi-definite everywhere.

Note that if f is quadratic, then, in view of Theorems 6.6 and 6.8, f is in H if and only if it is pseudoconcave. Hence a pseudoconcave quadratic function satisfies conditions (A) and (B).

We now return to discuss Examples 8.6 and 8.7 in view of the above conditions.

Example 8.9. Consider the function given by (8.41) and (8.42) in Examples 6.7 and 8.6. In fact, in Example 6.7 we have already established that Conditions (I.4) and (II.5) are satisfied. Hence $f \in H$.

To check conditions (I.2) and (II.2), we note that (8.52) reads

$$v^T \nabla f(x) = v_1 x_3 + v_3 x_1 = 0 \tag{8.86}$$

while (8.51) becomes

$$\nabla^2 f(x) v - \beta \nabla f(x) - \lambda v = \begin{pmatrix} 0 & 0 & 1 \\ 0 & 0 & 0 \\ 1 & 0 & 0 \end{pmatrix} \begin{pmatrix} v_1 \\ v_2 \\ v_3 \end{pmatrix} - \beta \begin{pmatrix} x_3 \\ 0 \\ x_1 \end{pmatrix} - \lambda \begin{pmatrix} v_1 \\ v_2 \\ v_3 \end{pmatrix} = 0 \tag{8.87}$$

together with the three equations

$$v_3 - \beta x_3 - \lambda v_1 = 0, \tag{8.88}$$

$$\lambda v_2 = 0, \tag{8.89}$$

$$v_1 - \beta x_1 - \lambda v_3 = 0. \tag{8.90}$$

If $v_1 = v_3 = 0$ then (8.86) holds, but we must have $v_2 = 1$, in this case implying $\lambda = 0$. Otherwise, (8.86) and (8.42) imply that both v_1 and v_3 are nonzero and possess opposing signs. If $\lambda > 0$ holds then (8.88) implies

$$v_3/v_1 - \beta x_3/v_1 > 0. \tag{8.91}$$

Hence, $\beta/v_1 < 0$ must hold. Similarly, (8.90) implies $\beta/v_3 < 0$. But then, v_1 and v_3 cannot have opposite signs and (I.2) holds. If $\lambda = 0$ then (8.88) and (8.90) imply $\beta = v_3/x_3 = v_1/x_1$, and since v_1, v_3 must have opposing signs, this can only hold if $\beta = 0$ holds. Hence (II.2) holds. The reader can verify that all other conditions of Theorem 8.17 are also satisfied for this function.

Example 8.10. We return to the pseudoconcave function given in Example 8.7. It was already shown that for this function there exists no $x \in C$ such that $H(x; \rho)$ is negative semidefinite for sufficiently large ρ. Since f is pseudoconcave, condition (I.1) must hold for every $x \in C$. We show, however, that (II.1) cannot hold. First note that $v^T\nabla f(x) = 0$, $v \neq 0$ imply $v_2/v_1 = x_2/x_1$. Moreover,

$$v^T\nabla^2 f(x)v = \frac{-2(v_1)^2 x_2/x_1 + 2v_1 v_2}{(x_1)^2} = 0 \tag{8.92}$$

for v_1, v_2 satisfying the above ratio. However,

$$\nabla^2 f(x)v = [1/(x_1)^2](-2x_2 v_1/x_1 + v_2, v_1)^T = [1/(x_1)^2](-v_2, v_1)^T \neq 0. \tag{8.93}$$

Thus, condition (II.1) does not hold, and $f \not\in H$.

Note that, in general, it suffices to find one point in C, for which the above condition does not hold, to conclude that $f \not\in H$.

There are other examples of functions that are quasiconcave [thus (I.1) holds], but are not in H:

$$f_1(x) = x_2/(1 + x_1) \tag{8.94}$$

$$f_2(x) = (x_1 - 1) + [(1 - x_1)^2 + 4(x_1 + x_2)]^{1/2}, \tag{8.95}$$

where

$$C = \{x \in R^2: x_1 > 0, x_2 > 0\}. \tag{8.96}$$

For both functions no point $x \in C$ satisfies (II.1). These functions are due to Kannai (1977) and Arrow and Enthoven (1961), respectively.

8.2.3. Some Expressions for $\rho_0(x)$

We shall now develop three different expressions for $\rho_0(x)$, given by (8.40), which are valid for functions in H. The first of these expressions is due to Fenchel (1951) and the others to Diewert (1978) and Schaible and Zang (1980). We recall that conditions (A) and (B) imply that for $x \in C$ satisfying $\nabla f(x) = 0$ the Hessian $\nabla^2 f(x)$ must be negative semidefinite. Thus, for these points $\rho_0(x)$ can have any value. Consequently, in the sequel we shall consider only points in C having a nonzero gradient.

First, we show that for functions in H it is possible to express $\rho_0(x)$ in terms of the coefficients of the characteristic equations of $\nabla^2 f(x)$, and of $\nabla^2 f(x)$ restricted to $\nabla f(x)^{\perp}$. The following derivation is based upon Fenchel (1951). To simplify the exposition we shall start by establishing an expression for the value of $\bar{r}$ that makes $H(r)$, given in (8.49), negative semidefinite for all $r \geq \bar{r}$. We assume that conditions (i.1) and (ii.2) of Proposition 8.13 are satisfied and thus the existence of $\bar{r}$ is ensured. The implication for f will eventually become straightforward, by setting $A = \nabla^2 f(x)$ and $b = \nabla f(x)$.

A characteristic equation of an $n \times n$ symmetric matrix A is given by

$$S_A(\lambda) = \det(A - \lambda I) = \prod_{i=1}^{n} (\lambda_i - \lambda)$$

$$= s_n - s_{n-1}\lambda + \cdots + (-1)^i s_{n-i}\lambda^i + \cdots + (-1)^n s_0 \lambda^n, \tag{8.97}$$

where $s_0 = 1$ and the roots λ_i of $S_A(\lambda)$ are the eigenvalues of A. Moreover, it is easy to show, using the signs of the roots of (8.97), that A is negative semidefinite if and only if $(-1)^j s_j \geq 0$ for $j = 1, \ldots, n$. It is also possible to show that the negative semidefiniteness of A implies that if for some $1 \leq l \leq n$, $s_l = 0$ holds, then $s_j = 0$ for all $j > l$, and that if q is the rank of A then $(-1)^j s_j > 0$ for $j \leq q$ and $s_j = 0$ for $j > q$.

Using Lemma 8.15 we obtain

$$S_{H(r)}(\lambda) = \det(A - \lambda I - rbb^T) = \det(A - \lambda I) + r \det\begin{pmatrix} 0 & b^T \\ b & A - \lambda I \end{pmatrix}$$

$$= S_A(\lambda) + r \det\begin{pmatrix} 0 & b^T \\ b & A - \lambda I \end{pmatrix}, \tag{8.98}$$

where $S_{H(r)}(\lambda)$ is the characteristic equation of $H(r)$, and the second determinant in (8.98) is the characteristic equation of A restricted to $b^{\perp}$.

To show this we note that, in view of (8.51)-(8.53), $v \neq 0$ is an eigenvector of A restricted to $b^{\perp}$ and λ is the corresponding eigenvector if and only if

$$\begin{pmatrix} 0 & b^T \\ b & A - \lambda I \end{pmatrix} \begin{pmatrix} -\beta \\ v \end{pmatrix} = 0. \tag{8.99}$$

That is,

$$\det \begin{pmatrix} 0 & b^T \\ b & A - \lambda I \end{pmatrix} = 0. \tag{8.100}$$

Equation (8.100) is a polynominal of degree n in λ. However, because of the zero entry, the coefficient of λ^n is zero. We shall now obtain an expression for the (normalized) characteristic equation of A restricted to $b^{\perp}$, denoted by $S_A^*(\lambda)$. For this, we have to divide (8.100) by the coefficient of λ^{n-1}. To find this coefficient, we divide the determinant in (8.100) by λ^{n-1} and let $\lambda \to \infty$. If we divide each of the n last rows of the matrix in (8.100) by λ, and then multiply the first column by λ, we obtain, letting $\lambda \to \infty$ and applying Lemma 8.14, that the coefficient of λ^{n-1} is

$$\det \begin{pmatrix} 0 & b^T \\ b & -I \end{pmatrix} = (-1)^n b^T b. \tag{8.101}$$

Dividing (8.100) by $(-b^T b)$, we obtain

$$\begin{aligned} S_A^*(\lambda) &= -\frac{1}{b^T b} \det \begin{pmatrix} 0 & b^T \\ b & A - \lambda I \end{pmatrix} \\ &= s_{n-1}^* - s_{n-2}^* \lambda + \cdots ; + (-1)^i s_{n-i-1}^* \lambda^i + \cdots + (-1)^{n-1} s_0^* \lambda^{n-1}, \end{aligned} \tag{8.102}$$

where $s_0^* = 1$. We can now write (8.98) as

$$S_{H(r)}(\lambda) = S_A(\lambda) - r b^T b S_A^*(\lambda). \tag{8.103}$$

Letting the characteristic equation of $H(r)$ be given by

$$S_{H(r)}(\lambda) = \bar{s}_n - \bar{s}_{n-1}\lambda + \cdots + (-1)^i \bar{s}_{n-i} \lambda^i + \cdots + (-1)^n \bar{s}_0 \lambda^n, \tag{8.104}$$

where $\bar{s}_0 = 1$, we obtain from (8.103), using (8.102) and (8.97), that

$$\bar{s}_j = s_j - r b^T b s_{j-1}^*, \qquad j = 1, \ldots, n. \tag{8.105}$$

Assume now that A restricted to $b^\perp$ is negative semidefinite, which implies that $(-1)^j s_j^* \geq 0$ for $j = 1, \ldots, n-1$, and suppose that its rank restricted to $b^\perp$ is $q - 1$. It follows that

$$(-1)^j s_j^* > 0, \qquad j = 0, \ldots, q-1; \qquad s_j^* = 0, \qquad j = q, \ldots, n-1. \tag{8.106}$$

Moreover, condition (ii.4) implies that the rank of A is at most q, and thus

$$s_j = 0, \qquad j = q+1, \ldots, n. \tag{8.107}$$

Returning now to (8.105), we have that $H(r)$ is negative semidefinite if and only if

$$(-1)^j \bar{s}_j \geq 0, \qquad j = 1, \ldots, n \tag{8.108}$$

or

$$(-1)^j s_j + rb^T b(-1)^{j-1} s_{j-1}^* \geq 0, \qquad j = 1, \ldots, n. \tag{8.109}$$

But since (8.106) and (8.107) hold, (8.109) must hold only for $j = 1, \ldots, q$. Consequently, $H(r)$ is negative semidefinite for every $r \geq \bar{r}$, where

$$\bar{r} = \max_{j=1,\ldots,q} \left\{ \frac{s_j}{b^T b s_{j-1}^*} \right\}. \tag{8.110}$$

Let the maximum in (8.110) be obtained for $j = j_0$; then by (8.105) and (8.109)

$$(-1)^j \bar{s}_j = (-1)^j s_j + \bar{r} b^T b(-1)^{j-1} s_{j-1}^* \geq 0, \qquad j = 1, \ldots, q. \tag{8.111}$$

Since $\bar{s}_{j_0} = 0$ and $H(\bar{r})$ is negative semidefinite, we must have

$$\bar{s}_j = s_j - \bar{r} b^T b s_{j-1}^* = 0, \qquad j_0 \leq j \leq q. \tag{8.112}$$

In particular, by letting $j = q$ in (8.112) we obtain the expression we sought:

$$\bar{r} = \frac{s_q}{b^T b s_{q-1}^*}. \tag{8.113}$$

We may now obtain our first expression for $\rho_0(x)$ for a function $f \in H$. Setting $A = \nabla^2 f(x)$, $b = \nabla f(x)$ and using the notation $s_j(x)$ and $s_j^*(x)$ for the coefficients of the characteristic equations of $\nabla^2 f(x)$, and of $\nabla^2 f(x)$ restricted to $\nabla f(x)^\perp$, respectively, we obtain

$$\rho_0(x) = \frac{s_{q(x)}(x)}{\nabla f(x)^T \nabla f(x) s^*_{q(x)-1}(x)}, \tag{8.114}$$

where $q(x) - 1$ is the rank of $\nabla^2 f(x)$ restricted to $\nabla f(x)^\perp$, or equivalently, $q(x) + 1$ is the rank of $B(x)$.

It should be mentioned that $\rho_0(x)$ is not necessarily a continuous function of x, even though f is a C^2 function and consequently, s_j and s_j^* are continuous functions of x. This is because $q(x)$ is an integer, and changes in q may cause discrete changes in ρ_0.

We now turn to obtain a second expression for $\rho_0(x)$, due to Schaible and Zang (1980). Again, first we find the value $\bar{r}$ for which $H(r)$ becomes negative semidefinite for all $r \geq \bar{r}$. Utilizing Proposition 8.16 and in particular equation (8.85), we obtain that it is necessary and sufficient for

$$r \geq -\det A_{\gamma,k}/\det D_{\gamma,k} \tag{8.115}$$

to hold for all $\gamma \in \Gamma_k$, $k = 1, \ldots, n$ satisfying $(-1)^k \det D_{\gamma,k} > 0$. Thus,

$$\bar{r} = \max_{\gamma,k} \{-\det A_{\gamma,k}/\det D_{\gamma,k} : (-1)^k \det D_{\gamma,k} > 0\}. \tag{8.116}$$

As above, this implies that for a function $f \in H$, we obtain a second expression for $\rho_0(x)$:

$$\rho_0(x) = \max_{\substack{\gamma \in \Gamma_k \\ k=1,\ldots,n}} \{-\det A_{\gamma,k}(x)/\det D_{\gamma,k}(x) : (-1)^k \det D_{\gamma,k}(x) > 0\}. \tag{8.117}$$

Note that in case $\nabla^2 f(x)$ is not negative semidefinite at x then for at least one principal minor $(-1)^k \det A_{\gamma,k}(x) < 0$ holds and, therefore, by Proposition 8.16, $(-1)^k \det D_{\gamma,k}(x) > 0$. Thus $\rho_0(x)$ in (8.117) is positive.

We shall now obtain our last expression for $\rho_0(x)$, due to Diewert (1978). Here $\rho_0(x)$ is represented in terms of the eigenvalues of $\nabla^2 f(x)$ restricted to $\nabla f(x)^\perp$. As before, we shall first derive an expression for $\bar{r}$ for a matrix A and a vector $b \neq 0$ satisfying (i.1) and (ii.1). We recall that V is the $n \times (n-1)$ matrix whose columns are the mutually orthogonal eigenvectors of A restricted to $b^\perp$, obtained by solving (8.51)-(8.53), and that Λ

is an $(n-1) \times (n-1)$ diagonal matrix of the corresponding restricted eigenvalues, which by (i.1) are nonpositive. Finally, let $N = (V, b)$. Then N is nonsingular, and $H(r)$ is negative semidefinite if and only if $N^T H(r) N$ is also negative semidefinite, that is, if and only if $N^T AN - rN^T bb^T N$ is negative semidefinite. However,

$$N^T AN = \begin{pmatrix} V^T \\ b^T \end{pmatrix} A(V, b) = \begin{pmatrix} V^T AV & V^T Ab \\ b^T AV & b^T Ab \end{pmatrix}. \tag{8.118}$$

Since $Av^i = \beta_i b + \lambda_i v^i$, by (8.51) and by using $V^T b = 0$, we obtain that $b^T AV = b^T b\beta$, where $\beta = (\beta_1, \beta_2 \ldots \beta_{n-1})^T$, and also $V^T AV = \Lambda$. Hence,

$$N^T AN = \begin{pmatrix} \Lambda & b^T b\beta \\ b^T b\beta^T & b^T Ab \end{pmatrix}. \tag{8.119}$$

Consequently, $H(r)$ is negative semidefinite if and only if the matrix given by

$$\begin{aligned} N^T H(r) N &= N^T AN - rN^T bb^T N \\ &= N^T AN - r\begin{pmatrix} V^T \\ b^T \end{pmatrix} bb^T (V, b) \\ &= N^T AN - r\begin{pmatrix} V^T bb^T V & V^T bb^T b \\ b^T bb^T V & (b^T b)^2 \end{pmatrix} \\ &= \begin{pmatrix} \Lambda & b^T b\beta \\ b^T b\beta^T & b^T Ab - r(b^T b)^2 \end{pmatrix} = \bar{B} \end{aligned} \tag{8.120}$$

is negative semidefinite, where the last equality follows from (8.119) and $V^T b = 0$. The $n \times n$ matrix P, given by

$$P = \begin{pmatrix} I & \bar{\beta} \\ 0 & 1 \end{pmatrix}, \tag{8.121}$$

where $\bar{\beta}_i = -b^T b\beta_i/\lambda_i$ if $\lambda_i < 0$ and $\bar{\beta}_i = 0$ otherwise $(i = 1, \ldots, n-1)$, is nonsingular. It follows that $\bar{B}$ is negative semidefinite if and only if $P^T\bar{B}P$ is also negative semidefinite. However,

$$P^T\bar{B}P = \begin{pmatrix} \Lambda & 0 \\ 0 & b^TAb - r(b^Tb)^2 - \sum_{\substack{i=1 \\ \lambda_i<0}}^{n-1} (b^Tb)^2\beta_i^2/\lambda_i \end{pmatrix}. \tag{8.122}$$

Thus, $H(r)$ is negative semidefinite for all values of r for which the (n, n)th entry of $P^T\bar{B}P$ is nonpositive. Consequently,

$$\bar{r} = \frac{b^TAb}{(b^Tb)^2} - \sum_{\substack{i=1 \\ \lambda_i<0}}^{n-1} \beta_i^2/\lambda_i \tag{8.123}$$

must hold. This implies that for a function $f \in H$

$$\rho_0(x) = \frac{\nabla f(x)^T\nabla^2 f(x)\nabla f(x)}{[\nabla f(x)^T\nabla f(x)]^2} - \sum_{\substack{i=1 \\ \lambda_i(x)<0}}^{n-1} \beta_i(x)^2/\lambda_i(x), \tag{8.124}$$

whether $\beta_i(x)$ and $\lambda_i(x)$ are the solutions of (8.51)-(8.53) for $A = \nabla^2 f(x)$ and $b = \nabla f(x)$.

Example 8.11. Consider the functions in Example 8.6. For this function (8.97) and (8.102) yield

$$S_{\nabla^2 f(x)}(\lambda) = -(-1)^1\lambda + (-1)^3(\lambda)^3 \tag{8.125}$$

and

$$S^*_{\nabla^2 f(x)}(\lambda) = -(-1)^1\frac{2x_1x_3}{(x_1)^2 + (x_3)^2}\lambda + (-1)^2(\lambda)^2. \tag{8.126}$$

Since $q(x) = 2$ for every $x \in C$ we have $s_2 = -1$ and $s_1^* = -2x_1x_3/[(x_1)^2 + (x_3)^2] = -2x_1x_3/\nabla f(x)^T\nabla f(x)$. Thus (8.114) implies that

$$\rho_0(x) = 1/(2x_1x_3). \tag{8.127}$$

Note that the same value of $\rho_0(x)$ is obtained from (8.117) and the values of $\det D_{\gamma,k}(x)$ and $\det A_{\gamma,k}(x)$ in Example 8.9.

To conclude this section we would like to comment on the relative efficiencies of computing $\rho_0(x)$ from (8.114) or (8.117). First we note that to compute $s_{q(x)}(x)$ it is necessary to compute all the principal minors of order $q(x)$ of $\nabla^2 f(x)$; that is, we must compute det $A_{\gamma,q(x)}(x)$ for all $\gamma \in \Gamma_{q(x)}$. Similarly, computing $s^*_{q(x)-1}(x)$ requires the computation of det $D_{\gamma,q(x)}(x)$ for all $\gamma \in \Gamma_{q(x)}$. Moreover, to carry out these operations we must check first whether (I.3) and (II.3) are satisfied at x. This process yields the value of $q(x)$ and involves $O(n^3)$ multiplications. However, to calculate (8.117), it would probably be more efficient to verify the existence of $\rho_0(x)$ by checking (I.4) and (II.5). In this case, the computation of $\rho_0(x)$ becomes relatively easy. However, the verification of (I.4) and (II.5) involves the computation of det $D_{\gamma,k}(x)$ for all $\gamma \in \Gamma_k$ and $k = 1, \ldots, n$, and the computation of det $A_{\gamma,k}(x)$ whenever det $D_{\gamma,k}(x) = 0$ holds. In case $q(x)$ is known, it is necessary to carry out these computations only for $k = 1, \ldots, q(x)$, since det $D_{\gamma,k}(x) =$ det $A_{\gamma,k}(x) = 0$ for $k > q(x)$. In any case, calculating $\rho_0(x)$ from (8.117) seems to be less efficient than calculating this value from (8.114). This comparison is only a rough one, and it does not consider condition (8.124). It is well known that the calculation of the eigenvalues of a matrix is a process requiring $O(n^3)$ multiplications. Thus, calculating $\rho_0(x)$ from (8.124) may prove to be more efficient than from (8.117). In conclusion, a more detailed and thorough examination of the computational aspects of expressions (8.114), (8.117), and (8.124) is needed to determine which is more efficient. However, the practical implications of the computability of $\rho_0(x)$ seem to be quite limited, since it may be necessary to compute its values for all possible values of $x \in C$.

8.2.4. The Global Conditions: Existence of *G*

In this section we consider functions that belong to the family H as characterized in the last sections. Our aim is to establish conditions that are necessary and sufficient for the existence of a concavifying function G for a given $f \in H$. Clearly, the existence of the function $\rho_0(x)$ such that $H(x; \rho_0(x))$ is negative semidefinite for every $x \in C$ is already ensured by assuming that $f \in H$. This, however, does not guarantee that f is G-concave for some G, even if $\rho_0(x)$ is continuous, because $\rho_0(x)$ may approach $+\infty$ as x approaches a boundary point of C. The following example, due to Schaible (1971), demonstrates such a situation.

Example 8.12. The function

$$f(x) = -(x_1)^3 - (x_2)^2 \tag{8.128}$$

is pseudoconcave on every open convex set contained in

$$Y = \{x \in R^2\colon x_1 < 0, -[-\tfrac{3}{4}(x_1)^3]^{1/2} < x_2 < [-\tfrac{3}{4}(x_1)^3]^{1/2}\}. \tag{8.129}$$

Consider an open convex subset $C \subset Y$ such that there exists a point $\bar{x}$ on the boundary of C that is also on the boundary of Y. For example, let

$$C = \{x \in R^2\colon x_1 > -4/3, x_2 > 0, 3x_1 + 3x_2 < -4/3\}. \tag{8.130}$$

Then $\bar{x} = (-4/3, 4/3)^T$. It can be shown that $f \in H$ with

$$\rho_0(x) = -\tfrac{1}{2}[\tfrac{3}{4}(x_1)^3 + (x_2)^2]^{-1}. \tag{8.131}$$

It is easy to see that $\rho_0(x) > 0$ for $x \in Y$ and that $\rho_0(x) \to \infty$ as x approaches a boundary point of Y. Consider all points $x \in C$ satisfying $f(x) = \bar{t} = 16/27$. This level curve is shown in Figure 8.2. Certainly since $\alpha < \bar{t} < \beta$ (see 8.37), the set $\{\rho_0(x)\colon f(x) = \bar{t}\}$ should be bounded from above if f is G-concave for some G. However, $\rho_0(x) \to \infty$ as $x \to \bar{x}$ along this level curve.

In view of the above example the following conditions should be satisfied for the function f to be G-concave (Fenchel, 1951):

(C) The function $g(t)$ defined on the open interval (α, β) by

$$g(t) = \sup\{\rho_0(x)\colon x \in C, f(x) = t\} \tag{8.132}$$

is finite for every $t \in (\alpha, \beta)$.

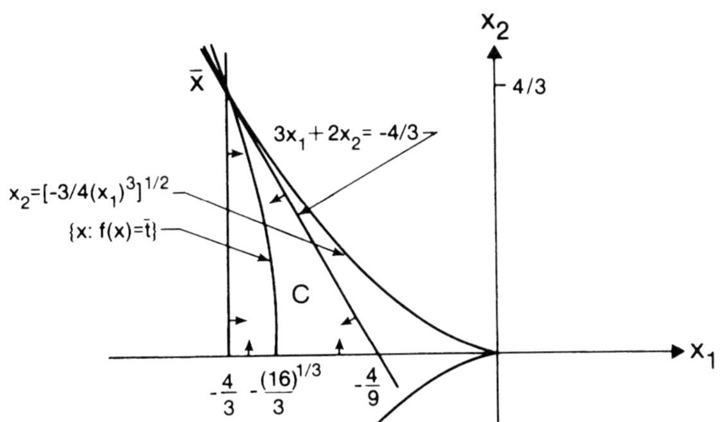

Figure 8.2. The level curve $f(x) = 16/27$.

By (8.114), (8.117), or (8.124) it is possible to obtain the following equivalent expressions:

$$g(t) = \sup\left\{\frac{s_{q(x)}(x)}{\nabla f(x)^T \nabla f(x) s^*_{q(x)-1}(x)} : x \in C, f(x) = t\right\}, \tag{8.133}$$

$$g(t) = \sup_{\substack{x \in C \\ f(x)=t}} \left[\max_{\substack{\gamma \in \Gamma_k \\ k=1,\ldots,n}} \{-\det A_{\gamma,k}(x)/\det D_{\gamma,k}(x) : (-1)^k \det D_{\gamma,k}(x) > 0\}\right] \tag{8.134}$$

and

$$g(t) = \sup_{\substack{x \in C \\ f(x)=t}} \left\{\frac{\nabla f(x)^T \nabla^2 f(x) \nabla f(x)}{[\nabla f(x)^T \nabla f(x)]^2} - \sum_{\substack{i=1 \\ \lambda_i(x)<0}}^{n-1} [\beta_i(x)]^2/\lambda_i(x)\right\}. \tag{8.135}$$

Note that (8.133) and the discussion following (8.114) imply that $g(t)$ is not necessarily continuous.

Example 8.13. Consider again the function in Examples 8.6 and 8.11. From (8.41), (8.127), and (8.132) we have that

$$g(t) = 1/(2t), \tag{8.136}$$

which is finite for $t \in (\alpha, \beta) = (0, \infty)$. Thus condition (C) holds.

Consider now a function f for which condition (C) is satisfied, that is, $g(t)$ is finite on (α, β). If G is a twice continuously differentiable function on (α, β) satisfying $G'(t) > 0$ and

$$-G''(t)/G'(t) \geq g(t) \tag{8.137}$$

then, in view of (8.48), f is G-concave over C. Equivalently, we may require the existence of a once continuously differentiable and positive function $h(t)$ on (α, β) satisfying

$$\frac{d \ln h(t)}{dt} = \frac{h'(t)}{h(t)} \leq -g(t). \tag{8.138}$$

This will imply that f is G-concave with $G'(t) = h(t)$. Hence condition C alone is not sufficient for G-concavity of f. The following example demonstrates such a case:

Example 8.14 (Avriel and Schaible, 1978). Let

$$f(x) = \begin{cases} -\int_0^x (\xi)^4[2 + \cos(1/\xi)]\,d\xi & \text{if } x > 0, \\ 0 \quad \text{if } x = 0, \\ \int_0^x (\xi)^4[2 + \cos(1/\xi)]\,d\xi & \text{if } x < 0. \end{cases} \tag{8.139}$$

This function is strictly pseudoconcave on R. To see this, consider

$$f'(x) = \begin{cases} -(x)^4[2 + \cos(1/x)] & \text{if } x > 0, \\ 0 \quad \text{if } x = 0, \\ (x)^4[2 + \cos(1/x)] & \text{if } x < 0, \end{cases} \tag{8.140}$$

and $f'(x) > 0$ for $x < 0$, $f'(0) = 0$, and $f'(x) < 0$ for $x > 0$. However, by computing the second derivative

$$f''(x) = \begin{cases} -(x)^2\{4x[2 + \cos(1/x)] + \sin(1/x)\} & \text{if } x > 0, \\ 0 \quad \text{if } x = 0, \\ (x)^2\{4x[2 + \cos(1/x)] + \sin(1/x)\} & \text{if } x < 0, \end{cases} \tag{8.141}$$

we can see that f'' changes sign in every neighborhood of the origin; thus f is not concave. It is easy to verify that conditions (i.1) and (ii.1) hold for this function, and using (8.40) and the symmetry of $f''(x)$ we have

$$\rho_0(x) = \frac{f''(x)}{[f'(x)]^2} = -\frac{4|x|[2 + \cos(1/x)] + \sin(1/|x|)}{x^6[2 + \cos(1/x)]^2} \tag{8.142}$$

if $x \neq 0$ and $\rho_0(0) = 0$; thus $f \in H$. By (8.132) we have

$$g(t) = \{\rho_0(x): f(x) = t\} \tag{8.143}$$

for every $t \leq 0$, where, without loss of generality, we consider only points satisfying $x \geq 0$. To show that f is G-concave, we should establish the existence of a positive and differentiable function $h(t)$ satisfying (8.138) for all $t \geq 0$, and then it will follow that f is G-concave with $G'(t) = h(t)$.

However, since the correspondence between $g(t)$ and $\rho_0(x)$ given by (8.143) is one-to-one for all $t \leq 0$ and $x \geq 0$, it is sufficient to establish the existence or nonexistence of a function $\tilde{h}(x)$ such that $\tilde{h}$ is differentiable for all $x \geq 0$ and

$$\frac{d \ln \tilde{h}(x)}{dx} = \frac{\tilde{h}'(x)}{\tilde{h}(x)} \leq -\rho_0(x) \tag{8.144}$$

for all $x \geq 0$. We shall show actually that such a function h cannot be differentiable at $x = 0$. For $k = 1, 2, \ldots$, let $x^k = [(3/2)\pi + 2k\pi]^{-1}$. Then

$$\rho_0(x^k) = -\frac{8x - 1}{4(x)^6}. \tag{8.145}$$

From (8.144) we have that

$$\frac{d \ln \tilde{h}(x)}{dx} \leq \frac{8x - 1}{4(x)^6} \tag{8.146}$$

must hold for $x = x^k$, which is equivalent to

$$\frac{d \ln \tilde{h}(x)}{dx} \leq \frac{d}{dx}\left[\frac{1 - 10x}{20(x)^5}\right]. \tag{8.147}$$

Thus, we have that

$$\tilde{h}(x) \leq \exp\left[(1 - 10x)/20(x)^5\right] \tag{8.148}$$

must hold for $x = x^k$. From (8.144) and (8.146) we have

$$\tilde{h}'(x) \leq \left[\frac{8x - 1}{4(x)^6}\right]\tilde{h}(x) \leq \left[\frac{8x - 1}{4(x)^6}\right] \exp\left[\frac{1 - 10x}{20(x)^5}\right], \tag{8.149}$$

for $x = x^k$, where the last inequality follows from (8.148). However, as $x \to 0$ the expression on the right-hand side of (8.149) goes to $-\infty$. Thus $\tilde{h}'(0)$ cannot exist.

In conclusion, the following condition should also be satisfied for f to be G-concave (Fenchel, 1956):

(D) There exists a differentiable function $h(t)$ on $t \in I_f(C)$ satisfying $h(t) > 0$ and

$$\frac{d}{dt} \ln [h(t)] \leq -g(t) \tag{8.150}$$

for $\alpha < t < \beta$, where $g(t)$ is given by (8.132) or by (8.133) through (8.135).

Example 8.15. Consider again the function f analyzed in Examples 8.6, 8.11, and 8.13. For this function we have that

$$h(t) = (t)^{-1/2} \tag{8.151}$$

satisfies condition (D).

It was shown by Crouzeix (1977) and Kannai (1977, 1981) that condition (C) can be stated in terms of Perron's integrability (see Natanson, 1960) or Lebesque integrability of $g(t)$. For further details the reader is referred to these references.

We conclude this section stating our main result. Certainly, because of their hierarchical nature, conditions (A)–(D) are sufficient for the G-concavity of f. In particular, condition (D) can be used to obtain an expression for G by letting $G' = h$. Moreover, these conditions are also necessary. To show this we note that in case f is G-concave then (A) trivially holds, and $G'(t) = h(t)$ is differentiable for $t \in I_f(C)$ and positive for $\alpha < t < \beta$. In addition, letting $\rho(x) = -G''(f(x))/G'(f(x))$ for nonmaximizing points $x \in C$ and $\rho(x) = 0$ for the maximizing ones, we have that $f \in H$ and (B) holds. To show that (C) holds, we note that for every $\alpha < t < \beta$

$$\begin{aligned} \tilde{g}(t) &= \sup\{-G''(f(x))/G'(f(x)) \colon x \in C, f(x) = t\} \\ &= -G''(t)/G'(t) < \infty \end{aligned} \tag{8.152}$$

holds, by the continuous differentiability of G. Since $g(t) \leq \tilde{g}(t)$, then (C) holds. Finally, $-g(t) \geq -\tilde{g}(t) = (d/dt)\{\ln[h(t)]\}$ and consequently (D) holds. To summarize, we can state the following theorem.

Theorem 8.18. Let f be a twice continuously differentiable function on the open convex set $C \subset R^n$. Then f is G-concave on C if and only if conditions (A)–(D) hold.

8.2.5. *G*-Concave Functions That Are *r*-Concave over Compact Sets

In this section we consider the family of twice continuously differentiable G-concave functions defined on the open convex set $C \subset R^n$, which are also r-concave over compact convex subsets $D \subset C$. That is, for such a G-concave function in every compact convex subset $D \subset C$, there exists a nonnegative number $r(D)$ such that f is $r(D)$-concave on D. A typical example is a G-concave function for which $G'(t) > 0$ for every $t \in I_f(C)$. We have already shown in Proposition 8.12 that such functions are $r(D)$-concave on every compact convex subset $D \subset C$. Consequently, once it is known that such a function is G-concave over a compact set, then it may be easier to obtain an expression for G in terms of an exponential function, where it is only necessary to determine the value of $r(D)$.

It is quite clear that, in general, f is r-concave on C with $r = \bar{r}$ (and consequently, for all $r \geq \bar{r}$) if $f \in H$ and $\sup_{x \in C} \{\rho_0(x)\}$ given by (8.40), is finite. Moreover, in case this supremum is nonnegative, $\bar{r}$ takes the value of the above supremum. Thus, f is r-concave if

$$\sup_{\substack{x \in C \\ z \in R^n \\ z^T \nabla f(x) \neq 0}} \left\{ \frac{z^T \nabla^2 f(x) z}{[z^T \nabla f(x)]^2} \right\} < \infty \tag{8.153}$$

holds. We have the following result:

Proposition 8.19 (Avriel, 1972). Let f be pseudoconcave on C. If (8.153) is satisfied and r^* is the supremal value, then f is $\bar{r}$-concave on C with $\bar{r} = \max(0, r^*)$.

The direct proof of the above proposition follows immediately from Corollary 8.11, using the negative semidefiniteness of $\nabla^2 f(x)$ restricted to $\nabla f(x)^\perp$, a property implied by the pseudoconcavity of f.

Note that in the above proposition, f is actually concave if $r^* \leq 0$, and if $\bar{r} > 0$, then it is the lowest possible value of r for a twice continuously differentiable pseudoconcave function. The above proposition is useful in the one-dimensional case, since for this case (8.153) becomes a very simple expression.

Example 8.16. Consider the function $f(x) = (x)^3 + x$ on $C = \{x \in R: -1 \leq x \leq 1\}$ presented in Example 8.5. For this function we have $f'(x) = 3(x)^2 + 1$, $f''(x) = 6x$ and

$$r^* = \sup_{0 \leq x \leq 1} \frac{6x}{[3(x)^2 + 1]^2}. \tag{8.154}$$

It can be shown that the above supremum is obtained for $x = 1/3$ and that $r^* = 9/8$, as was already pointed out in Example 8.5.

Example 8.17. Let $f(x) = -\log x$ and $C = R$. Then $f'(x) = -1/x$, $f''(x) = 1/(x)^2$, and

$$r^* = \sup_{x \in R} \frac{1/(x)^2}{(1/x)^2} = 1. \tag{8.155}$$

Thus, $-\log x$ is 1-concave.

In Proposition 8.7 we showed that a G-concave twice continuously differentiable function is also pseudoconcave, and this result applies, in particular, to r-concave functions. The converse statement, however, does not necessarily hold, as was shown by Examples 8.7 and 8.12, even in the one-dimensional case, as demonstrated by Example 8.14. We may mention here that a limited converse result was obtained in Avriel and Schaible (1978), where functions, termed *regular* were considered. A twice continuously differentiable function defined on a convex set $C \subset R^n$ is regular if whenever it obtains a maximum along some line segment at a point $\bar{x} \in C$, then along this line segment intersected with some open neighborhood of $\bar{x}$, the second derivative of f, as a function of one variable along the line segment, is nonpositive. The result obtained by Avriel and Schaible states that a regular pseudoconcave function is $r(I)$-concave as a function of one variable along every compact line segment I in C.

We return to discuss the main issue of this section, and note that in Example 8.14 we showed that a function belonging to the family H is not necessarily $r(D)$-concave on a compact convex set D. However, Avriel and Schaible (1978) considered a more restricted family of functions called H_c and defined as the family of functions in H for which $H(x; \rho(x))$ is negative semidefinite for some *continuous* function $\rho(x)$ on C. Furthermore, every function in H_c is pseudoconcave. Note that it follows from Example 8.14 that the inclusion $H_c \subset H$ is strict, that is, $H_c \neq H$. A different but equivalent characterization of the family H_c was given by Schaible and Zang (1980). They defined H_c as the family of functions in H satisfying the property that for every compact convex subset $D \subset C$, there exists an $\bar{r}(D)$ such that $H(x; \bar{r}(D))$ is negative semidefinite on D.

Certainly, the first definition implies the second by letting $\bar{r}(D) = \sup [\rho(x): x \in D]$, which is finite since ρ is continuous and D is compact. To show that the second definition implies the first, let $\{D_k\}$ be a sequence of compact convex sets in R^n such that $D_k \subset D_{k+1}$ for $k = 1, 2, \ldots$ and $C = \bigcup_{k=1}^{\infty} D_k$. Thus, there exist numbers $\{\bar{r}(D_k)\}$ such that $H(x, r(D_k))$ is negative semidefinite on D_k. Because $D_k \subset D_{k+1}$ we may assume that $\bar{r}(D_{k+1}) \geq \bar{r}(D_k)$. Since D_k are compact and convex, it follows (see Stoer and Witzgall, 1970, Theorem 3.8.3) that a continuous function ρ on C can be constructed such that $\rho(x) \geq \bar{r}(D_1)$ at $x \in D$, and $\rho(x) \geq \bar{r}(D_k)$ at $x \in D_k \backslash D_{k-1}$ for $k = 2, 3, \ldots$. It follows that $H(x; \rho(x))$ is negative semidefinite on C and that f satisfies the first definition.

In view of the above equivalence it is possible to characterize functions in H_c utilizing Theorem 8.17 and the expressions for $\rho_0(x)$ obtained in Section 8.2.3. Thus we have the following proposition.

Proposition 8.20 (Schaible and Zang, 1980). $f \in H_c$ if and only if $f \in H$ and for every compact convex subset $D \subset C$

$$r(D) = \max \{0, \sup_{x \in D} [\rho_0(x)]\} < \infty \tag{8.156}$$

holds, where $\rho_0(x)$ is given either by (8.114), (8.117), or (8.142).

For the function given in Example 8.6, we have, by Example 8.11, that $r(D) = \sup_{x \in D} \{1/(2x_1 x_3)\}$.

Thus we see that a pseudoconcave function f that belongs to H_c is also $r(D)$-concave, and that in case $r(D) > 0$ then it is the lowest possible bound for r. However, it is not necessarily r-concave or even G-concave on the whole of C, as shown by Example 8.12. The following proposition derives a necessary and sufficient condition for the G-concavity of an H_c function.

Proposition 8.21 (Avriel and Schaible, 1978). f is G-concave on C with $G'(t) > 0$ for $t \in I_f(C)$, if and only if $f \in H_c$ and the function $g(t)$ given by (8.132) is finite for all $t \in I_f(C)$.

Proof. If f is G-concave with $G'(t) > 0$ for $t \in I_f(C)$, then $f \in H_c$ with $\rho_0(x) = -G''(f(x))/G'(f(x))$, $g(t) \leq \tilde{g}(t) = -G''(t)/G'(t)$ for all $t \in I_f(C)$, and $g(t)$ is finite. Conversely, let $f \in H_c$ and suppose that $g(t)$ is finite for all $t \in I_f(C)$. Since f and ρ are continuous, so is g. Moreover, $H(x; g(t))$ is negative semidefinite for all $x \in C$ satisfying $f(x) = t$ and for all $t \in I_f(C)$. Finally, the continuity of g implies that the differential equation

$$-\frac{G''(t)}{G'(t)} = g(t) \tag{8.157}$$

has a solution

$$G(t) = \int^t \exp\left[\int^\eta - g(\tau)\, d\tau\right] d\eta \tag{8.158}$$

with $G'(t) > 0$. Hence, f is G-concave. $\square$

Note that a G-concave function for which $G'(t) = 0$ holds for $t = \beta$ does not necessarily belong to H_c since condition (D) of the last section does not necessarily require continuity of $g(t)$. As a result of the above proposition we have the following proposition.

Proposition 8.22. Let $f \in H_c$ and suppose that for every $t \in R$, the upper-level set

$$U(f, t) = \{x \in C: f(x) \geq t\} \tag{8.159}$$

is compact. Then f is G-concave.

Proof. The above hypotheses imply that $g(t)$, given by (8.132), is finite. The proof follows immediately by Proposition 8.21. □

Compactness of the level sets is not a necessary condition for functions in H_c to be concave transformable. For example, all nonconcave pseudoconcave quadratic functions have unbounded upper level sets, but they are concave transformable on solid convex sets by Theorem 6.3.

Let us consider now a more restricted family of $r(D)$-concave functions. These are functions for which a negative definite augmented Hessian exists at every $x \in C$. Following Avriel and Schaible (1972), we define H^s as the family of functions for which a negative definite augmented Hessian exists at every $x \in C$. That is, $f \in H^s$ if there exists a function $\rho(x)$ such that $H(x; \rho(x))$ is negative definite for every $x \in C$. We also let H_c^s be the family of all functions in H^s for which $H(x; \rho(x))$ is negative definite for some continuous function $\rho(x)$ on C. By using arguments similar to the one following the definition of H_c functions, we can equivalently define H_c^s functions as the family of functions in H^s satisfying the property that for every compact convex subset $D \subset C$, there exists an $\bar{r}(D)$ such that $H(x; \bar{r}(D))$ is negative definite on D (Schaible and Zang, 1980). Although $H_c \neq H$ we may obtain the following result, which will simplify the discussion to follow.

Proposition 8.23 (Avriel and Schaible, 1978). $H_c^s = H^s$.

Proof. Since $H_c^s \subset H^s$, we have to show that $H^s \subset H_c^s$. Let D be a compact set in C and let $\bar{x} \in D$. Since $f \in H^s$, there exists an $\varepsilon > 0$ such that $z^T \nabla^2 f(\bar{x}) z < 0$ on $Z = \{z: \|z\| = 1, |z^T \nabla f(\bar{x})| < \varepsilon\}$. Because of the continuity of ∇f and $\nabla^2 f$, there exists a neighborhood $N(\bar{x})$ of $\bar{x}$ such that for all $x \in N(\bar{x})$ we have $z^T\nabla^2 f(x)z < 0$ on Z and $|z^T\nabla f(x)| \geq \varepsilon/2 > 0$ for $z \notin Z$ satisfying $\|z\| = 1$. Let

$$r(\bar{x}) = \max\{0, \max[z^T\nabla^2 f(x)z/[z^T\nabla f(x)]^2 : x \in N(\bar{x}), z \notin Z, \|z\| = 1\}. \tag{8.160}$$

The right-hand side is finite and we see that $H(x; \bar{r})$ is negative definite on $N(\bar{x})$. Thus for every $x \in D$, $H(x; r)$ is negative definite on some neighborhood of x with finite r. Since D is compact we have by the Heine–Borel theorem (Apostol, 1974) that there exists a finite number of neighborhoods $N(x^k)$, $x^k \in D$ that cover D. Thus, $H(x, \bar{r}(D))$ is negative definite on D for $\bar{r}(D) = \max_k\{r(x^k)\}$. □

Thus, a function that belongs to H^s is actually $r(D)$-concave for $r(D) = \max\{0, \bar{r}(D)\}$. Moreover, $-\exp[-rf(x)]$ is strictly concave on D for every $r > r(D)$.

We now turn, in view of the above proposition, to derive necessary and sufficient conditions for a function f to belong to H^s_c. As we did for the family H, we first consider conditions that are necessary and sufficient for the negative definiteness of $H(r)$, given by (8.49), for all $r > \bar{r}$. Using the notation that preceded Proposition 8.16, and letting B be given by (8.56), we have the following proposition.

Proposition 8.24. The matrix $H(r)$ is negative definite for all $r > \bar{r}$ if and only if either condition (iii.1), (iii.2), or (iii.3) below holds.

(iii.1) $v \neq 0$, $v^Tb = 0$ implies $v^TAv < 0$.

(iii.2) For $k = 1, \ldots, n$, $(-1)^k \det D_k \geq 0$, and if $\det D_k = 0$ then $(-1)^k \det A_k > 0$ and $b_k = 0$.

(iii.3) A is negative definite whenever $b = 0$, and B is nonsingular and has exactly one positive eigenvalue whenever $b \neq 0$.

Proof. Condition (iii.1) is the well-known Finsler (1937) theorem. The proof of condition (iii.2) is similar to that of Proposition 8.16. That is, in view of Lemma 8.15, $H(r)$ is negative definite if and only if

$$(-1)^k[\det A_k + r\det D_k] > 0, \qquad k = 1, \ldots, n. \tag{8.161}$$

Thus, $(-1)^k \det D_k$ must be nonnegative and if it is zero then $(-1)^k \det A_k > 0$ must hold. However, since in the last case $\det A_k \neq 0$, we may apply Schur's formula (Gantmacher, 1959) given in Lemma 8.14 to obtain

$$\det D_k = \det A_k[-b_k^T A_k^{-1} b_k], \tag{8.162}$$

and since $\det D_k = 0$ we have

$$b_k^T A_k^{-1} b_k = 0. \tag{8.163}$$

We now show that A_k is negative definite, and thus $b_k = 0$ must hold. Condition (iii.1) implies that A has at most one positive eigenvalue. A well-known eigenvalues interlocking lemma (Stewart, 1973) implies that the same holds for A_k. Then since $(-1)^k \det A_k > 0$, we must have that all the eigenvalues of A_k are negative.

Now we have to show that condition (iii.3) is equivalent to the negative definiteness of $H(r)$. Clearly, if $b = 0$ then by (iii.1) we are done. Let $b \neq 0$ and assume that $H(r)$ is negative definite. Then (iii.1) holds and in view of (i.1) condition (i.3) holds. It is only left to show that B is nonsingular. Let (v_0, v^T), $v_0 \in R$, $v \in R^n$ be an eigenvector of B with zero eigenvalue, then

$$\begin{aligned} v^T b &= 0, \\ v_0 b + Av &= 0 \end{aligned} \tag{8.164}$$

holds. Clearly, $v \neq 0$, since otherwise, if $b \neq 0$ then $v_0 = 0$. Thus, multiplying the second equation in (8.164) by v^T, we obtain $v^T A v = 0$, contradicting (iii.1). Conversely, let $b \neq 0$ and assume that (iii.3) holds. Then (i.3) and consequently, (i.1) hold. To show that (iii.1) holds we establish that

$$v \neq 0, \qquad v^T b = 0, \qquad v^T A v = 0 \tag{8.165}$$

contradicts the nonsingularity of B and thus (iii.1) holds. Indeed, note that (8.165) and (i.1) imply that v is an eigenvector of A restricted to $b^\perp$ with zero eigenvalue. It follows that (8.51) holds with $\lambda = 0$, and consequently, (8.164) holds, which, in turn, implies that B is singular. □

Condition (iii.2) is due to Schaible and Zang (1980). Earlier versions of this condition can be found in Avriel and Schaible (1978), Bellman (1974), and Debreu (1952). Condition (iii.3) is due to Schaible (1981) and Zang (1981). Also note that if the matrix A satisfies the above conditions, then it cannot have more than one zero eigenvalue.

If we wish to compare the computational aspects of conditions (iii.1)–(iii.3), we immediately observe that it is impossible to check (iii.1) directly. In view of the discussion in Section 8.2.3, (iii.3) may be tested either by utilizing the Cholesky factorization of A in case $b = 0$ or Cottle's (1974) algorithm in case $b \neq 0$. Moreover, if $b \neq 0$, then checking (iii.1) is equivalent to testing negative definiteness of the matrix $Y^T A Y$ defined in condition (i.6), using the Cholesky factorization. As for (iii.2), certainly this condition is more compatible with (iii.3) than (i.4) and (ii.5) are with (i.3) and (ii.3); still, it seems to be inferior from the computational point of view.

Conditions that are necessary and sufficient for f to belong to H_c^s can now be derived as a straightforward application of Proposition 8.24. Letting $A = \nabla^2 f(x)$, $b = \nabla f(x)$, $b_k = (\nabla f(x))_k$ and letting $B(x)$, $D_k(x)$, $A_k(x)$ stand for B, D_k, and A_k, we obtain the following theorem.

Theorem 8.25. $f \in H_c^s$ if and only if for every $x \in C$ either condition (III.1), (III.2), or (III.3) below holds:

(III.1) $v \neq 0$, $v^T \nabla f(x) = 0$ implies $v^T \nabla^2 f(x) v < 0$.

(III.2) For $k = 1, \ldots, n$, $(-1)^k \det D_k(x) \geq 0$, and if $\det D_k(x) = 0$ then $(-1)^k \det A_k(x) > 0$ and $[\nabla f(x)]_k = 0$.

(III.3) $\nabla^2 f(x)$ is negative definite whenever $\nabla f(x) = 0$, and $B(x)$ is nonsingular and has exactly one positive eigenvalue whenever $\nabla f(x) \neq 0$.

Note that for the function $f(x) = x_1 x_3$ on $C = \{x \in R^3 : x_1 > 0, x_3 > 0\}$, considered in Examples 6.7, 8.6, and 8.9, $B(x)$ given in Example 6.7 is singular while $\nabla f(x) \neq 0$ for all $x \in C$. Consequently, (III.3) cannot hold [although (I.3) and (II.3) in Theorem 8.17 do hold], and $f \notin H_c^s$.

In view of Theorem 6.7, 6.9, or 6.11 we have that a nonconcave quadratic function belongs to H_c^s if and only if it is strictly pseudoconcave. For a general function note that condition (III.1) is actually the definition of strongly pseudoconcave functions introduced in Chapter 3 (see Definition 3.15 and Proposition 3.46). Therefore, we have the following corollary.

Corollary 8.26. $f \in H_c^s$ if and only if f is strongly pseudo-concave.

Thus, every strongly pseudoconcave function is $r(D)$-concave. Moreover, it is possible to show that strictly pseudoconcave functions become also $r(D)$-concave under some mild assumptions:

Corollary 8.27 (Schaible and Zang, 1980). $f \in H_c^s$ if and only if f is strictly pseudoconcave and for all $x \in C$ and $v \in R^n$, $v \neq 0$ the function $\bar{f}(\theta) = f(x + \theta v)$ has a local maximum at $\theta = 0$ only if $\bar{f}''(0) < 0$ and $\bar{f}'(0) = 0$.

Proof. The necessity is obvious since a strongly pseudoconcave function is strictly pseudoconcave. Conversely, if f is strictly pseudoconcave then $\bar{f}'(0) = v^T \nabla f(x) = 0$ implies that $\theta = 0$ is a local maximum of $\bar{f}$, and by the hypotheses $\bar{f}''(0) = v^T \nabla^2 f(x) v < 0$ holds. Thus (III.1) is satisfied. □

In view of the above corollary we have that every strictly pseudoconcave function for which the second derivative of a one-dimensional restriction is negative at a local maximum is $r(D)$-concave, since it is actually strongly pseudoconcave. We also obtain the following immediate result of Theorem 8.25, using (III.2):

Corollary 8.28 (Schaible and Zang, 1980). Assume that $\partial f/\partial x_1 \neq 0$ for every $x \in C$. Then $f \in H_c^s$ if and only if for all $x \in C$ and $k = 1, \ldots, n$, we have $(-1)^k \det D_k(x) > 0$.

Through interchanging variables, the above corollary remains true if $\partial f/\partial x_l \neq 0$ is assumed for some fixed l and for all $x \in C$. It is impossible, however, to drop this assumption, as can be seen from the following example:

Example 8.18. Let

$$f(x_1, x_2) = -\ln[(x_1)^2 + (x_2)^2], \tag{8.166}$$

and

$$C = \{x \in R^2: x_1 + x_2 > 1\}. \tag{8.167}$$

Then $\nabla f(x) = \alpha(-2x_1, -2x_2)^T$, where $\alpha = 1/[(x_1)^2 + (x_2)^2]$, and

$$\nabla^2 f(x) = 2\alpha^2 \begin{pmatrix} (x_1)^2 - (x_2)^2 & 2x_1x_2 \\ 2x_1x_2 & (x_2)^2 - (x_1)^2 \end{pmatrix}. \tag{8.168}$$

Furthermore, $v^T \nabla f(x) = 0$ implies that

$$v_1 x_1 = -v_2 x_2, \tag{8.169}$$

and

$$\begin{aligned} v^T \nabla^2 f(x) v &= 2(\alpha)^2 \{[(x_1)^2 - (x_2)^2](v_1)^2 + [(x_2)^2 - (x_1)^2](v_2)^2 + 4x_1x_2v_1v_2\} \\ &= -2(\alpha)^2 (x_1 v_2 + x_2 v_1)^2 < 0, \end{aligned} \tag{8.170}$$

where the second equality is a result of substituting (8.169), and the strict inequality follows since $v_2 \neq 0$, $v_1 = 0$ imply that $x_2 = 0$, which, in turn, imply that $x_1 \neq 0$. Thus $f \in H_c^s$. However, $\partial f/\partial x_1$ ($\partial f/\partial x_2$) becomes zero, say at $(0, 2) \in C$ [$(2, 0) \in C$].

We now proceed to derive expressions of $\rho_0(x)$ for functions in H_c^s. Note that in this case, $\rho_0(x)$ need not be continuous and $H(x; \rho_0(x))$ is not necessarily negative definite but only negative semidefinite. However, $\rho(x) > \rho_0(x)$ implies that $H(x; \rho(x))$ is negative definite, and there exists a continuous ρ on C. We refer first to (8.114) and note that for points $x \in C$ where $\nabla f(x) \neq 0$ we must have, in view of (III.3), that $q(x)$, the rank of $\nabla^2 f(x)$, is n. Thus,

$$\rho_0(x) = \frac{s_n(x)}{\nabla f(x)^T \nabla f(x) s_{n-1}^*(x)}. \tag{8.171}$$

From (8.97) we obtain by substituting $\lambda = 0$ that $s_n(x) = \det[\nabla^2 f(x)]$, whereas (8.102) implies that $s_{n-1}^*(x) = -\det[B(x)]/\nabla f(x)^T \nabla f(x)$. Therefore,

$$\rho_0(x) = -\det[\nabla^2 f(x)]/\det[B(x)] = -\det[A_n(x)]/\det[D_n(x)] \tag{8.172}$$

holds. This expression was also obtained by Gerencsér (1973), using a different approach. Moreover, if $\det A_n(x) \neq 0$ then Lemma 8.14 implies that

$$\det D_n(x) = \det A_n(x) \det[-\nabla f(x)^T \nabla^2 f(x)^{-1} \nabla f(x)], \tag{8.173}$$

and thus

$$\rho_0(x) = \begin{cases} \dfrac{1}{\nabla f(x)^T \nabla^2 f(x)^{-1} \nabla f(x)} & \text{if } \nabla f(x) \neq 0, \quad \det \nabla^2 f(x) \neq 0, \\ 0 \quad \text{if } \nabla f(x) \neq 0, & \det \nabla^2 f(x) = 0, \\ -\infty \quad \text{if } \nabla f(x) = 0, & \det \nabla^2 f(x) \neq 0. \end{cases} \tag{8.174}$$

Note that $\rho_0(x)$ given by the last expression is not continuous, if $\nabla f(x) = 0$ for some $x \in C$, and that $\nabla f(x) \to 0$ implies that $(-1)^n \det D_n(x) \to 0$. Thus in view of (III.2), $(-1)^n \det A_n(x)$ must have a positive limit as $\nabla f(x) \to 0$, and $-\det A_n(x)/\det D_n(x) = 1/\nabla f(x)^T \nabla^2 f(x)^{-1} \nabla f(x) \to -\infty$. Therefore, $\rho_0(x)$ as given by (8.174) is bounded from above on every compact subset of C. This agrees with the fact that $f \in H_c^s$, and it is possible to find a continuous function $\rho(x)$, satisfying $\rho(x) > \rho_0(x)$ for every $x \in C$, such that $H(x; \rho(x))$ is negative definite on C.

The expression in (8.174) is a specialization of Fenchel's (1951) formula (8.114) to H_c^s functions, and was obtained by Schaible and Zang (1980) utilizing a different approach. It was also obtained by Crouzeix (1977). It is possible to obtain another expression for $\rho_0(x)$, whenever $f \in H_c^s$, by specializing (8.124). The resulting expression will differ from (8.124) only by dropping the constraint $\lambda_i(x) < 0$ in the summation. However, (8.174) seems to be preferable from the computational point of view. We summarize the above discussion in the following proposition.

Proposition 8.29 (Schaible and Zang, 1980). Let $f \in H_c^s$. Then f is $r(D)$-concave on every compact convex subset $D \subset C$ for every $r(D)$ satisfying

$$r(D) > \max\{0, \sup_{x \in D} [\rho_0(x)]\} \tag{8.175}$$

where $\rho_0(x)$ is given by (8.174).

Note that $\sup\{\rho_0(x): x \in D\}$ may be negative, and thus expression (8.175) assures that $r(D)$ is nonnegative.

Example 8.19. Consider the quadratic function

$$Q(x) = \tfrac{1}{2}x^T A x + b^T x, \tag{8.176}$$

where $A \in R^{n \times n}$, $b \in R^n$. According to Theorems 8.25 and 6.9, $Q \in H_c^s$ if and only if Q is strictly pseudoconcave on C. This, in view of Proposition 6.5, implies that if Q is not concave then A is nonsingular, and a simple substitution in (8.174) gives

$$\rho_0(x) = \begin{cases} \dfrac{1}{2[Q(x) - \delta]} & \text{if } \nabla f(x) \neq 0 \\ -\infty & \text{if } \nabla f(x) = 0, \end{cases} \tag{8.177}$$

where δ is given by (6.46). In case Q is strictly concave then $\rho_0(x)$ is given by (8.177) if $\det A \neq 0$, and $\rho_0(x) = 0$ if $\det A = 0$.

Note that the discussion contained in the above example coincides with the results of Theorem 6.6.

8.3. Concave Transformable Functions: Domain and Range Transformations

In Sections 8.1 and 8.2 we considered nonconcave functions that can be transformed into concave functions by a one-to-one increasing transformation of their ranges. These functions were called G-concave functions.

Here we treat a more general family of concave transformable functions, which includes G-concave functions as a special subclass. These functions were introduced by Ben-Tal (1977) and discussed also by Avriel (1976) and Zang (1974). The idea that underlies the definition of these functions is that it is possible sometimes to apply a one-to-one transformation to the domain of a nonquasiconcave function so that its upper-level sets are transformed into convex sets, and in addition, to apply some monotone transformation (if needed) on the range of the transformed quasiconcave function to obtain a concave function.

To state the definition of this family of functions, we employ the concept of general mean-value functions, due to Hardy, Littlewood, and Polya (1952):

Suppose that f is a continuous function defined on a set $C \subset R^n$, where C is not necessarily convex, and let h be a continuous one-to-one and onto function defined on C with values in R^n. That is, $h : C \to R^n$. Similarly, let ϕ be a continuous increasing function defined on $I_f(C)$ with values in R. Note that both h and ϕ have one-to-one inverse functions h^{-1} and ϕ^{-1}, respectively, that satisfy $h^{-1}h(x) = x$, $hh^{-1}(\hat{x}) = \hat{x}$, $\phi^{-1}\phi(t) = t$ and $\phi\phi^{-1}(\hat{t}) = \hat{t}$ for every $x \in C$, $t \in I_f(C)$, and $\hat{x}$ and $\hat{t}$ are in the images of C and $I_f(C)$ under h and ϕ, denoted by $h(C)$ and $\phi(I_f(C))$, respectively. We also assume that $h(C)$ is a convex set.

Definition 8.3. The function $H(x^1, x^2; \theta) : [0, 1] \to C$ given by

$$H(x^1, x^2; \theta) = h^{-1}[(1 - \theta)h(x^1) + \theta h(x^2)] \tag{8.178}$$

is said to be an h-mean-value function. Similarly, $\Phi(f(x^1), f(x^2); \theta) : [0, 1] \to I_f(C)$ given by

$$\Phi(f(x^1), f(x^2); \theta) = \phi^{-1}[(1 - \theta)\phi(f(x^1)) + \theta\phi(f(x^2))] \tag{8.179}$$

is said to be a ϕ-mean-value function.

It can now be easily observed that the right-hand side of expression (8.1) is actually a ϕ-mean-value function of f with $\phi(t) = G(t)$. We now extend the concept of G-concavity to allow the possibility of taking h-generalized mean values of points in C instead of convex combinations as done in the left-hand side of (8.1). We have the following definition.

Definition 8.4 (Ben-Tal, 1977). A function f is said to be (h, ϕ)-concave on $C \subset R^n$, if for every $x^1 \in C$, $x^2 \in C$ and $0 \le \theta \le 1$ we have

$$f(H(x^1, x^2; \theta)) \ge \Phi(f(x^1), f(x^2); \theta). \tag{8.180}$$

It is said to be (h, ϕ)-convex if the inequality in (8.180) is reversed.

From (8.178) and (8.179) we have that (8.180) is equivalent to

$$f(h^{-1}[(1-\theta)h(x^1)+\theta h(x^2)]) \geq \phi^{-1}[(1-\theta)\phi(f(x^1))+\theta\phi(f(x^2))], \tag{8.181}$$

and that f is (h, ϕ)-convex if the inequality in (8.181) is reversed. It can be easily seen that (i) a concave function is (h, ϕ)-concave with $h(x) = x$, $\phi(t) = t$; (ii) a G-concave function is (h, ϕ)-concave with $h(x) = x$, $\phi(t) = G(t)$; (iii) an r-concave function is (h, ϕ)-concave with $h(x) = x$, $\phi(t) = -\exp(-rt)$.

We shall now see a few examples of (h, ϕ)-concave (convex) functions.

Example 8.20. Consider the following function on $C = R^2$ (Rosenbrock, 1960):

$$f(x) = -100[x_2 - (x_1)^2]^2 - (1 - x_1)^2. \tag{8.182}$$

This is a continuously differentiable nonconcave function having a unique maximum at $x^* = (1, 1)$. However, it does not belong to any family of generalized concave functions discussed in the preceding sections and chapters, such as pseudo- or quasiconcave functions. In fact, its upper-level sets are nonconvex "banana-shaped" sets, as can be seen in Figure 8.3.

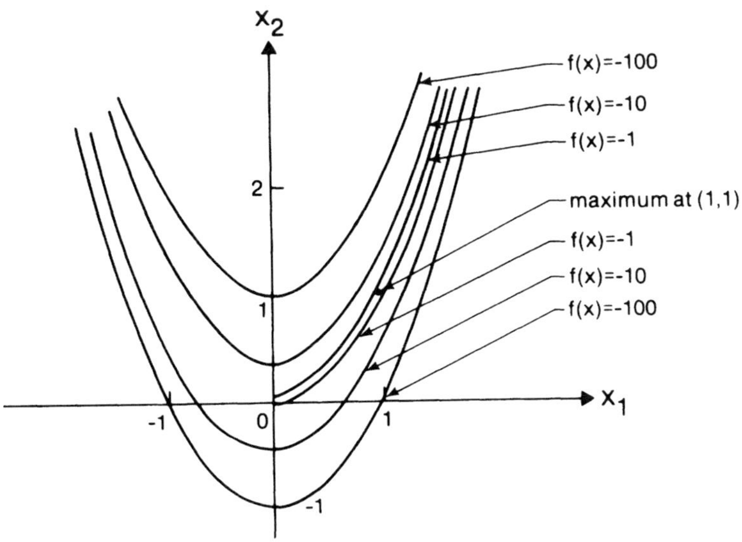

Figure 8.3. The function (8.182).

Letting

$$h(x) = \begin{pmatrix} 10[x_2 - (x_1)^2] \\ 1 - x_1 \end{pmatrix}, \tag{8.183}$$

$$h^{-1}(\hat{x}) = \begin{pmatrix} 1 - \hat{x}_2 \\ \frac{1}{10}\hat{x}_1 + (1 - \hat{x}_2)^2 \end{pmatrix}, \tag{8.184}$$

and $\phi(t) = t$, we obtain for every $x^1 \in R^2$, $x^2 \in R^2$

$$H(x^1, x^2; \theta) = \begin{pmatrix} (1-\theta)x_1^1 + \theta x_1^2 \\ (1-\theta)[x_2^1 - (x_1^1)^2] + \theta[x_2^2 - (x_1^2)^2] + [(1-\theta)x_1^1 + \theta x_1^2]^2 \end{pmatrix}, \tag{8.185}$$

and the reader can verify that for $0 \le \theta \le 1$,

$$f(H(x^1, x^2; \theta)) \ge (1-\theta)f(x^1) + \theta f(x^2). \tag{8.186}$$

In fact, from (8.182) and (8.184) we get

$$\phi f h^{-1}(x) = -(\hat{x}_1)^2 - (\hat{x}_2)^2, \tag{8.187}$$

a concave quadratic function.

We shall see below that a similar result holds for every (h, ϕ)-concave function; that is, they can be transformed into concave functions. In the last example we had a case where $\phi(t)$ was the identity function. In the next example neither h nor ϕ will be the identity function.

Example 8.21. Let f be a positive-valued function and let C be the positive orthant of R^n. Taking $h(x) = (\log x_1, \ldots, \log x_n)$ for $x > 0$ and $\phi(t) = \log t$ for $t > 0$, we can define f to be (log, log)-convex if for every two points x^1, x^2 in C and $0 \le \theta \le 1$, we have by (8.181) and after rearrangement,

$$f[(x_1^1)^{1-\theta}(x_1^2)^\theta, \ldots, (x_n^1)^{1-\theta}(x_n^2)^\theta] \le [f(x^1)]^{1-\theta}[f(x^2)]^\theta. \tag{8.188}$$

Posynomials that appear in a special branch of nonlinear programming called geometric programming (Duffin, Peterson, and Zener, 1967), are defined by

$$g(x) = \sum_{i=1}^{m} c_i \prod_{j=1}^{n} x_j^{a_{ij}}, \tag{8.189}$$

where $c_i > 0$ and $a_{ij} \in R$ are constants. Such functions are (log, log)-convex.

A detailed discussion regarding the implications of (h, ϕ)-convexity to geometric programming and, in particular, to geometric programming duality, can be found in Avriel (1976). The next example is concerned with (h, ϕ)-concavity [with $\phi(t) = t$] of concave fractional programs introduced in Chapter 7. First we note, however, that the most important feature of (h, ϕ)-concave functions is that they are concave transformable. In other words, they can be transformed into concave functions, as we shall see in the next theorem.

Theorem 8.30 (Ben-Tal, 1977). A function f is (h, ϕ)-concave on $C \subset R^n$ if and only if $\hat{f}$, given by

$$\hat{f}(\hat{x}) = \phi f h^{-1}(\hat{x}), \tag{8.190}$$

is concave on $h(C)$.

Proof. A function f is (h, ϕ)-concave on $C \subset R^n$ if and only if

$$f(h^{-1}[(1-\theta)h(x^1) + \theta h(x^2)]) \geq \phi^{-1}[(1-\theta)\phi f(x^1) + \theta\phi f(x^2)] \tag{8.191}$$

for every $x^1 \in C$, $x^2 \in C$, and $0 \leq \theta \leq 1$. Since ϕ is increasing, we obtain

$$\phi f(h^{-1}[(1-\theta)h(x^1) + \theta h(x^2)]) \geq (1-\theta)\phi f(x^1) + \theta\phi f(x^2). \tag{8.192}$$

Letting $h(x) = \hat{x}$ and substituting into the last inequality yields

$$\phi f h^{-1}((1-\theta)\hat{x}^1 + \theta\hat{x}^2) \geq (1-\theta)\phi f h^{-1}(\hat{x}^1) + \theta\phi h^{-1}(\hat{x}^2). \tag{8.193}$$

That is, $\phi f h^{-1}$ is concave. □

Example 8.22. Consider the concave fractional program

$$\sup_{x \in C} q(x) = f(x)/g(x) \tag{8.194}$$

subject to

$$h_j(x) \leq 0, \qquad j = 1, \ldots, m, \tag{8.195}$$

where $C \subset R^n$ is convex, f is nonnegative and concave, $g(x)$ is positive and convex, and $h_j(x)$, $j = 1, \ldots, m$ are convex on C. Applying Theorem 5.15, it follows that $q(x)$ is semistrictly quasiconcave on the convex feasible set of the above program. Introducing an additional variable $z \in R$, the above concave fractional program is equivalent to

$$\sup f(x)/z \tag{8.196}$$

subject to

$$h_j(x)/z \leq 0, \qquad j = 1, \ldots, m, \tag{8.197}$$

$$g(x)/z \leq 1, \tag{8.198}$$

$$x \in C, \qquad z > 0. \tag{8.199}$$

Consider now the following one-to-one domain transformation due to Charnes and Cooper (1962):

$$\begin{pmatrix} \hat{x} \\ \hat{z} \end{pmatrix} = h(x, z) = \begin{pmatrix} \dfrac{1}{z} x \\ \dfrac{1}{z} \end{pmatrix}; \tag{8.200}$$

that is,

$$\begin{pmatrix} x \\ z \end{pmatrix} = h^{-1}(\hat{x}, \hat{z}) = \begin{pmatrix} \dfrac{1}{\hat{z}} \hat{x} \\ \dfrac{1}{\hat{z}} \end{pmatrix}. \tag{8.201}$$

Substituting (8.201) in (8.196)-(8.199), we obtain the following equivalent transformed program:

$$\sup \hat{z} f\left(\frac{1}{\hat{z}} \hat{x}\right), \tag{8.202}$$

subject to

$$\hat{z}h_j\left(\frac{1}{\hat{z}}\hat{x}\right) \le 0, \qquad j = 1, \ldots, m, \tag{8.203}$$

$$\hat{z}g\left(\frac{1}{\hat{z}}\hat{x}\right) \le 1, \tag{8.204}$$

$$(\hat{x}, \hat{z}) \in \hat{C}, \tag{8.205}$$

where

$$\hat{C} = \{(\hat{x}, \hat{z}) \in R^n \times R: \hat{z} > 0, (1/\hat{z})\hat{x} \in C\}. \tag{8.206}$$

This problem is problem (7.24), and it was shown to be a concave program in Proposition 7.3 (see also the proof of Proposition 7.2). In particular, the objective function (8.202) is concave and the constraint functions (8.203) and (8.204) are convex on $\hat{C}$, which, in turn, is a convex set. That is, the functions in (8.196), (8.197), and (8.198) are (h, ϕ)-concave, convex, and convex, respectively, with $\phi(t) = t$ over all $x \in C$ and $z > 0$.

Many properties of (h, ϕ)-concave functions can be conveniently derived by introducing certain generalized operations of addition and multiplication, due to Ben-Tal (1977). Let h and ϕ be defined as above, and suppose further that $h(C) = R^n$ and $\phi(I_f(C)) = R$. Define the h-vector addition of $x \in C$ and $y \in C$ as

$$x \oplus y = h^{-1}(h(x) + h(y)) \tag{8.207}$$

and the h-scalar multiplication of $x \in C$ and $\lambda \in R$ as

$$\lambda \odot x = h^{-1}(\lambda h(x)). \tag{8.208}$$

Similarly, the ϕ-addition of two numbers, $\alpha \in I_f(C)$ and $\beta \in I_f(C)$, is given by

$$\alpha[+]\beta = \phi^{-1}(\phi(\alpha) + \phi(\beta)) \tag{8.209}$$

and the ϕ-scalar multiplication of $\alpha \in I_f(C)$ and $\lambda \in R$ as

$$\lambda[\cdot]\alpha = \phi^{-1}(\lambda\phi(\alpha)). \tag{8.210}$$

Finally, the (h, ϕ)-inner product of vectors $x \in C$, $y \in C$ is defined as

$$(x^T y)_{h,\phi} = \phi^{-1}([h(x)]^T h(y)). \tag{8.211}$$

The h-vector addition and scalar multiplication operations were suggested earlier in Craven (1975). A complete treatment of these operations from an algebraic point of view would require considerably more extensive background material. Interested readers are referred to Ben-Tal (1977).

Example 8.23. Let

$$h(x) = (\log x_1, \dots, \log x_n) \tag{8.212}$$

and

$$C = \{x: x \in R^n, x > 0\}. \tag{8.213}$$

Then

$$\begin{aligned} x \oplus y &= [\exp(\log x_1 + \log y_1), \dots, \exp(\log x_n + \log y_n)]^T \\ &= (x_1 y_1, \dots, x_n y_n)^T \end{aligned} \tag{8.214}$$

and for $\lambda \in R$

$$\begin{aligned} \lambda \odot x &= [\exp(\lambda \log x_1), \dots, \exp(\lambda \log x_n)]^T \\ &= [(x_1)^\lambda, \dots, (x_n)^\lambda]^T. \end{aligned} \tag{8.215}$$

It is easy to verify that mean-value functions can be written in this form as well—for example,

$$H(x^1, x^2; \theta) = [(1-\theta) \odot x^1] \oplus (\theta \odot x^2) \tag{8.216}$$

and

$$\Phi(\alpha^1, \alpha^2; \theta) = [(1-\theta)[\cdot]\alpha^1][+](\theta[\cdot]\alpha^2). \tag{8.217}$$

It follows then that the defining inequality (8.181) for (h, ϕ)-concave functions can be written as

$$f([(1-\theta) \odot x^1] \oplus (\theta \odot x^2)) \geq ((1-\theta)[\cdot]f(x^1))[+](\theta[\cdot]f(x^2)). \tag{8.218}$$

Thus (h, ϕ)-concave functions are "concave" under the foregoing generalized algebraic operations.

Using these operations, let us see how to generalize results presented in the preceding chapters for concave functions to (h, ϕ)-concave functions. For example, Proposition 2.15 states that if f_1 and f_2 are concave functions and λ is a nonnegative number, then $f_1 + f_2$ and λf_1 are also concave. The corresponding results for (h, ϕ)-concave functions f_1 and f_2 and for $\lambda \geq 0$ are that

$$f_1(x)[+]f_2(x) = \phi^{-1}[\phi(f_1(x)) + \phi(f_2(x))] \tag{8.219}$$

and

$$\lambda[\cdot]f_1(x) = \phi^{-1}(\lambda\phi(f_1(x))) \tag{8.220}$$

are also (h, ϕ)-concave. We have also seen that f is concave if and only if $(-f)$ is convex. Similarly, f is (h, ϕ)-concave if and only if

$$(-1)[\cdot]f(x) = \phi^{-1}(-\phi(f(x))) \tag{8.221}$$

is (h, ϕ)-convex.

In Theorem 2.12 we proved that a differentiable function on a convex set $C \subset R^n$ is concave if and only if

$$f(x^2) \leq f(x^1) + (x^2 - x^1)^T\nabla f(x^1) \tag{8.222}$$

for any two points $x^1 \in C$, $x^2 \in C$. Similarly, a differentiable function is (h, ϕ)-concave, with h and ϕ differentiable, if and only if

$$f(x^2) \leq f(x^1)[+][(x^2 \ominus x^1)^T\nabla^* f(x^1)]_{h,\phi}, \tag{8.223}$$

where

$$x^2 \ominus x^1 = x^2 \oplus (-1) \odot x^1 = h^{-1}(h(x^2) - h(x^1)) \tag{8.224}$$

and

$$\nabla^* f(x^1) = h^{-1}(\nabla \phi f h^{-1}(t)|_{t=h(x^1)}). \tag{8.225}$$

The inequality of (8.223) can be translated back into the more familiar form of ordinary algebraic operations, and then we obtain

$$\phi f(x^2) \leq \phi f(x^1) + \phi'(f(x^1)) \sum_{i=1}^{n} \sum_{j=1}^{n} \frac{\partial f(x^1)}{\partial x_j} \frac{\partial h_j^{-1}(h(x^1))}{\partial x_i}[h_i(x^2) - h_i(x^1)]. \tag{8.226}$$

The following is an immediate result of the above inequality.

Theorem 8.31 (Ben-Tal, 1977). Let f be a differentiable (h, ϕ)-concave function with h and ϕ differentiable on $C \subset R^n$, and let $x^* \in C$ satisfy $\nabla f(x^*) = 0$. Then x^* is a global maximum of f on C.

Proof. The result follows from (8.226), after substituting x^* for x^1. □

It is also possible to obtain a second-order characterization of twice differentiable (h, ϕ)-concave functions. For this property, as well as some applications of (h, ϕ)-convexity to statistical decision making, see Ben-Tal (1977). Because (h, ϕ)-concave functions are actually concave in some more general linear spaces, it is possible to obtain counterparts of many results that are obtained for concave optimization programs. For exmple, it is possible to develop a complete duality theory for "(h, ϕ)-concave" programs. See Avriel (1976) or Zang (1974) for further details.

References

APOSTOL, T. M. (1974), *Mathematical Analysis*, 2nd ed., Addison-Wesley, Reading, Massachusetts.

ARROW, K. J., and ENTHOVEN, A. C. (1961), Quasi-concave programming, *Econometrica* **29**, 779–800.

AUMANN, R. J. (1975), Values of markets with a continuum of traders, *Econometrica* **43**, 611–646.

AVRIEL, M. (1972), *r*-Convex functions, *Math. Programming* **2**, 309–323.

AVRIEL, M. (1973), Solution of certain nonlinear programs involving *r*-convex functions, *J. Optimization Theory Appl.* **11**, 159–174.

AVRIEL, M. (1976), *Nonlinear Programming: Analysis and Methods*, Prentice-Hall, Englewood Cliffs, New Jersey.

AVRIEL, M., and SCHAIBLE, S. (1978), Second order characterizations of pseudoconvex functions, *Math. Programming* **14**, 170–185.

AVRIEL, M., and ZANG, I. (1974), Generalized convex functions with applications to nonlinear programming, in *Mathematical Programs for Activity Analysis*, Edited by P. Van Moeseke, North-Holland, Amsterdam.

BELLMAN, R. (1974), *Introduction to Matrix Analysis*, 2nd ed., McGraw-Hill, New York.

BEN-TAL, A. (1977), On generalized means and generalized convexity, *J. Optimization Theory Appl.* **21**, 1–13.

CHABRILLAC, Y., and CROUZEIX, J. P. (1982), Definiteness and semi-definiteness of quadratic forms revisited, Mimeo, Department de Mathématiques Appliquées, Université de Clermont II, France.

CHARNES, A., and COOPER, W. W. (1962), Programming with linear fractional functionals, *Naval Res. Logistics Q.* **9**, 181–186.

COTTLE, R. W. (1974), Manifestation of the Schur complement, *Linear Alg. Appl.* **8**, 189–211.

CRAVEN, B. D. (1975), Converse duality in Banach spaces, *J. Optimization Theory Appl.* **17**, 229–238.

CROUZEIX, J. P. (1977), Contributions à l'étude des fonctions quasi-convexes, Thèse de Doctorat, U.E.R. des Sciences Exactes et Naturelles, Université de Clermont-Ferrand II, France.

CROUZEIX, J. P., and FERLAND, J. A. (1982), Criteria for quasi-convexity and pseudo-convexity: Relationships and numerical analysis, *Math. Programming* **23**, 193-205.

DEBREU, G. (1952), Definite and semidefinite quadratic firms, *Econometrica* **20**, 295-300.

DEBREU, G. (1976), Least concave utility functions, *J. Math. Economics* **3**, 121-129.

DE FINETTI, B. (1949), Sulle stratificazioni convesse, *Ann. Math. Pura Appl.* **30**, 123-183.

DIEWERT, W. E. (1978), Notes on transconcavity, unpublished manuscript.

DUFFIN, R. J., PETERSON, E. L., and ZENER, C. (1967), *Geometric Programming—Theory and Applications*, Wiley, New York.

FENCHEL, W. (1951), Convex cones, sets and functions, Mimeographed lecture notes, Princeton University, Princeton, New Jersey.

FENCHEL, W. (1956), Über konvexe Funktionen mit voreschriebenen Niveaumannigfaltigkeiten, *Math. Z.* **63**, 496-506.

FERLAND, J. A. (1981), Matrix-theoretic criteria for the quasi-convexity of twice continuously differentiable function, *Linear Alg. Appl.* **38**, 51-63.

FINSLER, P. (1937), Über das Vorkommen definiter und semidefiniter Formen in scharen quadratischer Formen, *Comment. Math. Helv.* **9**, 188-192.

GANTMACHER, F. R. (1959), *The Theory of Matrices*, Vol. I, Chelsea Publishing Company, New York.

GERENCSÉR, L. (1973), On a close relation between quasiconvex and convex functions and related investigations, *Math. Operationenforsch. Stat.* **4**, 201-211.

HARDY, G. H., LITTLEWOOD, J. E., and POLYA, G. (1952), *Inequalities*, 2nd ed., Cambridge University Press, Cambridge, England.

HORST, R. (1971a), Über mittelbar konvexe Optimierung, doctoral dissertation, Technische Hochschule Darmstadt.

HORST, R. (1971b), Mittelbare konvexe Funktionen and Optimierungsaufgaben, in *Methods of Operations Research*, Vol. XII, Edited by R. Henn, H. P. Künzi, and H. Schubert, Verlag Anton Hain, Meisenheim.

HURWICZ, L., JORDAN, J., and KANNAI, Y. (1984), On the demand generated by smooth and concavifiable preference ordering, University of Minnesota report.

KANNAI, Y. (1974), Approximation of convex preferences, *J. Math. Econ.* **1**, 101-106.

KANNAI, Y. (1977), Concavifiability and the constructions of concave utility functions, *J. Math. Econ.* **4**, 1-56.

KANNAI, Y. (1980), The ALEP definition of complementarity and least concave utility functions, *J. Econ. Theory* **22**, 115-117.

KANNAI, Y. (1981), Concave utility functions—existence, constructions and cardinality, in *Generalized Concavity in Optimization and Economics*, Edited by S. Schaible and W. T. Ziemba, Academic Press, New York.

KANNAI, Y. (1985), Engel curves, marginal utility of income, and concavifiable preferences, Working Paper, The Weizmann Institute of Sciences.

KANNAI, Y., and MANTEL, R. (1978), Non-convexifiable Pareto sets, *Econometrica* **46**, 571-575.

KATZNER, D. W. (1970), *Static Demand Theory*, Macmillan, New York.

KLINGER, A., and MANGASARIAN, O. L. (1968), Logarithmic convexity and geometric programming, *J. Math. Anal. Appl.* **24**, 388-408.

LINDBERG, P. O. (1981), Power convex functions, in *Generalized Concavity in Optimization and Economics*, Edited by S. Schaible and W. T. Ziemba, Academic Press, New York.

MANGASARIAN, O. L. (1969), *Nonlinear Programming*, McGraw-Hill, New York.

MARCUS, M., and MINC, H. (1964), *A Survey of Matrix Theory and Matrix Inequalities*, Allyn and Bacon, Boston.

MARTOS, B. (1966), Nem-linearis Programmozási módszerek hatóköre (The power of nonlinear programming methods), MTA Közgazdaságtudományi Intézetének Közleményei No. 20, Budapest (in Hungarian).

MARTOS, B. (1975), *Nonlinear Programming, Theory and Methods*, North-Holland, Amsterdam.

NATANSON, I. P. (1960), *Theory of Functions of a Real Variable*, Vol. II, Federick Ungar, New York (English translation),

ROSENBROCK, H. H. (1960), An automatic method for finding the greater or least value of a function, *Computer J.* **3**, 175–184.

SCHAIBLE, S. (1971), Beiträge zur quasikonvexen Programmierung, doctoral dissertation, University of Köln, Germany.

SCHAIBLE, S. (1972), Quasiconvex optimization in general real linear spaces, *Z. Oper. Res.* **16**, 205–213.

SCHAIBLE, S. (1977), Second-order characterizations of pseudoconvex quadratic functions, *J. Optimization Theory Appl.* **21**, 15–26.

SCHAIBLE, S. (1981), Generalized convexity of quadratic functions, in *Generalized Concavity in Optimization and Economics*, Edited by S. Schaible and W. T. Ziemba, Academic Press, New York.

SCHAIBLE, S., and ZANG, I. (1980), On the convexifiability of pseudoconvex C^2-functions, *Math. Programming* **19**, 289–299.

STEWART, G. W. (1973), *Introduction to Matrix Computations*, Academic Press, New York.

STOER, J., and WITZGALL, C. (1970), *Convexity and Optimization in Finite Dimensions I*, Springer, Berlin.

ZANG, I. (1974), Generalized convex programming, D.Sc. dissertation, Technion, Israel Institute of Technology, Haifa (In Hebrew),

ZANG, I. (1981), Concavifiability of C^2-functions: A unified exposition, in *Gernalized Concavity in Optimization and Economics*, Edited by S. Schaible and W. T. Ziemba, Academic Press, New York.

9

Additional Generalizations of Concavity

In this chapter we discuss some additional families of generalized concave functions. These families are considerably broader than those presented in the previous chapters.

First we discuss a generalization of concavity via support properties, due to Ben-Tal and Ben-Israel (1976). A well-known property that may serve to define a concave function over an open convex set is that its graph can be supported from above at each point by a hyperplane. This property is fundamental in the derivation of optimality, duality, and in the construction of algorithms for concave programming. A natural generalization of concavity is obtained by considering functions whose graphs are supported from above by supports that are not necessarily linear. This leads to the concept of F-concave functions, which is the subject of Section 9.1. We shall discuss properties of these functions and present several examples. One example ties F-concavity to a classical generalization of concavity developed by Beckenbach (1937); another example shows that the family of F-concave functions can become extremely large.

Further, in Section 9.2, following the work of Ortega and Rheinboldt (1970), and Avriel and Zang (1980), we present extensions of the families of generalized concave functions introduced in Chapter 3. The idea underlying the definitions of the new families is to examine the behavior of the function under consideration along continuous arcs instead of line segments as is done with the definitions of the ordinary classes of generalized concave functions. The resulting families of functions are more general than those introduced in Chapter 3 because the convexity of the upper-level sets which must hold for quasiconcave functions is replaced by a weaker property, namely, arcwise connectedness. After defining these functions and analyzing their properties, we shall show that the necessary local–global maximum properties that hold for the families of functions discussed in Chapter 3 also hold for the new families. Moreover, for the latter families of functions these conditions are shown, under some mild assumptions, to be sufficient.

9.1. *F*-Concave Functions

In this section generalization of concave functions via their support properties will be discussed. It is well known (see Theorem 2.12) that differentiable concave functions over open convex sets can be alternatively characterized by the "graph below the tangent" property. Namely, a differentiable function f is concave on an open convex set $C \subset R^n$ if and only if its hypograph, given by (2.33), is supported from above at the point $(x, f(x)) \in R^{n+1}$ by the hyperplane whose normal is $(-\nabla f(x), 1)$. This property is a special case of a general result, brought here without proof, concerning functions that are not necessarily differentiable.

Theorem 9.1 (Bazaraa and Shetty, 1979; Roberts and Varberg, 1973). Let f be a real function on the open convex set $C \subset R^n$. Then f is concave if and only if for every $x^0 \in C$ there exists a vector $z \in R^n$ depending on x^0 and such that

$$f(x) \leq f(x^0) + (x - x^0)^T z \tag{9.1}$$

for every $x \in C$.

The vector z satisfying (9.1) is called a *supergradient of f at* x^0 (Rockafellar, 1970; see also Definition 4.3). Note that at a point $x^0 \in C$ there may be more than one supergradient. However, a necessary and sufficient condition for the uniqueness of a supergradient at x^0 for a concave function f is that f is differentiable at x^0 and $z = \nabla f(x^0)$ (Rockafellar, 1970). Hence Theorem 2.12 is a special case of Theorem 9.1.

A natural generalization of concavity emanates from this "graph below the tangent property" via relaxing the linearity requirement of the support to the hypograph of the function under consideration. Ben-Tal and Ben-Israel (1976) considered families of generalized concave functions entitled *F-concave* by using a possibly nonlinear support in their definition.

Definition 9.1. Let $X \subset R^n$ and let F be a family of real functions on X. Let f be a real function on the open set $C \subset X$. Then f is said to be F-concave on C if for every $x^0 \in C$ there exists a function $g \in F$ such that

$$f(x^0) = g(x^0) \tag{9.2}$$

and

$$f(x) \leq g(x) \tag{9.3}$$

holds for all $x \in C$. f is called strictly F-concave on C if, in addition, (9.3) holds as a strict inequality for all $x \in C$, $x \neq x^0$. The function g satisfying (9.2) and (9.3) is called a support of f at x^0 in C.

The definition of (strict) F-convexity is carried out in a similar way by reversing the inequality in (9.3). It is easy to show that f is F-concave on C if and only if $(-f)$ is $(-F)$-convex on C where $(-F) = \{-g: g \in F\}$. The following example shows the relationship between concave and F-concave functions.

Example 9.1. Let F_a be the family of affine functions on $X = R^n$. That is,

$$F_a = \{g: g(x) = a^T x + b, a \in R^n, b \in R\}. \tag{9.4}$$

Then, according to the above definition and Theorem 9.1, a real function f on an open convex set $C \subset R^n$ is F_a-concave if and only if it is concave.

The next example considers functions with bounded Hessians.

Example 9.2. Let f be a twice continuously differentiable function on the open convex set $C \subset R^n$ and suppose that the Hessian of f is uniformly bounded from above on C. That is, there exists a number M such that for all $x \in C$

$$z^T \nabla^2 f(x) z \leq M z^T z \tag{9.5}$$

holds for all $z \in R^n$. Let $X = R^n$, and for a given $\alpha \in R$ define

$$F_\alpha = \{g: g(x) = \tfrac{1}{2}\alpha x^T x + a^T x + b, a \in R^n, b \in R\}. \tag{9.6}$$

Then f is F_α-concave provided

$$\alpha \geq M. \tag{9.7}$$

To show this, let $x^0 \in C$ and let $g \in F_\alpha$ be given by

$$a = \nabla f(x^0) - \alpha x^0 \tag{9.8}$$

and

$$b = f(x^0) - (x^0)^T \nabla f(x^0) - \tfrac{1}{2}\alpha (x^0)^T x^0. \tag{9.9}$$

Then

$$g(x) = f(x^0) + (x - x^0)^T \nabla f(x^0) + \tfrac{1}{2}\alpha (x - x^0)^T (x - x^0) \tag{9.10}$$

and (9.2) holds. Applying the Mean Value theorem, we obtain that for every $x \in C$ there exists a point $\bar{x} \in (x, x^0)$ such that

$$f(x) = f(x^0) + (x - x^0)^T \nabla f(x^0) + \tfrac{1}{2}(x - x^0)^T \nabla^2 f(\bar{x})(x - x^0). \tag{9.11}$$

In view of (9.10), (9.11), to establish (9.3) it is necessary to show that

$$(x - x^0)^T \nabla^2 f(\bar{x})(x - x^0) \le \alpha (x - x^0)^T (x - x^0) \tag{9.12}$$

holds. This, however, follows (9.5) and (9.7). Hence all functions with bounded from above Hessians are F_α-concave on open convex sets. Ben-Tal and Ben-Israel (1981) showed that these functions are supported by a different family F involving trigonometric functions. Note that if inequality (9.7) is strict, then f is strictly F_α-concave. If the Hessian of f is uniformly bounded from below on C, that is, there exists a number m such that for all $x \in C$

$$z^T \nabla^2 f(x) z \ge m z^T z \tag{9.13}$$

holds for all $z \in R^n$, then f is F_α-convex for $\alpha \le m$. Notice that an F_α-concave (convex) function with $\alpha \le 0$ ($\alpha \ge 0$) is concave (convex) on C.

As a special case, let f be a quadratic function given by

$$f(x) = \tfrac{1}{2}x^T Q x + c^T x + d, \tag{9.14}$$

where $c \in R^n$, $d \in R$, and Q is an $n \times n$ matrix. Then f is $F_{\lambda_{\max}}$-concave and $F_{\lambda_{\min}}$-convex where $\lambda_{\max}$ and $\lambda_{\min}$, respectively, denote the largest and smallest eigenvalues of the symmetric matrix $\tfrac{1}{2}(Q + Q^T)$.

The family of functions possessing uniformly bounded Hessians on convex sets is interesting in itself. It is a subset of a larger family of F_α-concave functions called ρ-concave. The concept of ρ-concavity was defined and discussed by Vial (1983) (see also Vial, 1982). A function f on a convex subset $C \subset R^n$ is said to be *ρ-concave* if there exists a real number ρ such that for every $x^1 \in C$, $x^2 \in C$

$$f(\lambda x^1 + (1-\lambda)x^2) \geq \lambda f(x^1) + (1-\lambda)f(x^2) + \rho\lambda(1-\lambda)\|x^2 - x^1\|^2 \quad (9.15)$$

holds for all $0 \leq \lambda \leq 1$. Note that a 0-concave function is concave, while a ρ-concave function with $\rho > 0$ is strongly concave (see Definition 2.7 and Proposition 2.19). Moreover, a twice continuously differentiable function satisfying (9.5) is ρ-concave for all $\rho \leq -M$. Vial (1983) showed that a function f, which is not necessarily twice differentiable, is ρ-concave if and only if it is F_α-concave with $\alpha = -\rho$. Vial (1983) also derived Karush–Kuhn–Tucker-type sufficient global optimality conditions for mathematical programs involving ρ-concave functions.

Example 9.3. Beckenbach (1937) defined *super F-functions* as follows: Let F_B be a family of continuous functions on $(l, u) \subset R$, satisfying the condition that for any two points $x^1 \in R$, $x^2 \in R$ satisfying $l < x^1 < x^2 < u$ and any two numbers $y^1 \in R$, $y^2 \in R$, there is a unique $g \in F_B$ satisfying

$$g(x^1) = y^1, \qquad g(x^2) = y^2. \quad (9.16)$$

Such a family will be called a *Beckenbach family on* (l, u). A function f on (l, u) is a *super F_B-function* if for any two points $l < x^1 < x^2 < u$, the member $g_{12} \in F_B$ defined by

$$g_{12}(x^i) = f(x^i), \qquad i = 1, 2, \quad (9.17)$$

also satisfies

$$g_{12}(x) \leq f(x) \quad (9.18)$$

for all $x_1 < x < x_2$. It was shown (see Ben-Tal and Ben-Israel, 1976, for necessity and Peixoto, 1948, 1949, for sufficiency) that a function on (l, u) is F_B-concave if and only if it is a super F_B-function. For a further discussion of super-F_B-functions (or sub-F_B-functions, their convex counterpart) see Beckenbach and Bellman (1965) and Roberts and Varberg (1973).

Example 9.4. Let

$$F = \{g: g(x) = b\cos(ax), a \in R, b \in R\} \quad (9.19)$$

be a family of real functions on $X = R$. The function

$$f(x) = \begin{cases} -1 & \text{if } x \neq 0 \\ 1 & \text{if } x = 0 \end{cases} \tag{9.20}$$

is F-concave on $C = R$. First note that every $g \in F$ with $b = 1$ satisfies (9.2) and (9.3) for $x^0 = 0$. Let $x^0 \neq 0$. Then $g(x) = \cos(\Pi x/x^0)$ (that is, $a = \Pi/x^0$, $b = 1$) satisfies (9.2), (9.3).

The above example is interesting since it implies that an F-concave function need not be continuous even if F is a family of continuous functions. Ben-Tal and Ben-Israel (1976) proved, however, that upper semicontinuity (see Definition 3.4) of all functions in F implies the same property for f.

Proposition 9.2. Let F be a family of upper semicontinuous functions on $X \subset R^n$ and let f be F-concave on the open set $C \subset X$. Then f is upper semicontinuous.

Proof. A function g on C is upper semicontinuous at $\bar{x} \in C$ if for every sequence $\{x^k\} \subset C$, $x^k \to \bar{x}$

$$\limsup_{k \to \infty} g(x^k) \leq g(\bar{x}) \tag{9.21}$$

holds (see Definition 3.4 and the ensuing discussion). Suppose that f is not upper semicontinuous. Then there exists a point $x^0 \in C$ and a sequence $\{x^k\} \subset C$, $x^k \to x^0$ such that

$$\limsup_{k \to \infty} f(x^k) > f(x^0) \tag{9.22}$$

holds. Suppose that $g \in F$ supports f at x^0. Then

$$g(x^0) = f(x^0) < \limsup_{k \to \infty} f(x^k) \leq \limsup_{k \to \infty} g(x^k) \leq g(x^0), \tag{9.23}$$

where the first inequality in (9.23) is (9.22), the second one follows the F-concavity of f, and the last one is a result of the upper semicontinuity of F. However, (9.23) means a contradiction. □

We now present some elementary properties of F-concave functions extending the properties of concave functions discussed in Chapter 2.

Proposition 9.3. Let F_i, $i \in I$, be a finite collection of families of real functions on X and suppose that $f_i(x)$, $i \in I$ are F_i-concave functions on $C \subset X$. Denote

$$\bar{F} = \left\{g: g = \sum_{i \in I} g_i, g_i \in F_i\right\}, \tag{9.24}$$

and

$$F = \bigcup_{i \in I} F_i. \tag{9.25}$$

Then

$$\bar{f}(x) = \sum_{i \in I} f_i(x) \tag{9.26}$$

is an $\bar{F}$-concave function and

$$f(x) = \inf_{i \in I} \{f_i(x)\} \tag{9.27}$$

is an F-concave function.

The proof of the above proposition is obvious and is left for the reader. Note that generally if f_i, $i \in I$ are F-concave (with the same F), then the function (9.27) is still F-concave, while (9.26) is F-concave only if the family F is closed under addition. The above result extends Propositions 2.15 and 2.18. The next proposition extends properties proved in Propositions 2.15, 2.16, and 2.17.

Proposition 9.4. Let f be F-concave on $C \subset X \subset R^n$ and let q be a function from R^m to R^n and ψ be a function from R to R. Let $\{\psi Fq\}$ be a family of functions from $q^{-1}(X) \subset R^n$ to R given by

$$\{\psi Fq\} = \{\tilde{g}: \tilde{g} = \psi gq, g \in F\}. \tag{9.28}$$

Then the function ψfq is $\{\psi Fq\}$-concave on $q^{-1}(C) \subset q^{-1}(X)$.

The proof of this proposition is also left for the reader.

Example 9.5. In Theorem 8.30 we showed that a function f is (h, ϕ)-concave on C if and only if $\phi f h^{-1}(x)$ is concave on $h(C)$. It follows that $\phi f h^{-1}(x)$ is F_a-concave (see Example 9.1) on $h(C)$. Consequently, substituting ϕ^{-1} for ψ and h for q in Proposition 9.4, we have that f is $\{\phi^{-1} F_a h\}$-concave or F-concave where

$$F = \{g: g(x) = \phi^{-1}(a^T h(x) + b), a \in R^n, b \in R\}. \tag{9.29}$$

We now turn to develop a characterization of differentiable F-concave functions, following the analysis of Ben-Tal and Ben-Israel (1976). In the sequel we shall confine ourselves to families F on $X \subset R^n$ characterized in terms of $n+1$ parameters. That is, each member $g \in F$ is a function on the open set $C \subset R^n$ of the form $g(a, b; \cdot)$, where $a \in A \subset R^n$ and $b \in B \subset R$. Such a family will be called an $(n+1)$-*parameter family*. The families of functions given by (9.4), (9.19), and (9.29) are examples of such $(n+1)$-parameter families. The family given by (9.6) is also of the same type, provided α is held constant. Note that for a given function f with a bounded from above Hessian on $C \subset R^n$, the family F_α can be determined by fixing the parameter α at a level that is not smaller than M in order to obtain F_α-concavity of f with respect to this $(n+1)$-parameter family.

To establish the main gradient type characterization of F-concave functions we impose some further assumptions on F and f:

(i) F is an $(n+1)$-parameter family of once-differentiable functions on $X \subset R^n$, such that the correspondence between (a, b) and $g(a, b; \cdot)$ is one to one.

(ii) Let $Z(F) \subset R \times R^n$ be denoted by

$$Z(F) = \bigcup \{\text{range}\,(g, \nabla_x g) \colon g \in F\} \tag{9.30}$$

where $\nabla_x g$ denotes the n-vector of partial derivatives of g with respect to its (vector) last argument. Then for every $x \in X$ and $(w, y) \in Z(F)$, the system

$$w = g(a, b; x), \tag{9.31}$$

$$y = \nabla_x g(a, b; x), \tag{9.32}$$

has a unique solution $(a, b) \in A \times B$.

(iii) The function f is once differentiable on the open set $C \subset X$ and $(f(x), \nabla f(x)) \in Z(F)$ for all $x \in C$.

If assumptions (i)–(iii) are satisfied, then for every $x \in C$ we can denote by $(a(x), b(x))$ the unique solution of the system

$$f(x) = g(a, b; x), \tag{9.33}$$

$$\nabla f(x) = \nabla_x g(a, b; x). \tag{9.34}$$

That is, for every $x \in C$

$$f(x) = g(a(x), b(x); x) \tag{9.35}$$

$$\nabla f(x) = \nabla_x g(a(x), b(x); x) \tag{9.36}$$

holds. We are now able to fully characterize support functions of f in cases where they exist.

Lemma 9.5. Under assumptions (i)–(iii), if f is supported by $g \in F$ at $x^0 \in C$, then $g(a(x^0), b(x^0); x)$ is the unique support function of f at x^0.

Proof. Suppose that $g(a^0, b^0; x)$ supports f at $x^0 \in C$. That is, the function v given by

$$v(x) = g(a^0, b^0; x) - f(x) \tag{9.37}$$

is, by (9.3), nonnegative and

$$v(x^0) = g(a^0, b^0; x^0) - f(x^0) = 0 \tag{9.38}$$

by (9.2). Hence x^0 is a local minimum of v and since C is open

$$\nabla v(x^0) = \nabla_x g(a^0, b^0; x^0) - \nabla f(x^0) = 0 \tag{9.39}$$

must hold. It follows from (9.35), (9.36), (9.38), and (9.39) that $a^0 = a(x^0)$, $b^0 = b(x^0)$ must hold, and g is the only support of f at x^0. □

We can now prove a characterization of differentiable F-concave functions:

Theorem 9.6 (Ben-Tal and Ben-Israel, 1976). Suppose that assumptions (i)–(iii) hold. Then f is F-concave on C if and only if for every $x^0 \in C$, $x \in C$

$$f(x) \le g(a(x^0), b(x^0); x) \tag{9.40}$$

is satisfied. Furthermore, f is strictly F-concave if and only if a strict inequality holds in (9.40) whenever $x^0 \neq x$.

Proof. Note that, in view of (9.35), at $x = x^0$ equality in (9.40) must hold. It follows that if (9.40) holds then $g(a(x^0), b(x^0); x)$ is a support of f at x^0 and that x^0 is the unique point of support if a strict inequality in (9.40) holds for $x = x^0$. Conversely, if f is F-concave, then by Lemma 9.5, g is the unique support of f at x^0 and (9.40) holds. If f is strictly F-concave, then since it is also F-concave it follows again that g is the unique support of f at x^0 and (9.40) then holds with a strict inequality for $x \neq x^0$. □

Note that although (9.40) does not employ the derivatives of f explicitly, it is still a first-order condition since $a(x^0)$ and $b(x^0)$ are obtained from the system (9.35) and (9.36).

Example 9.6. Consider the family of affine functions F_a given by (9.4). Then, at x^0 (9.33) and (9.34) take the form

$$f(x^0) = g(a, b; x^0) = a^T x^0 + b, \tag{9.41}$$

$$\nabla f(x^0) = \nabla_x g(a, b; x^0) = a, \tag{9.42}$$

and the solution to this system is

$$a(x^0) = \nabla f(x^0), \tag{9.43}$$

$$b(x^0) = f(x^0) - (x^0)^T \nabla f(x^0). \tag{9.44}$$

Hence we obtain

$$g(a(x^0), b(x^0); x) = a(x^0)^T x + b(x^0) = f(x^0) + (x - x^0)^T \nabla f(x^0), \tag{9.45}$$

and (9.40) takes the form

$$f(x) \leq f(x^0) + (x - x^0)^T \nabla f(x^0), \tag{9.46}$$

which is the well-known gradient inequality (2.70) of concave functions.

Example 9.7. Consider now the family F_a of Example 9.2 defined by (9.6). In this case, at x^0 (9.33) and (9.34) take the form

$$f(x^0) = \tfrac{1}{2}\alpha (x^0)^T x^0 + a^T x^0 + b, \tag{9.47}$$

$$\nabla f(x^0) = \alpha x^0 + a, \tag{9.48}$$

from which

$$a(x^0) = \nabla f(x^0) - \alpha x^0, \tag{9.49}$$

$$b(x^0) = f(x^0) - (x^0)^T \nabla f(x^0) + \tfrac{1}{2}\alpha (x^0)^T x^0, \tag{9.50}$$

and hence

$$g(a(x^0), b(x^0); x) = f(x^0) + (x - x^0)^T \nabla f(x^0) + \tfrac{1}{2}\alpha (x - x^0)^T (x - x^0) \tag{9.51}$$

is obtained. Thus, f is F_α-concave if and only if

$$f(x) \le f(x^0) + (x - x^0)^T \nabla f(x^0) + \tfrac{1}{2}\alpha(x - x^0)^T(x - x^0) \tag{9.52}$$

holds for all $x \in C$, $x^0 \in C$. In case f is twice continuously differentiable on C, then using the Mean Value theorem as was done in Example 9.2, we have that f is F_α-concave if and only if f has a bounded from above Hessian on C and $\alpha \ge M$.

It is also possible to derive second-order characterizations of F-concave functions. See Ben-Tal and Ben-Israel (1976) for this analysis as well as for a further elaboration on first-order conditions. To conclude this section we present some further properties of differentiable F-concave functions due to Ben-Tal and Ben-Israel (1981). We have shown in Proposition 9.2 that upper semicontinuity is a property inherited by F-concave functions from the family F. The next proposition discusses similar induction of unimodality and pseudoconcavity properties.

Proposition 9.7. Let F be an $(n + 1)$-parameter family of differentiable functions on $X \subset R^n$, each having the property that a stationary point in X is a global maximum. Then a differentiable F-concave function on $C \subset X$ inherits the same property. Moreover, if every $g \in F$ is pseudoconcave then so is f.

Proof. Let $x^0 \in C$ satisfy

$$\nabla f(x^0) = 0. \tag{9.53}$$

Then (9.36) implies

$$\nabla f(x^0) = \nabla_x g(a(x^0), b(x^0); x^0) = 0. \tag{9.54}$$

However, (9.54) implies

$$g(a(x^0), b(x^0); x) \le g(a(x^0), b(x^0); x^0) \tag{9.55}$$

for all $x \in C$. Moreover, from (9.40) we get

$$f(x) \le g(a(x^0), b(x^0); x) \le g(a(x^0), b(x^0); x^0) = f(x^0), \tag{9.56}$$

where the second inequality in (9.56) follows from (9.55) and the equality from (9.35). The proof of the pseudoconcave part is similar. □

The fact that F-concave functions possess nonlinear supports everywhere makes it possible to derive a duality theory for such functions. See Ben-Tal and Ben-Israel (1981) for a complete discussion.

9.2. Generalized Arcwise Connected Functions

In this section we further extend the notions of generalized concavity presented in Chapter 3. If we examine the definitions of families of generalized concave functions that use only function values, such as Definitions 3.8, 3.11, and 3.14 and Theorem 3.1, we can see that the behavior of the function under consideration is always tested along line segments. If we were to move between each pair of points along continuous arcs and were to still require the same properties to hold, we could obtain new classes of functions. These families of functions, presented below, are more general than those introduced in Chapter 3, because, as we shall show, the convexity of the upper-level sets which must hold for quasiconcave functions is replaced by a weaker property, namely, arcwise connectedness.

The general approach to this types of function was introduced by Ortega and Rheinboldt (1970) and further extended by Avriel and Zang (1980). We shall adopt here the notation of Avriel and Zang, but it is important to note that here we generalize pseudoconcave and quasiconcave functions, whereas in Avriel and Zang (1980), the convex counterparts of these classes of functions were extended under the same terminology.

In this section we first introduce definitions and some properties of the new families of functions. Next we show that the necessary local-global maximum properties which hold for the families of functions introduced in Chapter 3 also hold for the new families. Moreover, for the latter families of functions these conditions are shown under some mild assumptions to be sufficient.

Let us first introduce the definition of an arcwise connected set (see, for example, Apostol, 1974).

Definition 9.2. A set $C \subset R^n$ is said to be arcwise connected (AC) if for every pair of points $x^1 \in C$, $x^2 \in C$ there exists a continuous vector valued function $H(x^1, x^2; \theta)$, called an arc, defined on the unit interval $[0, 1] \subset R$ and with values in C such that

$$H(x^1, x^2; 0) = x^1, \qquad H(x^1, x^2; 1) = x^2. \tag{9.57}$$

We use the same notation here for arcs as we used for h-mean value functions in Definition 8.3, because (8.178) actually defines an arc connecting x^1 and x^2 in C. In the sequel, $H(x^1, x^2; \theta)$ will denote a continuous arc connecting x^1 and x^2. Note that every convex set in R is AC, since the function

$$H(x^1, x^2; \theta) = (1-\theta)x^1 + \theta x^2 \tag{9.58}$$

is an arc in the sense of the above definition. Thus in view of Definition 2.1, the concept of arcwise connected sets is a generalization of convex sets. Following Ortega and Rheinboldt (1970), we now extend the concept of quasiconcavity.

Definition 9.3. A real function defined on the AC set $C \subset R^n$ is called quasiconnected (QCN) if for every $x^1 \in C$, $x^2 \in C$ there exists an arc $H(x^1, x^2; \theta)$ in C satisfying

$$f(H(x^1, x^2; \theta)) \geq \min[f(x^1), f(x^2)] \tag{9.59}$$

for every $0 \leq \theta \leq 1$. Alternatively, f is QCN if and only if for every $\alpha \in R$ the upper-level set $U(f, \alpha)$ [see (2.38)] is arcwise connected.

The reader can easily verify that these two definitions of QCN functions are equivalent. Moreover, Definition 3.1 and Theorem 3.1 imply that a quasiconcave function is QCN with $H(x^1, x^2; \theta)$ given by (9.58).

Example 9.8. Let $x \in R^2$ and let f be defined by

$$f(x) = \begin{cases} -(x_1x_2)^2 & \text{if } x_1x_2 \leq 5 \\ -25 & \text{otherwise.} \end{cases} \tag{9.60}$$

Level curves of this function are shown in Figure 9.1. It is clear that this function is not quasiconcave, but one can find arcs that will satisfy the definition of quasiconnectedness. For example, we can connect each pair of points through the origin, that is,

$$H(x^1, x^2; \theta) = \begin{cases} (1-2\theta)x^1, & 0 \leq \theta \leq \frac{1}{2} \\ (2\theta-1)x^2, & \frac{1}{2} \leq \theta \leq 1 \end{cases} \tag{9.61}$$

for every $x^1 \in R^2$, $x^2 \in R^2$.

Following the same idea as above, we introduce additional families of functions extending the concepts of strictly and semistrictly quasiconcave functions.

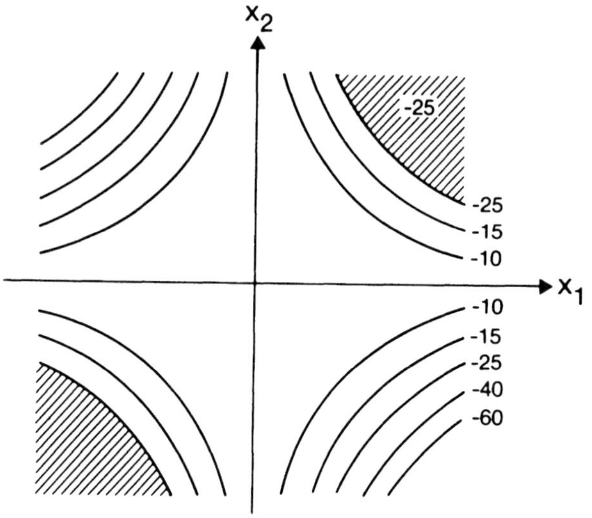

Figure 9.1. A quasiconnected function.

Definition 9.4. A real function f defined on the AC set $C \subset R^n$ is called strictly quasiconnected (STQCN) if for every $x^1 \in C$, $x^2 \in C$, $x^1 \neq x^2$, such that $f(x^1) \geq f(x^2)$ there exists an arc $H(x^1, x^2; \theta)$ in C satisfying

$$f(H(x^1, x^2; \theta)) > f(x^2) \tag{9.62}$$

for $0 < \theta < 1$. Under the same assumptions f is called semistrictly quasiconnected (SSTQCN) if (9.62) is satisfied for every $x^1 \in C$, $x^2 \in C$ such that $f(x^1) > f(x^2)$.

The class of STQCN functions was first introduced by Ortega and Rheinboldt (1970).

Example 9.9. Let $x \in R^2$ and let f_1 be defined by

$$f_1(x) = \begin{cases} -(x_1)^2 + (x_2)^2 & \text{if } x_2 \leq 0 \\ -(x_1)^2 - (x_2)^2 & \text{if } x_2 > 0. \end{cases} \tag{9.63}$$

The level curves of this function are shown in Figure 9.2. This is an STQCN function with arcs given by (9.58) if both $x_2^1 \leq |x_1^1|$, and $x_2^2 \leq -|x_1^2|$, and by (9.61) otherwise.

Take now

$$f_2(x) = -(x_1 x_2)^2. \tag{9.64}$$

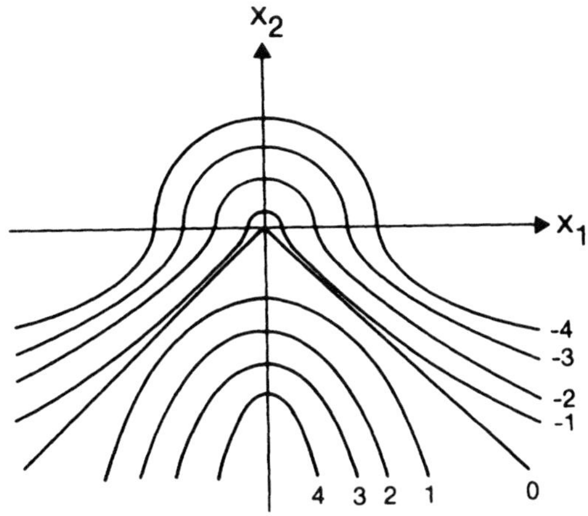

Figure 9.2. The STQCN function defined by (9.63).

This function is similar to the one given in Example 9.8. It is SSTQCN with arcs given by (9.61). It is also QCN, but not STQCN, as can be seen by taking, for example, $x^1 = (-1, 0)$, $x^2 = (1, 0)$.

A comparison of Definitions 3.8 and 3.11 with Definition 9.4 shows that strictly and semistrictly quasiconcave functions are STQCN and SSTCQN, respectively. It follows from the above definitions that an STQCN function is SSTQCN and also QCN. There are, however, SSTQCN functions that are not QCN. An example of such a function is given below.

Example 9.10. Let f be defined on R^2 by

$$f(x_1, x_2) = \begin{cases} -e^{x_1} & \text{if } -1 \le x_2 \le 1 \\ -(2 - |x_2|)e^{x_1} & \text{if } 1 \le x_2 \le 2 \quad \text{or} \quad -2 \le x_2 \le -1 \\ 0 & \text{if } x_2 \ge 2 \quad \text{or} \quad x_2 \le -2. \end{cases} \tag{9.65}$$

This function is continuous and SSTQCN. The required arcs can be constructed, for example, in the following way: Whenever $f(x^1) > f(x^2)$, start at x^2 and proceed in a straight line parallel to the negative x_1 axis until a point $\bar{x} = (\bar{x}_1, x_2^2)$ is reached that satisfies $\bar{x}_1 < \min\{x_1^1, x_2^2, \ln[-f(x^2)]\}$. Make a right-angle turn and proceed in the $x_2^1 - x_2^2$ direction to the point $\bar{\bar{x}} = (\bar{x}_1, x_2^1)$. Then connect $\bar{\bar{x}}$ with x^1 by a straight line. To see that f is not QCN, take $\alpha = 0$ and observe that

$$U(f, 0) = \{x: x \in R^2, x_2 \ge 2 \text{ or } x_2 \le -2\} \tag{9.66}$$

is not an AC set.

The above example is the SSTQCN counterpart of Example 3.4, which shows that a semistrictly quasiconcave function is not necessarily quasiconcave. In that case, however, we have established a sufficient condition (Proposition 3.30) for a semistrictly quasiconcave function to be quasiconcave. We generalize this result here by proving a necessary and sufficient condition for a SSTQCN function to be QCN. Let

$$G(f) = \{\alpha \in R: U(f, \alpha) \neq \varnothing\}, \tag{9.67}$$

and let

$$\bar{U}_f = \begin{cases} U(f, \alpha_{\text{sup}}) & \text{if } G(f) \text{ is bounded from above} \\ \varnothing & \text{otherwise,} \end{cases} \tag{9.68}$$

where

$$\alpha_{\text{sup}} = \sup\{\alpha \in R: \alpha \in G(f)\}. \tag{9.69}$$

That is, $\bar{U}_f$ is the upper-level set corresponding to the supremal value of f, in case such a supremum exists, and it is the empty set otherwise. Then we have the following proposition.

Proposition 9.8. Let f be a real SSTQCN function defined on the AC set $C \subset R^n$. Then f is QCN if and only if $\bar{U}_f$ is AC.

Proof. If f is QCN then by Definition 9.3, $\bar{U}_f$ is AC. Assume now that $\bar{U}_f$ is AC. We only have to show that $x^1 \in C$, $x^2 \in C$, and $f(x^1) = f(x^2)$ imply the existence of an arc satisfying (9.59), since the case $f(x^1) > f(x^2)$ is covered by Definition 9.4. If $\bar{U}_f \neq \varnothing$ and $x^1 \in \bar{U}_f$, $x^2 \in \bar{U}_f$, then we are done by the hypothesis. Otherwise, there exists a point $\bar{x} \in C$ satisfying $f(\bar{x}) > f(x^1) = f(x^2)$. Since f is SSTQCN, we have $H(\bar{x}, x^1; \theta)$ and $H(\bar{x}, x^2; \theta)$ such that

$$f(H(\bar{x}, x^1; \theta)) > F(x^1), \qquad 0 < \theta < 1, \tag{9.70}$$

$$f(H(\bar{x}, x^2; \theta)) > f(x^2), \qquad 0 < \theta < 1. \tag{9.71}$$

Take

$$H(x^1, x^2; \theta) = \begin{cases} H(\bar{x}, x^1; 1 - 2\theta), & 0 \le \theta \le \frac{1}{2} \\ H(\bar{x}, x^2; 2\theta - 1), & \frac{1}{2} \le \theta \le 1. \end{cases} \tag{9.72}$$

Then the arc given by (9.72) satisfies (9.59). □

Relating this proposition to the preceding example, we observe that $G(f)$ is bounded above, $\alpha_{\sup} = 0$ and $\bar{U}_f$ is not AC.

The following definition extends the notion of pseudoconcavity.

Definition 9.5. A real function f defined on the AC set $C \subset R^n$ is called pseudoconnected (PCN) if for every $x^1 \in C$, $x^2 \in C$ such that $f(x^1) > f(x^2)$ there exists an arc $H(x^1, x^2; \theta)$ in C and a positive number $\beta(x^1, x^2)$ satisfying

$$f(H(x^1, x^2; \theta)) \ge f(x^2) + \theta(1 - \theta)\beta(x^1, x^2) \tag{9.73}$$

for $0 < \theta < 1$. Under the same assumptions f is called strictly pseudoconnected (STPCN) if (9.73) is satisfied whenever $x^1 \ne x^2$ and $f(x^1) \ge f(x^2)$.

A (strictly) pseudoconcave function is (STPCN) PCN with $H(x^1, x^2; \theta)$ given by (9.58). A PCN function is also SSTQCN, and a STPCN function is STQCN, hence QCN.

Let us turn now to some differential properties of the functions under discussion.

Definition 9.6. The arc $H(x^1; x^2; \theta)$ is said to have a directional derivative at $\theta = 1$ if there exists a vector $\nabla^- H(x^1, x^2; 1) \in R^n$ and a vector-valued function α: $[0, 1] \to R^n$ satisfying $\lim_{t \to 0^+} \alpha(t) = 0$ such that

$$H(x^1, x^2; \theta) = H(x^1, x^2; 1) + (\theta - 1)\nabla^- H(x^1, x^2; 1) + (1 - \theta)\alpha(1 - \theta) \tag{9.74}$$

is satisfied for $0 \le \theta \le 1$. The vector $\nabla^- H(x^1, x^2; 1)$, given by

$$\nabla^- H(x^1, x^2; 1) = \lim_{\theta \to 1^-} \frac{H(x^1, x^2; \theta) - x^2}{\theta - 1} \tag{9.75}$$

is called the directional derivative of $H(x^1, x^2; \theta)$ at $\theta = 1$.

Let f be differentiable on the AC set $C \subset R^n$ and let $x^1 \in C$, $x^2 \in C$. Suppose that $H(x^1, x^2; \theta)$ is an arc in C having a directional derivative at $\theta = 1$. Then we can write

$$f(H(x^1, x^2; \theta)) = f(x^2) + (\theta - 1)[\nabla^- H(x^1, x^2; 1)]^T \nabla f(x^2) + (1 - \theta)\bar{\alpha}(1 - \theta) \tag{9.76}$$

for $0 \le \theta \le 1$, where $\bar{\alpha}: [0, 1] \to R$ satisfies $\lim_{t \to 0^+} \bar{\alpha}(t) = 0$. The real number $f^-(H(x^1, x^2; 1))$, given by

$$f^-(H(x^1, x^2; 1)) = [\nabla^- H(x^1, x^2; 1)]^T \nabla f(x^2) = \lim_{\theta \to 1^-} \frac{f(H(x^1, x^2; \theta)) - f(x^2)}{\theta - 1} \tag{9.77}$$

is called the *directional derivative of f with respect to the arc* $H(x^1, x^2; \theta)$ *at* $\theta = 1$ and note that it exists whenever $\nabla^- H(x^1, x^2; 1)$ exists.

Definition 9.7. A real differentiable function f defined on an AC set $C \subset R^n$ is said to be differentially quasiconnected (DQCN) if it is QCN and for every $x^1 \in C$, $x^2 \in C$ there exists an arc $H(x^1, x^2; \theta)$ possessing a directional derivative at $\theta = 1$ and satisfying (9.59).

We similarly define differential subclasses of the other classes of functions introduced in this section, that is, we have families of DSTQCN, DSSTQCN, DPCN, and DSTPCN functions. We now present an example demonstrating that a family of functions introduced previously and its differential subfamily are not necessarily equivalent.

Example 9.11. Let

$$f(x) = (x)^2, \qquad C = \{x: x \in R, x \ge 0\}. \tag{9.78}$$

For every $x^1 \in C$, $x^2 \in C$ define

$$H(x^1, x^2; \theta) = [\theta(x^2)^2 + (1 - \theta)(x^1)^2]^{1/2} \tag{9.79}$$

and

$$\beta(x^1, x^2) = f(x^1) - f(x^2). \tag{9.80}$$

We observe that f is PCN. Suppose now that f is DPCN. Taking $x^1 > 0$, $x^2 = 0$ we must have in view of (9.77) and (9.73)

$$f^-(H(x^1, x^2; 1)) = [\nabla^- H(x^1, x^2; 1)]^T \nabla f(x^2) \le -\beta(x^1, x^2) < 0, \tag{9.81}$$

a contradiction since $\nabla f(x^2) = 0$.

We shall now show that it is possible to derive a definition of DPCN and DSTPCN functions that is similar to Definition 3.13. The proof of the next proposition, due to Avriel and Zang (1980), is similar to that of Proposition 3.38 and is therefore omitted.

Proposition 9.9. Let f be a real differentiable function on the AC set $C \subset R^n$. Then f is DPCN if and only if for every $x^1 \in C$, $x^2 \in C$, such that $f(x^1) > f(x)$ there exists an arc $H(x^1, x^2; \theta)$ in C satisfying

$$f(H(x^1, x^2; \theta)) > f(x^2) \tag{9.82}$$

for $0 < \theta < 1$ and

$$f^-(H(x^1, x^2; 1)) < 0. \tag{9.83}$$

Similarly f is DSTPCN if and only if (9.82) and (9.83) are satisfied for every $x^1 \in C$, $x^2 \in C$, $x^1 \neq x^2$, and $f(x^1) \geq f(x^2)$.

For arcs $H(x^1, x^2; \theta)$ given by (9.58), it can be seen that $\nabla^- H(x^1, x^2; 1) = x^2 - x^1$. Therefore, in view of Definition 3.13, differentiable (strictly) pseudoconcave functions are (DSTPCN) DPCN.

We shall now present several propositions that show that the local-global properties, established for the generalized concave functions introduced in Chapter 3, also hold for the extended families of functions presented here. Since the proofs of Theorems 9.10–9.13 are quite simple and similar to their counterparts in Chapter 3, we omit them. The first result generalizes Proposition 3.3.

Theorem 9.10 (Ortega and Rheinboldt, 1970). Let f be a real QCN function, defined on the AC set $C \subset R^n$. If $x^* \in C$ is a strict local maximum of f then x^* is a strict global maximum of f over C. The set of points at which f attains its global maximum over C is AC.

Our next result extends the "only if" part of Theorem 3.37.

Theorem 9.11 (Avriel and Zang, 1980). Let f be a real SSTQCN function defined on the AC set $C \subset R^n$ and let $x^* \in C$ be a local maximum of f. Then x^* is a global maximum of f over C.

The extension of Proposition 3.29 is given by the following theorem.

Theorem 9.12 (Ortega and Reinboldt, 1970). Let f be a real STQCN function defined on an AC set $C \subset R^n$ and let $x^* \in C$ be a local maximum of f. Then x^* is a unique global maximum of f over C.

Finally, we extend Theorem 3.39.

Theorem 9.13 (Avriel and Zang, 1980). Let f be a real differentiable function on an open set containing the AC set $C \subset R^n$ and let $x^* \in C$ be a point where $\nabla f(x^*) = 0$. If f is DPCN, then x^* is a global maximum of f over C. If f is DSTPCN, then x^* is the unique global maximum of f over C.

We now turn to establish converse results to Theorems 9.10–9.13, obtained by Avriel and Zang (1980). Since the families of functions introduced here do not necessarily have convex upper-level sets, it is possible to obtain results stating that a function satisfying some local global maximum properties must belong to one of the new classes. The assumptions needed to derive these results are quite mild; and as we show in an example, functions that do not satisfy these assumptions are quite pathological in nature.

In the sequel we shall make use of the strict upper-level set mapping of f, given by (3.112):

$$U^0(f, \alpha) = \{x \in C: f(x) > \alpha\}. \tag{9.84}$$

We also denote by cl the closure operation of sets in R^n. Then we have the following definition.

Definition 9.8. Let f be a real function defined on the AC set $C \subset R^n$. Then f is said to be a well-behaved (WB) function if for every $\alpha \in R$ such that $U^0(f, \alpha)$ is AC, it follows that $U^0(f, \alpha) \cup \{\hat{x}\}$ is AC for every $\hat{x} \in cl\, U^0(f, \alpha)$.

In the sequel we shall deal mainly with well-behaved functions. In fact, most of our sufficient local–global maximum properties hold only for WB functions. However, as we shall later demonstrate, restricting ourselves to WB functions only is a very weak assumption. The next results deal with conditions under which a QCN function is also SSTQCN. First we have the following lemma.

Lemma 9.14. Let f be a well-behaved function defined on the AC set $C \subset R^n$ and assume that $U^0(f, \alpha)$ is AC for every $\alpha \in R$. Suppose that $x^1 \in C$, $x^2 \in C$, $f(x^1) > f(x^2)$, and x^2 is not a local maximum of f. Then there exists an arc $H(x^1, x^2; \theta)$ satisfying

$$f(H(x^1, x^2; \theta)) > f(x^2) \tag{9.85}$$

for every $0 \leq \theta \leq 1$.

Proof. Let $\hat{\alpha} = f(x^2)$. Then $x^1 \in U^0(f, \hat{\alpha})$ and we can find a sequence $\{x^i\} \subset U^0(f, \hat{\alpha})$, converging to x^2 such that $x^i \in C \cap N_{\delta_i}(x^2)$, where

$$N_{\delta_i}(x^2) = \{x: \|x - x^2\| < \delta_i\} \tag{9.86}$$

and

$$\delta_i = 1/i, \qquad i = 2, 3, \ldots \tag{9.87}$$

Thus x^2 is a cluster point of $U^0(f, \hat{\alpha})$. Since f is a WB-function, the set $U^0(f, \hat{\alpha}) \cup \{x^2\}$ is AC and there exists an arc $H(x^1, x^2; \theta)$ contained in $U^0(f, \hat{\alpha})$ for $0 \leq \theta < 1$. □

The following lemma will also be useful in the sequel.

Lemma 9.15. Let f be a real QCN function defined on the AC set $C \subset R^n$. Then $U^0(f, \alpha)$ is AC for every $\alpha \in R$.

Proof. Let f be QCN. Then the sets $U(f, \alpha)$ are AC for every $\alpha \in R$. Let $\bar{\alpha} \in R$ be such that $U^0(f, \bar{\alpha})$ is not empty. Then, for every $x^1 \in U^0(f, \bar{\alpha})$, $x^2 \in U^0(f, \bar{\alpha})$ we have

$$f(x^1) > \bar{\alpha}, \qquad f(x^2) > \bar{\alpha}. \tag{9.88}$$

Assume that $f(x^1) \geq f(x^2)$. Since f is QCN there exists an arc $H(x^1, x^2; \theta)$ satisfying

$$f(H(x^1, x^2; \theta)) \geq f(x^2) > \bar{\alpha} \tag{9.89}$$

for every $0 \leq \theta \leq 1$. Hence $H(x^1, x^2; \theta)$ is in $U^0(f, \bar{\alpha})$ and $U^0(f, \bar{\alpha})$ is AC. □

Now we can state and prove a result analogous to the "if" part of Theorem 3.37.

Theorem 9.16 (Avriel and Zang, 1980). Let f be a quasiconnected (QCN) well-behaved (WB) function defined on the AC set $C \subset R^n$. If every local maximum of f is a global one over C then f is semistrictly Q-connected (SSTQCN).

Proof. Suppose that $x^1 \in C$, $x^2 \in C$, and $f(x^1) > f(x^2)$. It follows from the hypotheses that x^2 is not a local maximum of f. By Lemma 9.15 $U^0(f, \alpha)$ is AC for every $\alpha \in R$. Taking $\alpha = f(x^2)$ we conclude, by Lemma 9.14, that there exists an arc $H(x^1, x^2; \theta)$ satisfying (9.85) for every $0 \le \theta < 1$. Hence f is SSTQCN. □

We shall now demonstrate by an example that the assumption of f being a WB-function is necessary for the preceding theorem to hold.

Example 9.12. For $k = 1, 2, \ldots$, let $A^k \subset R^2$ be defined by

$$A^k = \{x: (2)^{-k} < x_1 < 3(2)^{-(k+1)}, 0 \le x_2 \le 1\} \tag{9.90}$$

and let

$$B^1 = \{x: x \in R^2, 0 < x_1 < 1, -1 < x_2 < 0\}, \tag{9.91}$$

$$B^2 = \{x: x \in R^2, x_1 = 0, 0 \le x_2 \le 1\}, \tag{9.92}$$

$$B^3 = \{(1, 0)\}. \tag{9.93}$$

Then let

$$C = \left(\bigcup_{k=1}^{\infty} A^k\right) \cup B^1 \cup B^2 \cup B^3. \tag{9.94}$$

The set C is shown in Figure 9.3. It is AC with the following arcs: For a given pair of points $x^1 \in C$, $x^2 \in C$ with $x_1^1 \ge x_1^2$ use a straight line segment

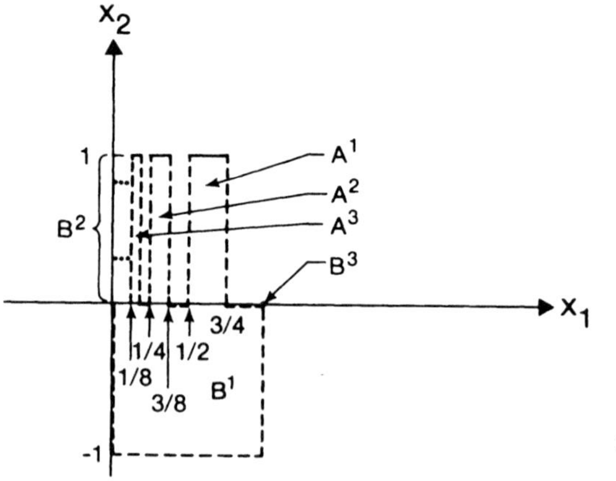

Figure 9.3. The set C in Example 9.12.

if possible, otherwise use piecewise linear arcs $H(x^1, x^2; \theta)$ (of no more than three segments) such that $[H(x^1, x^2; \theta^1)]_1 \geq [H(x^1, x^2; \theta^2)]_1$ for $\theta^1 < \theta^2$. Let

$$f(x) = x_1 - 1. \tag{9.95}$$

Then f is a continuous function over C having a unique local maximum at $x^* = (1, 0)$ which is also global. It is easy to verify that f is QCN, using the above-described arcs. However, it is not SSTQCN since if we take $x^1 = (\frac{1}{2}, -\frac{1}{2})$, $x^2 = (0, 1)$ we obtain

$$f(x^1) = -1/2 > f(x^2) = -1, \tag{9.96}$$

and every arc connecting x^1 to x^2 must have a line segment in B^2, contradicting (9.62). In this case, f is not a WB-function on C, because we can take $\alpha = -1$ and then

$$U^0(f, -1) = C \backslash B^2. \tag{9.97}$$

The point $x^2 = (0, 1)$ is in cl $U^0(f, -1)$ since there are sequences in A^k that converge to x^2, but clearly $U(f, -1) \cup \{(0, 1)\}$ is not AC.

It is quite evident that the above example is a pathological one. In fact, all functions that are not well-behaved functions are of the same nature, and consequently, restricting ourselves to well-behaved functions only is a rather mild assumption.

We shall now derive some additional general results in the form of conditions under which a continuous real function on R^n is SSTQCN. Here we utilize the concept of a connected set (Apostol, 1974): A subset D of R^n is said to be *connected* if there exist no nonempty disjoint sets $S^1 \subset R^n$, $S^2 \subset R^n$, such that neither contains cluster points of the other, satisfying $D = S^1 \cup S^2$. We recall that every arcwise connected set is connected, while the opposite does not necessarily hold. However, every open connected set is arcwise connected. We then have

Theorem 9.17 (Avriel and Zang, 1980). Let f be an upper semicontinuous real function defined on a closed set $C \subset R^n$. Suppose that $U(f, \alpha)$ are compact for all $\alpha \in R$. Also suppose that every local maximum of f is a global one over C and that the set of maxima is connected. Then the upper-level sets $U(f, \alpha)$ are connected for every $\alpha \in R$.

Proof. Suppose that the hypotheses hold and there exists an $\bar{\alpha} \in R$ such that $U(f, \bar{\alpha})$ is disconnected. We can, therefore, write $U(f, \bar{\alpha}) = S^1 \cup S^2$, where S^1 and S^2 are nonempty disjoint sets such that neither contains cluster points of the other, hence both are closed relative to $U(f, \bar{\alpha})$. It follows then that S^1 and S^2 are closed subsets of R^n. Moreover, since $U(f, \bar{\alpha})$ is compact, S^1 and S^2 are also compact and there exist points $x^1 \in S^1$ and $x^2 \in S^2$ that are maxima of f over S^1 and S^2, respectively (Mangasarian, 1969). Since x^1 is not a cluster point of S^2, there is an open ball such that $N_\delta(x^1) \cap S^2 = \varnothing$. Hence $x \in N_\delta(x^1) \cap C$ implies either $x \in S^1$ or $x \notin U(f, \bar{\alpha})$. In both cases $f(x) \leq f(x^1)$ and x^1 is a local maximum of f. The same argument holds for x^2. But one of these points, say x^1, must also be a global maximum of f over C and then x^2 is either a local maximum that is nonglobal or it is also a global maximum, disconnected from x^1, contradicting the hypotheses. □

To prove our next theorem we need the following lemma.

Lemma 9.18. Let f be a real function defined on a subset $C \subset R^n$. If the upper-level sets $U(f, \alpha)$ are connected for every $\alpha \in R$, then the strict upper-level sets $U^0(f, \alpha)$ are also connected for every $\alpha \in R$.

Proof. Suppose that the hypotheses hold; $\bar{\alpha} \in R$ and $U^0(f, \bar{\alpha})$ is disconnected, hence it is nonempty. Then there exist open subsets $A \subset R^n$ and $B \subset R^n$ whose intersections with $U^0(f, \bar{\alpha})$, are disjoint nonempty sets and

$$U^0(f, \bar{\alpha}) = [A \cap U^0(f, \bar{\alpha})] \cup [B \cap U^0(f, \bar{\alpha})]. \tag{9.98}$$

Take now any $\tilde{\alpha} > \bar{\alpha}$ such that $U(f, \tilde{\alpha})$ is nonempty. Then $U(f, \tilde{\alpha}) \subset U^0(f, \bar{\alpha})$ and $U(f, \tilde{\alpha})$ is connected by the hypotheses. Hence $U(f, \tilde{\alpha})$ is in A or in B but not in both. Suppose $U(f, \tilde{\alpha}) \subset A$. Hence,

$$\bigcup_{\tilde{\alpha} > \bar{\alpha}} U(f, \tilde{\alpha}) = U^0(f, \bar{\alpha}) \subset [A \cap U(f, \bar{\alpha})] \tag{9.99}$$

contradicting that $B \cap U^0(f, \bar{\alpha})$ is nonempty. □

Now we can state and prove the following theorem.

Theorem 9.19 (Avriel and Zang, 1980). Let f be a continuous real well-behaved (WB) function defined on R^n and suppose that for every $\alpha \in R$ the upper-level sets $U(f, \alpha)$ are compact. Also, suppose that every local maximum of f is a global one over R^n and the set of maxima is connected. Then f is an SSTQCN function on R^n.

Proof. By continuity of f we have that $U^0(f, \alpha)$ are open for every $\alpha \in R$. By Theorem 9.17, $U(f, \alpha)$ are connected for every $\alpha \in R$ and by Lemma 9.18 also the strict level sets are connected for every $\alpha \in R$. The strict level sets $U^0(f, \alpha)$ are therefore open and connected for all $\alpha \in R$, hence they are arcwise connected. Let $x^1 \in R^n$, $x^2 \in R^n$ satisfying $f(x^1) > f(x^2)$. By the hypotheses x^2 is not a local maximum of f and by Lemma 9.14 there exists an arc $H(x^1, x^2; \theta)$ satisfying (9.85). □

The hypotheses in the last theorem are sufficient for SSTQCN functions. They are, however, not necessary. In Example 9.10, we have a SSTQCN function whose set of global maxima is given by

$$U(f, 0) = \{x: x \in R^2, x_2 \geq 2 \text{ or } x_2 \leq -2\}, \tag{9.100}$$

which is disconnected. Hence, if we restrict ourselves to continuous WB-functions on R^n with compact upper-level sets, we have that the family of SSTQCN functions is the family of functions whose local maxima are global and form a connected set.

Let us now derive sufficient local-global extremum properties for STQCN functions, that will be similar to those presented in Theorems 9.16 and 9.19.

Theorem 9.20 (Avriel and Zang, 1980). Let f be a quasiconnected (QCN) well-behaved (WB) function defined on the AC set $C \subset R^n$. Assume that f has a unique local maximum which is also global over C. Then f is strictly quasiconnected (STQCN).

Proof. By Theorem 9.16 f is SSTQCN. It follows that for every $x^1 \in C$, $x^2 \in C$ such that $f(x^1) > f(x^2)$ we have an arc $H(x^1, x^2; \theta)$ satisfying

$$f(H(x^1, x^2; \theta)) > f(x^2) \tag{9.101}$$

for $0 \leq \theta < 1$. Let $x^1 \neq x^2$ and suppose that $f(x^1) = f(x^2) = \hat{\alpha}$. Clearly, neither x^1 nor x^2 is a global maximum of f, and $U^0(f, \hat{\alpha})$ is nonempty. Let $\tilde{x} \in U^0(f, \hat{\alpha})$. Since f is SSTQCN, there exist arcs $H(\tilde{x}, x^1; \theta)$ and $H(\tilde{x}, x^2; \theta)$ satisfying

$$f(H(\tilde{x}, x^i; \theta)) > f(x^i), \qquad i = 1, 2 \tag{9.102}$$

for $0 \leq \theta < 1$. Let

$$H(x^1, x^2; \theta) = \begin{cases} H(\tilde{x}, x^1; 1 - 2\theta), & 0 \leq \theta \leq \frac{1}{2} \\ H(\tilde{x}, x^2; 2\theta - 1), & \frac{1}{2} \leq \theta \leq 1. \end{cases} \tag{9.103}$$

Then

$$f(H(x^1, x^2; \theta)) > f(x^2) \tag{9.104}$$

for $0 < \theta < 1$ and f is STQCN. □

We may now obtain a more general result.

Theorem 9.21 (Avriel and Zang, 1980). Let f be a continuous well-behaved (WB) function defined on R^n and suppose that for every $\alpha \in R$ the upper-level sets $U(f, \alpha)$ are compact. Also suppose that f has a unique local maximum which is also global over R^n. Then f is an STQCN function on R^n.

Proof. By Theorem 9.19, f is SSTQCN. The rest of the proof is the same as in Theorem 9.20. □

It is possible to derive similar sufficient local-global extremum properties for DPCN, PCN, DSTPCN, and STPCN functions as well. We state these results without proofs. For more details see Avriel and Zang (1980).

Theorem 9.22. Let f be a continuously differentiable quasiconnected (QCN) WB-function defined on the open AC set $C \subset R^n$. Assume that every stationary point of f in C is a global maximum of f over C. Then f is a differentially pseudoconnected (DPCN) and hence a pseudoconnected (PCN) function over C.

Theorem 9.23. Let f be a continuously differentiable WB-function defined on R^n and suppose that for every $\alpha \in R$ the upper-level sets $U(f, \alpha)$ are compact. Also suppose that every stationary point of f is a global maximum over R^n and the set of maxima is connected. Then f is a DPCN and hence a PCN function on R^n.

Theorem 9.24. Let f be a continuously differentiable QCN WB-function defined on the open AC set $C \subset R^n$. Assume f has a unique stationary point that is also the global maximum of f over C. Then f is a DSTPCN, and hence an STPCN function on C.

Theorem 9.25. Let f be a continuously differentiable WB function defined on R^n and suppose that for every $\alpha \in R$ the upper-level sets $U(f, \alpha)$ are compact. Also suppose that f has a unique stationary point that is also the global maximum of f over R^n. Then f is a DSTPCN, and hence a STPCN function on R^n.

Characterizations of functions having local-global extremum properties were also derived by Martin (1981, 1982), utilizing the concept of connectedness instead of arcwise connectedness. Employing maximizing sets, rather than maximizing points as in Avriel and Zang (1980), Martin (1982) characterized functions having connected upper-level sets and obtained a result that is in the spirit of Theorem 9.17. Horst (1982) defined a new family of functions containing the family of SSTQC functions, called *limes arcwise strictly quasiconcave.* His definition employs sequences of arcs connecting every pair of points, rather than a single arc as in the definitions of this section. It is also shown there that for such a function, defined over a compact AC set, a local maximum must be global, while a function over a compact convex set whose local maxima are global, and form an AC set, must belong to the above family. These results do not impose assumptions on the upper-level set (i.e., compactness and well-behaved function assumptions) as in Theorem 9.19.

For more general characterizations of functions whose local extrema, or stationary points, are global extrema, see Gabrielsen (1985), Zang and Avriel (1975), and Zang, Choo, and Avriel (1976, 1977),

Finally, we note that the concept of concavity can be extended too by allowing movements along arcs instead of line segments. See Avriel and Zang (1980) and Singh (1983),

References

APOSTOL, T. M. (1974), *Mathematical Analysis*, 2nd ed., Addison-Wesley, Reading, Massachusetts.

AVRIEL, M., and ZANG, I. (1980), Generalized arcwise connected functions and characterizations of local-global minimum properties, *J. Optimization Theory Appl.* **32**, 407-425.

BAZARAA, M. S., and SHETTY, C. M. (1979), *Nonlinear Programming, Theory and Algorithms*, Wiley, New York.

BECKENBACH, E. F. (1937), Generalized convex functions, *Bull. Am. Math. Soc.* **43**, 363-371.

BECKENBACH, E. F., and BELLMAN, R. (1965), *Inequalities*, Springer-Verlag, New York.

BEN-TAL, A., and BEN-ISRAEL, A. (1976), A generalization of convex functions via support properties, *J. Australian Math. Soc.* **21**, 341-361.

BEN-TAL, A., and BEN-ISRAEL, A. (1981), *F*-Convex functions: Properties and applications, in *Generalized Concavity in Optimization and Economics*, Edited by S. Schaible and W. T. Ziemba, Academic Press, New York.

GABRIELSEN, G. (1986), Global maxima of real valued functions, *J. Optimization Theory Appl.*, to appear.

HORST, R. (1982), A note on functions whose local minima are global, *J. Optimization Theory Appl.* **36**, 457-463.

MANGASARIAN, O. L. (1969), *Nonlinear Programming*, McGraw-Hill, New York.

MARTIN, D. H. (1981), Connectedness of level sets as a generalization of concavity, in *Generalized Concavity in Optimization and Economics*, Edited by S. Schaible and W. T. Ziemba, Academic Press, New York.

MARTIN, D. H. (1982), Connected level sets, minimizing sets and uniqueness in optimization, *J. Optimization Theory Appl.* **36**, 71-91.

ORTEGA, J. M., and RHEINBOLDT, W. C. (1970), *Iterative Solution of Nonlinear Equations in Several Variables*, Academic Press, New York.

PEIXOTO, M. M. (1948), On the existence of a derivative of generalized convex functions, *Summa Brasiliensis Math.* **2**, No. 3.

PEIXOTO, M. M. (1949), Generalized convex functions and second order differential inequalities, *Bull. Am. Math. Soc.* **55**, 563-572.

ROBERTS, A. W., and VARBERG, D. W. (1973), *Convex Functions*, Academic Press, New York.

ROCKAFELLAR, R. T. (1970), *Convex Analysis*, Princeton University Press, Princeton, New Jersey.

SINGH, C. (1983), Elementary properties of arcwise connected sets and function, *J. Optimization Theory Appl.* **41**, 377-387.

VIAL, J.-PH. (1982), Strong convexity of sets and functions, *J. Math. Econ.* **9**, 187-205.

VIAL, J.-PH. (1983), Strong and weak convexity of sets and functions, *Math. Oper. Res.* **8**, 231-259.

ZANG, I., and AVRIEL, M. (1975), On functions whose local minima are global, *J. Optimization Theory Appl.* **16**, 183-190.

ZANG, I., CHOO, E. U., and AVRIEL, M. (1976), A note on functions whose local minima are global, *J. Optimization Theory Appl.* **18**, 555-559.

ZANG, I., CHOO, E. U., and AVRIEL, M. (1977), On functions whose stationary points are global minima, *J. Optimization Theory Appl.* **22**, 195-208.

10

Supplementary Bibliography

After completion of the manuscript the authors became aware of the following additional references that are relevant for this book:

ANEJA, Y. P., and NAIR, K. P. K. (1984), Ratio dynamic programs, *Oper. Res. Lett.* **3**, 167-173.

BECTOR, C. R., and BHATIA, B. L. (1985), Duality for a multiple objective nonlinear nonconvex program, *Utilitas Math.* **28**, 175-192.

BENSON, H. P. (1985), Finding certain weakly efficient vertices in multiple objective linear fractional programming, *Manage. Sci.* **31**, 240-245.

BONCOMPTE PONS, M. (1985), Programacion fraccional generalizada, thesis, Universidad di Barcelona, Facultad de Matematicas.

BORDE, J. (1985), Quelques aspects theoriques et algorithmiques en quasiconvexité, doctoral thesis, Université de Clermont II, Departement de Mathematiques Appliquées.

BORWEIN, J. M. (1981), Direct theorems in semi-infinite convex programming, *Math. Programming* **21**, 301-318.

CAMBINI, A., and MARTEIN, L. (1986), On the Fenchel-like and Lagrangian duality in fractional programming, *Methods Oper. Res.* **53**, 21-32.

CAMBINI, A., and MARTEIN, L. (1986), A modified version of Martos's algorithm for the linear fractional problem, *Methods Oper. Res.* **53**, 33-44.

CASTAGNOLI, E., and MAZZOLENI, P. (1986), Generalized convexity for functions and multifunctions and optimality conditions, Dipartimento di Matematica, Universita di Pisa, Technical Report No. 134.

CHANDRA, S., CRAVEN, B. D., and MOND, B. (1985), Symmetric dual fractional programming, *Z. Oper. Res.* **29**, 59-64.

CHANDRASEKARAN, R., and TAMIR, A. (1984), Optimization problems with algebraic solutions: Quadratic fractional programs and ratio games, *Math. Programming* **30**, 326-339.

CHARNES, A., COOPER, W. W., GOLANY, B., and SEIFORD, L. (1985), Foundations of data envelopment analysis for Pareto-Koopmans efficient empirical production functions, *J. Econometrics* **30**, 91-107.

CHARNES, A., COOPER, W. W., and THRALL, R. M. (1986), Classifying and characterizing efficiencies and inefficiencies in data envelopment analysis, *Oper. Res. Lett.* **5**, 105-110.

CHOO, E. U., SCHAIBLE, S., and CHEW, K. P. (1985), "Connectedness of the efficient set in three criteria quasiconcave programming, *Cah. Centr. Etud. Rech. Oper.* **27**, 213-220.

CROUZEIX, J. P., and LINDBERG, P. O. (1986), Additively decomposed quasiconvex functions, *Math. Programming* **35**, 42-57.

DEUMLICH, R., and ELSTER, K. H. (1980), Duality theorems and optimality conditions for nonconvex optimization problems, *Math. Operationenforsch. Stat. Ser. Optimization* **11**, 181-219.

DEUMLICH, R., and ELSTER, K. H. (1983), ϕ-Conjugation and nonconvex optimization, a survey (Part I), *Math. Operationenforsch. Stat. Ser. Opt.* **14**, 125-149.

DOMBI, P. (1985), On extremal points of quasiconvex functions, *Math. Programming* **33**, 115-119.

FERLAND, J. A., and POTVIN, J. Y. (1985), Generalized fractional programming: Algorithms and numerical experimentation, *Eur. J. Oper. Res.* **20**, 92-101.

FLACHS, J. (1985), Generalized Cheney-Loeb-Dinkelbach-type algorithms, *Math. Oper. Res.* **10**, 674-687.

HACKMAN, S. T., and PASSY, U. (1986), Semi-convexity, Mimeograph Series No. 388, Faculty of Industrial Engineering and Management, Technion, Haifa.

HIRCHE, J. (1983), Verallgemeinerte Konvexitaet bei Summen und Produkten linearer und gebrochen linearer Funktionen, *Wiss. Z. Univ. Halle* **32**, 91-99.

HIRCHE, J. (1984), On programming problems with a linear-plus-linear-fractional objective function, *Cah. Centr. Etud. Rech. Oper.* **26**, 59-64.

IVANOV, E. H., and NEHSE, R. (1983), Relations between generalized concepts of convexity and conjugacy, *Math. Operationenforsch. Stat. Ser. Optimization* **13**, 1-9.

JAGANNATHAN, R. (1985), An algorithm for a class of nonconvex programming problems with nonlinear fractional objectives, *Manage. Sci.* **31**, 847-851.

JAHN, J., and SACHS, E. (1986), Generalized quasiconvex mappings and vector optimization, *SIAM J. Control Optimization* **24**, 306-322.

JEYAKUMAR, V. (1985), First and second order fractional programming duality, *Opsearch* **22**, 24-41.

KAWOHL, B. (1985), Rearrangements and convexity of level sets in partial differential equations, Lecture notes in Mathematics 1150, Springer Verlag, Heidelberg.

KAWOHL, B. (1986), Geometrical properties of level sets of solutions to elliptic problems, *Proc. Symp. Pure Math.* **45**, Part 2, 25-36.

KOMLOSI, S. (1983), Some properties of nondifferentiable pseudo-convex functions, *Math. Programming* **26**, 232-237.

KORNBLUTH, J. S. H. (1986), On the use of multiple objective linear programming algorithms to solve problems with fractional objectives, *Eur. J. Oper. Res.* **23**, 78-81.

LUHANDJULA, M. K. (1984), Fuzzy approaches for multiple objective linear fractional optimization, *Fuzzy Sets Syst.* **13**, 11-23.

MARTINEZ LEGAZ, J. E. (1981), Un concepto generalizado de conjugacion, applicacion a las funciones quasiconvexas, doctoral thesis, Universidad de Barcelona.

MARTINEZ LEGAZ, J. E. (1983), Exact quasiconvex conjugation, *Z. Oper. Res.* **27**, 257-266.

MOND, B., and SCHECHTER, M. (1980), Duality in homogeneous fractional programming, *J. Inf. Optimization Sci.* **1**, 271-280.

NYKOWSKI, I., and ZOLKIEWSKI. Z. (1985), A compromise procedure for the multiple objective linear fractional programming problem, *Eur. J. Oper. Res.* **19**, 91-97.

PASSY, U., and PRISMAN, E. Z. (1984), Conjugacy in quasi-convex programming, *Math. Programming*, **30**, 121-146.

PASSY, U., and PRISMAN, E. Z. (1985), A convex-like duality scheme for quasi-convex programs, *Math. Programming* **32**, 278-300.

ROTHBLUM, U. G. (1985), Ratios of affine functions, *Math. Programming* **32**, 357-365.

SCOTT, C. H., and JEFFERSON, T. R. (1980), Fractional programming duality via geometric programming duality, *J. Australian Math. Soc. Ser. B* **21**, 398-401.

SCOTT, C. H., and JEFFERSON, T. R. (1981), Conjugate duality for fractional programs, *J. Math. Anal. Appl.* **84**, 381-389.

SINGER, I. (1984), Generalized convexity, fractional hulls and applications to conjugate duality in optimization, in: *Selected Topics in Operations Research and Mathematical Economics*, Lecture Notes in Economics and Mathematical Systems, Vol. 226, Springer-Verlag, New York, pp. 49-79.

SINGH, C. (1986), A class of multiple-criteria fractional programming problems, *J. Math. Anal. Appl.* **115**, 202-213.

SINGH, C., and HANSON, M. A. (1986), Saddlepoint theory for nondifferentiable multiobjective fractional programming, *J. Inf. Optimization Sci.* **7**, 41-48.

SNIEDOVICH, M. (1986), Fractional programming revisited, NRIMS Technical Report, TWISK 459, Pretoria.

STANCU-MINASIAN, I. M. (1985), A third bibliography of fractional programming, *Pure Appl. Math. Sci.* **22**, 109-122.

WOLF, H. (1986), Solving special nonlinear fractional programming problems via parametric linear programming, *Eur. J. Oper. Res.* **23**, 396-400.

ZALMAI, G. J. (1986), Duality for a class of continuous-time homogeneous fractional programming problems, *Z. Oper. Res.* **30**, A43-A48.

Author Index

Subject Index